SCIENCE DEIFIED
& SCIENCE DEFIED

SCIENCE DEIFIED
& SCIENCE DEFIED

The Historical Significance of Science in Western Culture

Volume 2: From the Early Modern Age through the Early Romantic Era, ca. 1640 to ca. 1820

Richard Olson

University of California Press

Berkeley Los Angeles Oxford

University of California Press
Berkeley and Los Angeles, California

University of California Press
Oxford, England

Copyright © 1990 by The Regents of the University of California

Library of Congress Cataloging-in-Publication Data

(Revised for vol. 2)
Olson, Richard, 1940–
  Science deified & science defied.

  Includes bibliographical references and indexes.
  Contents: [1] From the Bronze Age to the
beginnings of the modern era, ca. 3500 B.C. to
ca. A.D. 1640—v. 2. From the early modern age
through the early romantic era, ca. 1640 to ca.
1820.
  1. Science—History. 2. Science and civilization.
I. Science deified and science defied. II. Title.
Q125.O44 1982   303.4'83   82–40093
ISBN 0–520–04621–8 (v. 1)   AAY8579
ISBN 0–520–06846-7 (v. 2: alk. paper)

Printed in the United States of America

1 2 3 4 5 6 7 8 9

The paper used in this publication meets the minimum requirements of
American National Standard for Information Sciences—Permanence of
Paper for Printed Library Materials, ANSI Z39.48–1984 ∞

To the memory of Penny, who helped
keep my weight down and spirits
up for twelve good years

Contents

Acknowledgments

Support for this work has been gratefully received from a National Endowment for the Humanities Research Division Grant and a National Science Foundation Summer Grant. In addition, I have received valuable advice and encouragement from all of my colleagues in the Humanities and Social Sciences Department at Harvey Mudd College and from the members of the History Department at the University of California, Riverside, who served as my gracious and helpful hosts during a sabbatical leave in 1984–1985. Mark Smith and Roger Ransom were particularly generous with their time and advice.

Richard Westfall and James Force read all or parts of the manuscript at various stages, providing welcome encouragement and suggestions, and both Stan Holwitz and Elizabeth Knoll of the University of California Press have been magnificently patient and supportive.

The typing of a long manuscript and seemingly endless revisions were cheerfully accomplished by Susan Riley and Barbara Graham. All illustrations were created by Sally Yost of *Saragraphics*, and the plates were produced by the Audio Visual Department at the University of California, Riverside, and by the Smithsonian Institution.

Any clarity of expression which the text might show must be credited largely to my wife, Kathleen, whose passionate antagoism to pompous prose has saved the reader from unnecessary additional perplexity and to G. Patton Wright, who copyedited the manuscript.

Richard G. Olson

Introduction

Most social critics of the last fifty years are agreed that scientific attitudes, ideas, and institutions are tremendously important in our lives—indeed they seem to play for modern social-analysts a role as central as that which sex played for early twentieth-century psychoanalysts. Sandra Harding, author of *The Science Question in Feminism*, expresses a widely shared feeling when she writes:

> We are a scientific culture.—Scientific rationality has permeated not only the modes of thinking and acting in our public institutions, but even the ways we think about the most intimate details of our private lives. . . . Neither God nor tradition is privileged with the same credibility as scientific rationality in modern cultures.[1]

If what Harding says is true—and I am convinced that it is—then it is tragic that so few persons in our culture are aware of how, why, with what consequences, and to what extent scientific influences have become so pervasive. Indeed, our ignorance diminishes our capacity to be active agents in shaping our own destinies. In this context, it seems to me that historians share with psychoanalysts some very central aims and obligations. Both groups seek to make us self-conscious about aspects of our past which influence our lives without our knowledge and thus in ways which are beyond our control. Both groups understand that awareness does not guarantee that wise decisions will be made; but they are equally committed to the assumption that without awareness, intelligent choices are not possible.

The present volume is the second of a proposed three-part investigation of ways in which science has interacted with other features of Western cultures. It seeks to develop its themes not through encyclopedic coverage, but through relatively detailed explorations of selected illustrative episodes. And it concentrates on interactions between the traditional foci of humanistic study (religion, political thought, philosophy, literature, and the arts) and the scientific tradition, emphasizing how the former have been shaped by the latter.

I neither deny nor doubt that the interactions between science and other aspects of culture are bi-directional; to be sure, the scientific tradition is itself transformed through its interactions with other elements of culture. In fact, my first major historical study, *Scottish Philosophy and British Physics, 1750–1880: A Study in the Foundations of the Victorian Scientific Style*, sought to show how a tradition of moral philosophy within the Scottish educational system shaped the study of physics in Britain during the nineteenth century.[2] Furthermore, I am quite willing to admit that many of the ways in which science shapes our lives are mediated through technological developments rather than through the less tangible media of ideas and attitudes. In chapter 8 I argue that the significance of science for European economic activities became important rather earlier than most historians of technology and economic historians generally acknowlege. Yet, in what follows, I choose to emphasize the unidirectional impact of science on other aspects of culture and to concentrate on those aspects that we associate broadly with terms like "beliefs," "values," "tastes," and "ideologies," for three basic reasons.

First, although the vision of science which was dominant from the early nineteenth century until the mid-twentieth century—the so-called "positivist" vision of science—denied that true science could be influenced by social and cultural factors, trends in the social studies of science since the mid-1930s have led to a sophisticated and extensive scholarship relating to the various social interests that shape the growth of science. The need for one more general analysis in this area is thus marginal. Second, significant as science has been in stimulating technological and consequent economic and social changes, technologically mediated impacts of science on Western culture have also been extensively and intelligently studied. Third, and most important, the nontechnological influences of science on other aspects of culture have been deeply obscured or distorted by the one major tradition of scholarship which has continually asserted their importance for at least two hundred years—the positivist tradition.

From the end of the eighteenth century, when Jean-Antoine-Nicholas de Caritat de Condorcet wrote his *Esquisse d'un tableau historique des progrès de l'esprit humain* (Sketch of the Historical Progress of the Human Mind), until at least the third decade of the twentieth century and the publication of John Herman Randall's *The Making of the Modern Mind,*[3] the history of the Western world has been viewed predominantly in terms of a progressive unfolding of truth and freedom grounded in the constant advance of scientific knowledge. For our purposes, however, there are two sets of assumptions central to positivist historiography, each of which has served to interfere with any attempt to understand adequately the cultural implications of the sciences.

First, the universe studied by scientists is assumed to be inhabited solely by objects whose character and behavior are completely independent of any attempt to study them and whose behavior can be discovered and then described in inerrant and invariable "positive" statements. One limited but absolutely critical implication of this perspective has been explored in a particularly important way by B. F. Skinner, who asserts that "no theory changes what it's a theory about."[4] There is a serious question about whether such a statement can be interpreted in a way that could be true even if the objects of the theory were inanimate; but there is no question that it is *false* when it is applied to humans and other living organisms and when its terms are understood in their ordinary meanings. One of the most important and useful discoveries of mid-twentieth-century psychology is the so-called "pygmalion effect" or the "self-fulfilling prophecy." Research has unambiguously shown that the expectations which an observer has about the behavior of human or animal subjects affect the behavior being studied; thus, one of the most fundamental premises of positivist belief is undermined.[5] Furthermore, since expectations are almost always a consequence of holding some formal or informal theory about the objects of study, in an extremely important way theories can and often do consequently change what they are theories about. Indeed, that human actions are direct responses to *beliefs* or theories about the world rather than responses to a world apart from our beliefs makes a historical study of science a central aspect of Western culture; for to an extremely high degree our beliefs and our consequent behaviors are *shaped* and not merely mirrored by our sciences.

The second positivistic claim that has distorted approaches to the cultural consequences of the sciences is closely related to the first claim that science is the sole source of positive, value-neutral, and culture-independent knowledge (otherwise known as *truth*). According to this second claim, already implicit in Condorcet's writings, "the acquisition and systematization of positive knowledge are the only human activities which are truly cumulative and progressive," from which it follows that, "if we wish to explain the progress of mankind, then we must focus our attention on the development of science and its applications."[6] This perspective grants scientific knowledge a curious and unwarrantable privileged position. Attachment or commitment to any belief not sanctioned by science demands explanation in terms of the economic, social, psychological, theological, or other interests that it serves; but once known, a scientific statement is presumed to command assent simply because it is obvious and true. On the one hand, then, the appropriation of scientific knowledge by any group becomes natural, and we are relieved of investigating conditions that might encourage or discourage the progressive spread of scientific knowledge and attitudes. On the other hand, the failure to use scientific knowledge or, even worse, the deliberate opposition to the spread of scientific knowledge and activities becomes unnatural and a sign of irrationality, perversion, or conscious malevolence. The consequence of these assumptions is that serious historical investigation of the process of assimilation and rejection of scientific ideas and attitudes has been blocked.

Though it does not follow logically from the claim that science alone produces progress, it is also generally accepted by believers in the progressive nature of science that any "regressive" or otherwise undesired aspect of culture could not depend upon genuine science. Thus, when some trend presumed by an author to be undesirable is obviously linked to scientific developments, it must be interpreted—as is frequently the case with analysts of the conservative uses of social Darwinism or the use of Einstein's theory of Special Relativity to justify cultural relativism—as a perversion or misinterpretation of the relevant science. In those cases where scientists themselves obviously and immediately participate in the undesirable activities—as was the case when anthropologists, psychologists, and sociologists used scientific arguments to support racist and class-biased policies during the nineteenth century—the relevant scientific activities are automatically presumed to involve bad-science, pseudo-science, or mistaken-science rather than ordinary "positive" science.[7] Once again, by prejudging the nature of scientific impacts on other aspects of our life, such assumptions preclude anything but the most biased historical analyses.

In what follows, though I may not be able to avoid showing my admiration for science as an impressive stage for the exercise of the human imagination and intelligence, I do hope to be able to avoid many of those positivistic assumptions that have perverted most of the traditional attempts to understand interactions between science and other institutions within Western culture. Science will not be presumed to be the sole source of truth. It will not be presumed to be automatically progressive. It will not be presumed to be self-evident or self-justifying. And it will not be presumed incapable of producing attitudes and beliefs that I find morally repugnant. Put in a more positive way, science *will be* presumed to be a human cultural institution whose interactions with other cultural institutions are subject to all of the kinds of questions that might be asked of any other: "What interests does it legitimately serve?" and "What interests does it legitimately threaten?" From this point of view it becomes possible both to ask why some group or individual within a particular historical context should selectively accept elements of the scientific tradition and to understand that other groups or persons might have legitimate and rational grounds for rejecting them.

The way we understand how science and other aspects of culture interact with one another is very sensitive to our precise understanding of the meaning of the term "science"; so before going further, I would like to summarize the definition of science and the general interpretive framework that was established in greater detail in the first volume. For purposes of this work *science* is not taken to be a particular kind of knowledge but a cultural institution characterized for each particular time and place by *a set of activities and habits of mind aimed at contributing to an organized, universally valid, and testable body of knowledge about phenomena.* These characteristics are usually embodied in systems of concepts, rules of procedure, theories, or model investigations that are generally accepted by groups of practitioners—the scientific specialists.

As Thomas Kuhn has clearly demonstrated, there may be times and places in which virtually all persons we might define as scientists in a culture agree upon nearly all of the activities and habits of mind, or attitudes, which define science; but frequently there are substantial disagreements among members of different subspecialties within the scientific community about what activities and attitudes are legitimately scientific. During much of the eighteenth century and again in the early twentieth century, many psychologists accepted some introspective activities as a legitimate source of scientific knowledge, whereas neither chemists nor physicists would do so, for example. There are even important occasions when members of the same subspecialty disagree about critical defining characteristics. Thus, among mid-to-late nineteenth-century physicists, William Thomson and James Clerk Maxwell argued that the activity of constructing "mechanical" models of physical phenomena was a key feature of physics, whereas Ernst Mach and Henri Poincaré violently opposed this activity in favor of a more general "mathematical" modeling procedure.

The points to keep in mind for what follows is that *science* involves both some features that are relatively uniform, over time and from one scientific group to another, and others that vary substantially. When we try to evaluate the impact of scientific activities and habits of mind upon other aspects of culture, we need to consider both those that are highly stable across time and subdisciplines and those that are transient or that emerge only in limited portions of the scientific community.

When we think of science as a cultural institution characterized by certain activities and habits of mind often embodied in concepts and theories, then we can imagine a number of ways in which science might interact both positively and negatively with other cultural institutions, including religion and politics. Concepts or theories derived from scientific activity may be incorporated into the subjects dealt with by other cultural specialists because they meet some perceived needs of those specialists, or they may be attacked and rejected by other cultural specialists because they seem to challenge the conceptual structures or practical claims of their specialties. Similarly, activities or attitudes that first become important in connection with science may be incorporated by or be renounced by other specialist groups. The integration or rejection of scientific elements will generally be highly differential—i.e., some attitudes or concepts developed within the scientific community may be adopted by other groups because they serve the current interests of those groups at the same time that other elements are ignored as irrelevant or actively attacked and rejected because they are perceived to be threatening. Moreover, the degree of integration of scientific elements into initially nonscientific institutions is likely to be highly sensitive to specific historical circumstances.

At almost all times and places, members of any cultural specialty will simultaneously seek to expand the domain of influence of their own specialty and to protect it from outside pressures. This combination of imperialist and defensive postures is natural in part because cultural specialties offer means for the acquisition and exercise of power and status, and because few humans are immune to the desire to be obeyed and admired. Perhaps more important, it is natural because cultural specialists almost invariably become so committed to the importance of the values, activities, and aims of their own specialties that they honestly believe that the good of the entire culture which they serve can best be advanced by the subordination or abandonment of other specialists' perspectives—hence, the positivist philosophy of history discussed above. When any given specialty manages to expand its influence, it is also almost inevitable that opposition to that expansion will build up within other specialties that will attack many of the fundamental aims, values, and activities of the perceived aggressor specialty.

The present work seeks to establish briefly the activities and habits of mind which dominated science during the period from ca. 1640 to ca. 1820. It then seeks to explore in some detail the attempts of members of the scientific specialty to expand the influence of science into religious, social, political, economic, and even artistic domains. It pays special attention to the specific historical circumstances that encouraged non-scientists to incorporate scientific elements into their domains of interest, making it possible for science to become perhaps *the* dominant cultural specialty of the modern Western world—one that shapes all aspects of Western culture as extensively as religion shaped those of the European Middle Ages. Finally, it seeks to explore the emergence of critical pockets of opposition to the cultural imperialism of science.

With few exceptions, the topics dealt with in this work are drawn from English and French sources. In part this selection reflects my own background, which is heavily oriented to the British, and in part it reflects particularly dynamic scientific cultures in Britain and France during the early modern period. But I openly admit that a study of the interactions between science and other aspects of European culture between 1640 and 1820 with any pretention to completeness would also have to deal with developments in the

Netherlands, Scandinavia, Russia, the German-speaking regions, Eastern Europe, Spain, and Italy as well.

The first great efflorescence of Western science occurred in Classical Greece. In that setting, however, those who sought to extend the methods, attitudes, and conceptual systems of pre-Socratic natural philosophers to religious, ethical, political, and economic domains failed miserably to integrate their enterprise with any other significant institutions of contemporary culture. It is true that Greek educational institutions were transformed with the emergence of the Academy, the Lyceum, the Porch, and the Garden (vol. I, chap. 4), and it is true that the scientific conceptual systems developed by Plato, Aristotle, and their followers had a long-term impact on Western intellectual life. But the activities of the scientific intellectual elites were almost totally irrelevant to the ongoing social, economic, political, and religious activities of the bulk of the Greek populace.

We can hardly be surprised at the lack of immediate impact that scientific attitudes and activities had on Greek life when we consider the context in which they appeared. No coherent and substantial or powerful group within Greek society had any reason to promulgate science or scientism (the wholesale extension of scientific ideas, attitudes, and activities to other domains). On the contrary, scientific developments offered severe challenges to the only established Hellenic institutions to which they were clearly related at all—the religious cults (vol. I, chap. 3). Thus scientism was not merely unintegrated with dominant religious and political institutions, it was usually seen as absolutely opposed to them. Under these circumstances it is surprising that there was so little effective suppression of science or scientists—though attempts were made to discourage naturalistic speculations.

There were, moreover, few links between science and economic institutions in the ancient world. Some pre-Socratics—most notably the Milesian, Thales—were personally connected with commercial concerns. But early Greek science neither had nor sought significant productive implications. If any early science had practical implications, it was medicine, and even in that case the role of science vis-à-vis common sense and ritual was problematic. By the time of Aristotle, scientific philosophers were proudly proclaiming the total divorce of their main concerns from those of the mundane worlds of crafts, trades, and even social life.

Finally, competition among early Greek intellectual schools

over claims about the nature of the world and the proper ways of seeking knowledge about it led both to mutual antagonisms that discredited all attempts to establish scientific knowledge and to widespread skepticism among the populace about the worth of any of the methods or conclusions of the new sciences.

Scientific elements did have a substantial impact on early Christianity, in part because the early Church fathers had to differentiate their beliefs from those of other Near Eastern cults that had incorporated important astronomical elements, and in part because Greek philosophical perspectives—especially Platonism— were useful for Christians attempting to defend themselves against pagan critics.

When the center of intellectual life moved into the new universities developing in medieval Europe during the twelfth and thirteenth centuries, science and Christianity again came into close contact in a context which forced a mutual accommodation manifested in the great synthesis of religious and philosophical thought by St. Thomas Aquinas during the mid-thirteenth century. There was still relatively little direct interaction among science and major social and economic activities, although ideas and attitudes derived from Platonic and Islamic science began to influence how Medieval and Renaissance thinkers envisioned the structure of social and political life.

Once again in early modern Europe, most notably first in Italy and the Holy Roman Empire, then in England, France, and Holland, there was a spectacular flowering of science—the so-called scientific revolution. With this scientific activity came innumerable attempts to extend scientific techniques, ideas, and attitudes to artistic, religious, political, social, and economic concerns. But this time circumstances were vastly more favorable to the integration of science and other cultural domains than they had been in ancient Greece. Scientific investigations had been firmly established for three centuries within the universities that produced Europe's leaders in all fields related to intellectual or bureaucratic life. Moreover, by the early seventeenth century, virtually all important Christian sects had intense commitments to some form of the study of natural philosophy.

At one end of the religious spectrum, the Council of Trent (1563) had recommitted the Catholic Church to the Thomistic reconciliation between science and religious faith. As a consequence, by the early seventeenth century, the Dominican order, led by Cardinal Cajetan, had resuscitated Thomistic theology; moreover, the Jesuit order within the Church had both reinvigorated St. Thomas's Aristotelian approach to scientific issues and had produced a large and able cadre of men capable of teaching and working in the new traditions of Copernican mathematical astronomy and mathematical engineering. One recent first-rate scholar has claimed that "the single most important contributor to the support of the study of experimental physics in the seventeenth century was the Catholic Church, and within it, the Society of Jesus";[8] and if one considers the number of textbooks produced and students taught, the Jesuits must also be counted as extremely important contributors to the support of what was then called "mixed mathematics," including its applications to such "speculative" sciences as astronomy and to such practical arts as engineering.

At the other end of the religious spectrum, some millenarian Protestant sects interpreted the discovery of comets, supernovae, and "new stars" by late sixteenth- and early seventeenth-century astronomers as signs of the decay of the present world and of the imminent second coming of Christ.[9] These groups also often viewed Hermetic, Andreaean, and Baconian claims of the reform of knowledge (vol. I, chap. 9) linked to the recent voyages of discovery, trade, and colonization, as a fulfillment of the prophecy of *Daniel* 12:4, which stated that near the time of judgment, "many shall run to and fro, and knowledge shall be increased."[10] In line with their expectation of the coming of Christ's reign on earth, these groups accepted responsibility for preparing the way by seeking knowledge to apply to the amelioration of man's condition and to reestablishing man's dominion over nature. John Milton expressed the hopeful millenarian sentiment, writing:

> . . . the spirit of man, no longer confined within the dark prison house, will reach out far and wide till it fills the whole world and the spaces far beyond with the expansion of its divine greatness. Then at last, most of the chances and changes of the world will be so quickly perceived to him who holds this stronghold of wisdom hardly anything can happen in his life which is unforeseen or fortuitous. He will indeed seem to be one whose rule and dominion the stars obey, to whose command earth and sea harken, and whom winds and tempests serve; to whom, lastly, mother nature herself has surrendered as if indeed some God had abdicated the throne of the world and intrusted its rights, laws, and administration to him as governor.[11]

Between the Catholic right and millenarian left among Christian sects, moderate groups like the Anglican Latitudinarians became, if anything, more committed to scientific ways of thinking than those of either extreme[12] (chap. 3); for they hoped that the techniques of scientific argumentation, the certainty of scientific conclusions, and the evidence of God's activity called to man's attention by natural science would provide a way of discovering a common ground for religion and a way of eliminating sectarian disputes. As Joseph Glanvill put the moderates' case, scientific inquiry, "dispose[s] men's Spirits to more *calmness* and *modesty, clarity* and *prudence* in the Differences of *Religion,* and even silence[s] *Disputes* there." Moreover, by "open Inquiry in the great field of nature—[men] will find themselves disposed to mere *indifference* toward those *petty notions,* in which they were before apt to place a great deal of Religion; and to reckon that *all that* will signify lies in [a] *few, certain operative* Principles. . . ."[13]

Except, then, for small groups of religious conservatives—especially among Dominicans in the Catholic Church and Laudians in the High Anglican Church and a few radical sectarians like the anti-intellectual Anabaptists, virtually all seventeenth-century religious groups had developed a direct and active interest in scientific activities and attitudes of one kind or another. By the same token, many seventeenth-century political groups linked their ideologies to scientific and cosmological arguments.

Macrocosm-microcosm analogies associated with Aristotelian cosmology and pre-Copernican astronomy provided a significant set of arguments for supporting hierarchial ecclesiastical and political structures during the sixteenth century (vol. I, chap. 8). In the early seventeenth century, they emerged as the primary intellectual support for theories of absolute monarchy and the divine right of kings.[14] On the one hand, James I of England claimed in addressing Parliament that "kings are compared to the head of this macrocosm of body of man,"[15] and Thomas Hobbes managed to link the macrocosm-microcosm analysis with the latest mathematical and mechanical traditions of the scientific revolution in a much misunderstood, much maligned, and tremendously influential defense of absolute monarchy, *The Leviathan* (chap. 1). On the other hand, Copernican astronomy, the empirical scientific traditions oriented to Renaissance artisans and engineers, and the alchemical tradition associated with Paracelsus were associated with anti-Aristocratic and antihierarchic sentiments and with a growing sense of worth and value on the part of the producers of the material goods of society.

In the early seventeenth century, republican or democratic ideologies were not so often and so clearly grounded in scientific concepts and techniques as theories of absolute monarchy. Political reformers, nonetheless, both emphasized the need for science in their proposals and exploited empirical analyses of social conditions in formulating their ideas. Thus, the radical "digger," Gerard Winstanley, envisioned a Christian communist society in which natural history and natural science were major public undertakings. The supporter of "independents" in Parliament, Samuel Hartlib, forwarded to the Long Parliament on 25 October 1641 a proposal for a commonwealth that emphasized "social improvement by means of economic planning and scientific research."[16] The more respectable William Petty sought to create a science of "Political Arithmetic" (see chaps. 2 and 4) to guide public policy choices. And the republican theorist James Harrington grounded his proposals in a self-conscious and extremely quantitative "political anatomy"[17] of society, patterned on William Harvey's anatomical studies of the heart and blood (chap. 2).

Thus, just as most seventeenth-century religious sects were centrally concerned with scientific and scientistic arguments, most political groupings drew in some way from scientific ideas and supported scientific activity. In fact, as we shall see, the terms of political discourse were dramatically scientized during the seventeenth century in ways that continue to inform the dominant traditions of Western political thought even today.

Even if science and scientism had not been so thoroughly integrated into religious and political discussions, however, the place of science in seventeenth- and eighteenth-century culture would have been central because of its real and presumed integration into productive economic activities (chap. 8). Expectations continued to exceed the actual fruits of applied science as they had in the sixteenth century. But there is no doubt that mathematics in particular played a central role in navigation, which was essential to the extensive foreign trade of all European nations, in the critically important fields of civil and military engineering, and in accounting, which was essential for merchants and which stimulated the rise of a quantitatively formulated science of economics. Furthermore, chemistry, as crude as it might have been, played an increasingly important role in fields like iron-making, textile-finishing, and salt-manufacturing for a variety of purposes. What is perhaps even more important, in England and Germany especially, Baconian empirical attitudes promoted innumerable agricultural and industrial "projects" that provided a constant stimulus to the economy.[18]

No more than in ancient Greece were all scientists agreed about methods or results during the seventeenth and eighteenth centuries. Within natural philosophy, mutually exclusive explanations of celestial and terrestrial phenomena were variously offered in terms of Aristotelian causes: material, efficient, formal, and final; in terms of Hermetic occult sympathies and antipathies; in terms of Nominalist mathematical descriptions; and—increasingly as the seventeenth century progressed—in terms of purely "corpuscular" mechanisms. Even supporters of the same basic explanatory systems often violently disagreed regarding the character of scientific knowledge and the proper methods of attaining it. Gottfried W. Leibniz and Robert Boyle were both committed to offering corpuscular explanations of phenomena, for example; but the rationalist Leibniz retained the age-old demand that scientific causes be philosophically necessary. Boyle, the empiricist, was convinced that nothing more than contingent knowledge obtained by empirical method was attainable by man. Of Boyle's masterful experiments, Leibniz commented that

> he does nothing but draw from an infinity of beautiful experiments a conclusion which he should have taken as his first principle; that is, that everything in nature is done mechanically. This principle can be rendered certain only by reason, and never by experiments, no matter how many one does.[19]

As in ancient Greece, disputes among scientists in the seventeenth century did intensify skeptical feelings. Responding to this skepticism, John Donne lamented that the "new philosophy calls all in doubt" and compared his sense of intellectual disorientation with the emotional upheaval of his world at the death of Elizabeth Drury: "'Tis all in pieces, all coherence gone."[20] And Donne was by no means alone in his state of confusion. Unlike the case two thousand years earlier, however, conflicts among groups of seventeenth-century scientists were not able to discredit the entire enterprise among most influential audiences; for it had become too central to religious, political, and economic life. Instead of rejecting all of the conflicting claims of scientists, each group with a theological or political commitment tended to adopt and exploit certain scientific theories, methods, and attitudes while rejecting others.

Late in the seventeenth century and throughout the eighteenth century, the diversity of scientific systems and methods began to give way to a diffuse consensus often loosely associated with the various writings of Isaac Newton. After a short lag, the ideological diversity and sectarian bitterness of the seventeenth century was replaced by greater religious tolerance and a set of attitudes that we now associate with the term "liberalism." The new ideology emphasized competitive *laissez-faire* economic activity, equality of opportunity, and individual rights. It should be no surprise to the reader by now that European liberalism was closely linked with the dominant visions of science. It will be one of the chief aims of this work to argue that this linkage was not simply incidental and that many of the features of liberal social, political, and economic doctrines owe important characteristics to the scientistic arguments used to formulate and legitimize them (chaps. 4 and 6). We shall see that the growth of socialist doctrines in the eighteenth century was also closely tied to certain scientific developments (chap. 6) and that modern conservatism often opposed liberal and socialist developments because of the scientific trends that spawned them (chap. 9). At the same time, we will see that there was a strong "natural religion" movement associated with the new scientific synthesis and that while this may have led initially toward piety and a lessening of sectarian infighting, it opened the way both for modern "fundamentalism" and for a bland "Deism" that ultimately led to a significant undermining of the long-standing centrality of religion within European culture (see chap. 3).

Finally, we will see that even in matters of aesthetics (artistic and literary criticism), scientific attempts to develop an understanding of human intellectual and emotional life transformed our ideas during the seventeenth and eighteenth centuries (see chap. 7), producing a "Romantic" aesthetic theory that was used to develop fundamental criticisms not only of science-based religion and politics but also of the scientific enterprise itself (chap. 9).

The Extension of Mechanical and Mathematical Philosophies to the Study of Man and Society

During the early decades of the seventeenth century, two new emphases emerged among natural philosophers to supplement and partially supplant the fifteenth- and sixteenth-century intensification of concerns with experimentation and applied, or "mixed," mathematics. A nearly obsessive emphasis on empiricism had emerged largely among Paracelsan alchemists; and mixed mathematics had become extremely important not only because of its centrality to the dynamic science of astronomy but also because it was indispensable in such domains as cartography, navigation, and both civil and military engineering. But in the first three decades of the seventeenth century, mathematics became almost more important for its conceptual structure than for the applicability of its results. Moreover, the machines constructed by engineers became increasingly important not just for their direct physical uses but also for their service to the scientific community as the source of conceptual models to explain a wide range of natural phenomena.

Both the new and the revised emphasis on hypothetico-deductive systems of thought patterned on Euclidean geometry and the new search for mechanical explanations of phenomena were promoted by René Descartes, whose followers were said to be imbued with the *esprit géométrique* as well as the *esprit du méchanisme.*

DESCARTES AND THE GEOMETRICAL SPIRIT

Cartesian philosophy emerged in connection with detailed works in natural science—especially Descartes's *Essays on Dioptrics,*

Meteors, and Geometry which appeared in 1637—but it is best understood as a response to philosophical and religious crises brought on by the Protestant Reformation and by the recovery of the ancient Greek writings of Sextus Empiricus by late Renaissance humanists.[1] Sextus posed an awesome problem: before one can accept that some proposition has been established beyond a doubt, one must have some set of established criteria by which absolute certainty can be judged. Such criteria, however, can be established beyond doubt only by appeal to a previously established set of criteria. As a consequence, no proposition can be proven to be true or necessary, and—according to skeptics like Sextus—one must suspend judgment about everything. Absolutely nothing can be known with the kind of certainty Plato and Aristotle hoped to attain.

This peculiar intellectual puzzle assumed great importance during the Reformation. Salvation depended upon acceptance of the true doctrines of Christian religion, but this was a time when many different sects claimed to be sole possessors of spiritual truths. Similarly—though at first far less critical for most human lives— skepticism seemed to preclude the possibility of justifying scientific claims to natural truths.

This was the situation faced by the young French nobleman, René Descartes. Born in 1596, Descartes attended the Jesuit college of La Flèche near Paris from 1606 to 1615, absorbing not only the revived Aristotelian doctrines of the Jesuit natural philosophers, but the mathematical tradition associated with engineering as well. He found his formal education unsatisfying in light of the newly revived skeptical attitudes. As he later wrote, "I found myself embarrassed with so many doubts and errors, that it seemed to me that the effort to instruct myself had no effect other than the increasing discovery of my own ignorance."[2] The study of law and extensive travels throughout Europe demonstrated to Descartes the existence of a tremendous variety of beliefs and customs, belying the assumption that there is a fund of common human knowledge which lay beyond the bounds of possible doubt. Clearly there was almost *no* belief so absurd that somebody, somewhere did not accept it; conversely, there was virtually no belief so certain that somebody did not doubt it. Mathematics alone seemed to Descartes to approach certitude in the character of its demonstrations; and the young scholar's studies of analytic geometry in 1619 led him to believe that he had created not simply a new branch of mathematics, but

Fig. 1. Places associated with the cultural impact of science during the seventeenth and eighteenth centuries.

a new and powerful science, applicable to virtually all topics of intellectual concern.

Reflecting on why mathematics was so certain, Descartes became convinced that the only two possible ways of gaining factual knowledge were by experience or by deduction[3] and that "inferences from experience are frequently fallacious,"[4] but that deduction *cannot* be erroneous when done by a normal, rational person. Of course, he recognized that the conclusions of deductive arguments *are* often just plain wrong or incompatible with the conclusions of other equally legitimate deductive arguments. But this could happen only when one of the mutually exclusive consequences was deduced from improperly grounded premises.[5]

Just like Plato and Aristotle before him, Descartes had pushed the locus of the problem of scientific certainty back to the establishment of initial premises—Aristotle would have called them *archai*. But Descartes took a tack quite different from that of Aristotle. He argued in *Rules for the Direction of the Mind* (1632) that since the premises of most sciences are dependent on sensory experience and since sensory experience is inherently fallible, most sciences are unavoidably fallible. Arithmetic and geometry, however, "deal with an object so pure and uncomplicated, that they need make no assumptions at all which experience renders uncertain."[6] Descartes goes on to suggest that even traditional geometry deals with figures as they are tangibly or visually experienced, whereas algebra deals with the purest abstractions, independent even of this limited connection with the potentially tainted realm of experience.[7]

The special priority given to *intelligible* rather than *sensory* objects linked Descartes with the long Platonic and neo-Platonic tradition of Western philosophy, and it had important implications for continental mathematics and mechanics. Continental mathematicians, including Malebranche, d'Alembert, and Joseph Louis Lagrange—all Cartesians on this issue—insisted that algebraic, or analytic, formulations of mathematics were to be preferred to geometry for their certainty. This same notion was carried into mechanics and given its clearest expression in Lagrange's great treatise on rational mechanics, the *Méchanique Analytique* (1788). In this work Lagrange boasted:

> One will find no figures in this work. The method which I have presented demands neither constructions, nor geometrical or mechanical reasoning, but only algebraic operations subject to a regular and uniform order.[8]

The Cartesian arguments so far presented illustrate the radical methodological opposition between the Cartesian notion of science and that of the empiricist school represented by Bacon in England and by such revivers of ancient Epicureanism as Pierre Gassendi in France. For the Cartesian "rationalists," *nothing* dependent on sensory experience was to be trusted. For the empiricists, all worthwhile knowledge was ineluctably contingent. For the rationalists *no* contingent knowledge was worthy of the designation "scientific."

Descartes's emphasis on the certainty of deductive arguments left him with much the same problem faced by Plato and Aristotle much earlier. How can one establish the first principles from which necessary scientific consequences can be generated? Once again Descartes suggested that there are only two possibilities: *induction,* which he dismissed out-of-hand, and *intuition,* which must be understood in a special sense. He wrote:

> By intuition I understand not the fluctuating testimony of the senses, not the misleading judgment that proceeds from the blundering constructions of the imagination, but the conception which an unclouded and attentive mind gives so readily and distinctly that we are wholly freed from doubt about that which we understand. . . . [It] springs from the light of reason alone; it is more certain than deduction itself, for it is simpler, though deduction, as we have noted above, cannot by us be erroneously conducted. Thus each individual can mentally have intuition of the fact that he exists, and that he thinks; that the triangle is bounded by three lines only, the sphere by a single superficies, and so on.[9]

Throughout the remainder of his writings Descartes replaced the term "intuitive knowledge" with what seemed to him its equivalent: "*clear* and *distinct* knowledge"; for it was the clarity and distinctness of intuitions which guaranteed their truth.

The secret of discovering truths and escaping skepticism thus becomes adherence to a method which begins in intuitive, or clear and distinct, principles and which proceeds to deduce consequences from them. That is, the secret lies in casting all sciences in the form familiar from Euclidean geometry. It is for this reason that the French spoke of the Cartesian *esprit géométrique.* And it is for this reason that treatises on topics from mechanics (e.g., J. L. L. Lagrange's *Méchanique Analytique*), optics (e.g., Christian Huyghens' *A Treatise on Light*), and anatomy (e.g., Giovanni Borelli's *De Montum Animalum*), to music (e.g., Marin Mersenne's *Harmonie Universelle*), ethics (e.g., Benedictus Spinoza's *Ethics*), and politics (e.g., Thomas Hobbes's *Leviathan*) were presented in the form of definitions, postulates, axioms, and theorems with purportedly formal, deductive demonstrations. It is for this reason too that Bernard Fontenelle, perpetual secretary of the *Académie Fran-çaise,* could write at the end of the seventeenth century:

> A work on politics, on morals, a piece of criticism, even a manual on the art of public speaking would, other things being equal, be all the better for having been written by a geometrician. The order, the clarity, the precision, and the accuracy which have distinguished

the worthier kind of books for some time past now, may well have been due to the geometrical method which has been continuously gaining ground, and which somehow or other has an effect on people who are quite innocent of geometry. . . .[10]

The continental fad for mathematics became so extreme by the early eighteenth century, in fact, that the gossip columnist of the Parisian *Mercure Galant* parodied the mood by writing that

quite recently there were two young ladies in Paris whose heads had been so turned by this branch of learning that one of them declined to listen to a proposal of marriage unless the candidate for her hand undertook to learn how to make telescopes, so often talked of in the *Mercure Galant*; while the other young lady positively refused a perfectly eligible suitor because he had been unable, within a given time, to produce any new idea about "squaring the circle."[11]

Descartes's methodological discussions involved another emphasis in addition to those on deduction and intuition. According to the method of "resolution" and "composition" developed by Paduan Aristotelian philosophers in the sixteenth century,[12] phenomena were to be resolved, or analyzed, step by step into their component parts until some irreducible and well-established principles were reached. After this resolution, the phenomena were to be recomposed, or synthesized, by demonstrating that the original phenomena followed logically from the principles discovered in the resolutive stage. Today this method is best illustrated in chemistry in which we claim to understand a new compound only after it has been analyzed, or decomposed into its constituent elements, and then synthesized, or reconstituted from those elements. According to Descartes's formulation, "we reduce obscure and involved propositions step by step to those that are simpler and then starting with the intuitive apprehension of all those that are absolutely simple, attempt to ascend to the knowledge of all others by precisely similar steps."[13] Recalling that, for Descartes, the more general and abstract propositions are the simpler ones, we see that this method amounts to showing that particular phenomena can be deduced from abstract principles directly intuited. From the empiricist's viewpoint, on the contrary, particulars alone are directly apprehended, and abstract principles must be painstakingly approached through inductive processes.

One further aspect of Descartes's variant of the resolutive—compositive technique demands consideration. The presumption that all complex phenomena are reducible to simple, self-explana-tory ones whose logical combination obviously "accounts for" the former is at least gratuitous and may be false. But the presumption of resolvability is a powerful tactical device, one that reinforced and was reinforced by the other key tendency within Cartesianism: that toward "mechanical" explanations of phenomena.

The link between Descartes's reductive version of the analytic method and mechanical explanations was particularly well articu-lated by Benedictus Spinoza in *A Short Treatise of God, Man, and His Well Being* (ca. 1662). Here Spinoza claims that anything

> composed of different parts, must be such that the parts thereof, taken separately, can be conceived and understood without one another. Take, for instance, a clock which is composed of many dif-ferent wheels, cords, and other things; in it, I say each wheel, cord, etc. can be conceived and understood separately, without the com-posite whole being necessary thereto.[14]

No claim could be farther from the guiding principles of Aristote-lian analysis, in which neither the formal nor the final causes of living entities could be given meaning independent of the whole or-ganisms to which they were applied. In fact, the Cartesian analytic approach to knowledge (as opposed to earlier Aristotelian and Hermetic approaches or to later organic or holistic approaches) limits itself by the presumption that knowledge must be somehow atomic—isolatable into small unitary and self-contained chunks. As Spinoza's illustration suggests, mechanical systems are among the clearest examples of complex entities created out of simple, separately meaningful elements. Whether *all* complex phenomena or objects are of that kind remains an open question even today. Many philosophers and respectable scientists continue to insist that some parts derive meaning only from their place in a larger whole; but proponents of all versions of mechanical philosophy in-sist that all wholes are of the kind described by Spinoza above.

THE MECHANICAL PHILOSOPHY

When Descartes came to develop an explanation of natural phenomena grounded in his newly articulated method of philosophizing, the physics he created was mechanistic. That is, (1) it sought to understand all phenomena in terms of particles of matter in motion, and (2) it insisted that the only way that the motion of any particle could be changed was by direct contact with some other particle. In the words of the German philosopher and mathematician Gottfried W. Leibniz, "a body is never moved naturally except by another body which touches it and pushes it; after that it continues until it is prevented by another body which touches it. Any other kind of operation on bodies is either miraculous or imaginary."[15] Any explanatory scheme that shares the demand that phenomena be explained by and *only* by particle impacts is said to be mechanical. Any which does not cannot be mechanical though it might well be materialistic.

According to Descartes, matter is defined by its spatial extension. That which is extended is matter. Given this definition, there can be no space which is not filled with matter because there is no space without extension. It thus follows that when any identifiable particle moves, it must set in motion a closed loop or vortex of matter of some kind. To see how such vortical motions could be used to explain ordinary natural phenomena, we shall briefly consider Descartes's explanation of how bits of iron are oriented in the presence of a magnet.[16]

A magnet gives off closed streams of tiny screw-shaped particles (see fig. 2). These move through space, spinning as they advance, much like rifle bullets. Iron contains small spiral pores or tubes filled only with a subtle, easily moved matter and shaped so that the magnetic particles can pass through when the pores and particles are properly aligned. When a piece of iron is introduced into a region in which a loop of magnetic particles is flowing, the particles impinge on the openings of the pores in the iron, rotating the iron until the magnetic particles can flow through without impediment.

Although Descartes was probably the most influential of the seventeenth-century mechanical philosophers, mechanical philosophy originated outside the Cartesian school and extended far beyond it—even into the enemy empiricist camps of Bacon and Gassendi and their adherents. How and why did one kind of theory become so pervasive in natural philosophy that its principles should have seemed intuitive to those who demanded intuition and induced from experience to those who demanded empirical evidence for their explanatory premises? The seventeenth-century vogue for mechanical explanations had two clearly identifiable kinds of cause. One, a revival of ancient atomic theories, was linked most

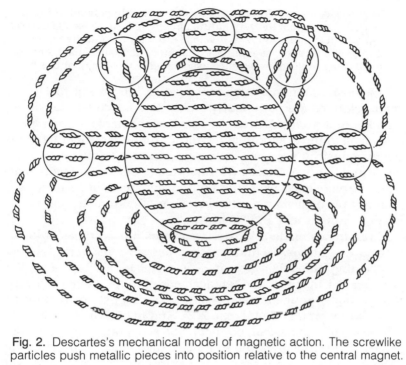

Fig. 2. Descartes's mechanical model of magnetic action. The screwlike particles push metallic pieces into position relative to the central magnet.

directly with the humanistic intellectual elite.[17] The second and vastly more important cause was artifactual. It resulted from a growing general familiarity with mechanical devices produced by engineers, clockmakers, and creators of automata.

In order to appreciate the impact of mechanical devices upon the early modern mind, we must briefly comment on the use of metaphor and analogy in the creation of theoretic schemes. The African cultural anthropologist Robin Horton has addressed this issue in a way that is particularly illuminating:

> In evolving a theoretical scheme, the human mind seems constrained to draw inspiration from analogy between the puzzling observations to be explained and certain already familiar phenomena. . . . In the genesis of a typical theory the drawing of an analogy between the familiar and the unfamiliar is followed by making a model in which something akin to the familiar is postulated as the reality underlying the unfamiliar. . . . What do we mean here by "familiar phenomena"? Above all, we mean phenomena strongly associated in the mind of the observer with order and regularity.

> That theory should depend on analogy with things familiar in this sense follows from the very nature of explanation. Since the overriding aim of explanation is to disclose order and regularity underlying apparent chaos, the search for explanatory analogies must tend toward those areas of experience associated with such qualities.[18]

In most traditional African societies, according to Horton, human relations are particularly stable and hence familiar in his sense. For that reason Africans sought their explanations of natural phenomena anthropomorphically and in highly personal terms. In modern industrial societies, however, personal relations are more chaotic and our experiences with inanimate objects are more regular and stable; hence, we pattern our explanations of social as well as natural phenomena upon the behavior of inanimate objects and seek to be impersonal and "objective."

Even in pre-modern Western societies during times of unstable social conditions, we find that theoretic schemes were grounded in something other than personal or social terms. In the ancient Near East, for instance, the replacement of anthropomorphic with astral accounts of phenomena occurred during periods of political and social instability (vol. I, chap. 2). And in classical Greece during a period of rapid social, political, and economic change, the pre-Socratic philosophers moved to replace theological with naturalistic theories to account for both the natural and social worlds (vol. I, chaps. 3 and 4). Once again in early modern Europe, a time of intense religious, social, and political instability,[19] there was a renewed tendency to seek theoretical explanations in impersonal terms. At this point, the Renaissance tradition of engineering and the earlier Medieval rise of mechanical technologies which had harnessed wind, water, and animal power for productive activities, provided a suggestive source of "familiar" analogs to draw upon. Mechanical devices of all sorts—grindstones, water wheels, locks of canals, pumps, and above all, clockworks, and the closely related automata (simulated persons or animals moved by elaborate hidden mechanisms of strings, springs, weights, rollers, and hydraulic devices)—suggested themselves as analogs for natural phenomena. Ancient automata had already been used to illuminate the principles of animal motion by Aristotle;[20] but the dramatic changes that led directly to the seventeenth-century dominance of mechanistic theories began in the fourteenth century in response to the creation of planetarium clocks—weight-driven mechanisms which combined the functions of ordinary time indicators with those of later orreries (models simulating the motions of the heavenly bodies).

As early as 1232, the Sultan of Damascus had presented to Frederick II of Hohenstaufen a clock that imitated the motions of the sun and the moon; by 1271, Robert the Englishman reports that throughout Europe "clockmakers are trying to make a wheel or disk, which will move exactly as the equinoctial circle does."[21] And about 1364, an Italian clockmaker, Giovanni de Dondi, created a magnificent mechanical clock that included seven movements simulating the motions of the sun, moon, and all five known planets, and reproducing their varying velocities and distances from the earth. In addition it boasted an ordinary hour dial and a dial of movable feasts which kept saints' days properly correlated with dates from year to year.[22] (See plate 1.)

Clocks like those of de Dondi provided metaphors for a variety of purposes. The French courtier Jean Froissart, for example, became one of the first authors to exploit the possibilities of the clock mechanism in a poem, *Li Orloge amoureus* ("The Clock of Love"), of 1369. The poem praises mechanical clocks, calling them "beautiful and noteworthy instruments, as pleasing as they are useful";[23] then it goes on to present a long and elaborate comparison of the attributes of courtly love to the movements in a weight-driven clock. John Huizinga has unkindly but perhaps appropriately called this poem the ultimate example of the decadence of Medieval symbolism.[24] Vastly more important was the "clock of wisdom" tradition which appeared in fourteenth- and fifteenth-century illuminated manuscripts. In this artistic and literary tradition, the harmonious arrangement of machinery and bells in a striking mechanical clock was called upon to proclaim the wisdom of God and to inspire devotion.[25] The "clock of wisdom" tradition appealed to the sense of familiarity discussed by Horton in justifying its metaphoric use of the clock; for it referred directly to the order, regularity, and harmonious characteristics of clockworks as those features which bring to mind God's wisdom.

The clock of love and the clock of wisdom illustrate that the clock was making its presence felt as a powerful metaphor during the fifteenth century; but it was in connection with astronomy that, in Lynn White's words, "the metaphor became a metaphysics."[26] Beginning in the thirteenth century, clock mechanisms had been designed to illustrate celestial motions. As clockworks and other

Pl. 1. Model of de Dondi's Astronomical Clock of 1364
(Smithsonian Institution)

mechanical devices became increasingly familiar and impressive, a conceptual inversion, or gestalt switch, took place; and scholars began to explain heavenly motions in terms of their better-understood mechanical analogs rather than vice versa.

Far and away the most important Medieval expression of the "clock-universe" analogy which dominated early seventeenth-century cosmology was that of Nicole Oresme in his *Livre du ciel et du Monde* (*Book of the Heavens and the Earth*) of 1370. This work was written shortly after Oresme returned from Italy where he had the opportunity to inspect de Dondi's planetarium clock. Speaking of a Christianized Aristotelian cosmology in which intelligences or angels were deputized by God to move the various heavenly spheres, Oresme said:

> When God created them [the heavens], he placed in them qualities and moving powers just as he put weight in terrestrial things and put in them resistances against the moving powers. . . . And these powers are so moderated, tempered, and ordered against their resistances that the movements are made without violence. But except for the violence it is similar when a man has made a clock and he lets it go and be moved by itself. Thus God let the heavens be moved continually according to the proportions that the moving powers have to their resistances and according to the established order.[27]

Here Oresme has exploited two elements of the clock-universe analogy which would become important during the seventeenth century: first, the notion that the universe resembles a mechanical artifact in its ability to run by itself once set in motion; second, the notion that such a universe must be the creation of a God who can be understood principally as a superior artisan or craftsman.

The clock-universe analogy did not play a major role in Renaissance astronomical theory, in part because planetarium clocks declined dramatically in sophistication after de Dondi until the work of Joost Burgi in the 1590s.[28] The nascent imagery was thus not sustained and enriched during this period. In addition, Copernican developments pre-empted other astronomical considerations during the later sixteenth century. But in the early seventeenth century, particularly in the works of Johannes Kepler, the mechanical philosophy, symbolized in the clock-universe, reentered cosmological theory to stay. Kepler had begun his scientific career as an opponent of mechanistic theories. In his *Mysterium Cosmographicum* of 1596, he expressed the animistic vision of the world which infused Hermetic and neo-Platonic theories—referring, for example, to the souls (*animae*) which move the planets.[29] But sometime before 1610, when he published the second edition of this work, he became a convert to the mechanical philosophy. Here he corrected his earlier references to souls: "if we substitute for the word soul the word force, then we get just the principle which underlies my new physics of the skies. . . . I have come to the conclusion that this force must be substantial."[30] Kepler is now concerned with inanimate rather than animate causes for celestial motions.

Kepler's new perspective began to develop around 1600 when he came into contact with William Gilbert's magnetic studies and began to think of magnetism as causing the motion of the moon around the earth.[31] Kepler's first unambiguous statement of commitment to a mechanical philosophy, however, appeared in a letter to a friend in 1605, and was phrased in terms of the clock-universe analogy:

> I am now much engaged in investigating physical causes. My goal is to show that the celestial machine is *not* the likeness of the divine being, but is the likeness of a clock (he who believes the clock is animate attributes the glory of the maker to the thing made). In this machine nearly all the variety of movements flow from one very simple magnetic force just as in a clock all the motions flow from a simple weight.[32]

This statement goes much further toward a fully developed mechanical philosophy than the influence of Gilbert would have warranted; for Gilbert never changed his view of world as animate, whereas Kepler insisted that to call the world alive was to detract from its artisan-like creator. When we ask what stimulated Kepler's transformation from neo-Platonic animist to mechanical theorist, we seem to come back to the impact of sophisticated planetarium clocks. The first European maker of planetarium clocks to rival de Dondi was Joost Burgi. Burgi was not only one of Europe's premier instrument makers, he was also the first-class observational astronomer who initiated the use of accurate clocks to measure right ascension of stars (the angular distance along the celestial equator). In 1592 Emperor Rudolph, whose court at Prague Kepler was to join in 1600, bought one of Burgi's masterpieces, a planetarium clock that simulated the motions of sun, moon, and the five planets.[33] Not only was Kepler familiar with the Burgi planetarium clock, but when Burgi entered Rudolph's employment, he and Kepler became close friends, observing together during 1604.[34] Later, Burgi even developed a form of logarithms to simplify the kinds of astronomical calculations that Kepler had to do. Given the timing of Kepler's conversion to the mechanical philosophy and the way he first phrased his commitment, it is almost certain that a principal stimulus was the "Emperor's mechanic," Burgi, and his creations.

Kepler's new mechanical theorizing had important and immediate implications for mathematical astronomy. Once Kepler recognized that physical forces emanating from heavenly bodies caused celestial motions, he saw that the lines of apsides of all planetary orbits (i.e., lines of bilateral symmetry connecting the points of least and greatest distance from the sun) must pass through the physical body of the sun and not through the empty center of the earth's orbit. This recognition not only forced a break with the mathematical details of the Copernican system but also played a key role in the mathematical theorizing that led Kepler to discover his famous laws of planetary motion: (1) that the planets move in ellipses with the sun at one focus, (2) that the line drawn from the sun to each planet sweeps out equal areas in equal times, and (3) that the periods of planetary motion, T_i, are proportional to $r_i^{3/2}$ where r_i is the mean distance from the planet i. These three laws, in turn, were indispensable ingredients in Isaac Newton's reformulation of theoretical astronomy in the late seventeenth century.[35]

Shortly after Kepler's revival of mechanical cosmological theorizing, mechanistic theories came to dominate accounts of astronomical phenomena. Descartes, for example, accounted for the phenomena of the solar system by positing that it was a huge whirlpool or vortex composed predominantly of small particles which have been turned into smooth spheres through their constant bumping and sliding over one another. In the interstices between particles of this material, which he called the "second element," was the extremely fine or subtle "first element" composed of the dust ground away from the second element during the process which made it spherical. Within this whirlpool of first and second matter float larger particles merged into large bodies called planets, just as clumps of debris are carried along in eddies of water. The large bodies tend to move outward as they are carried around; but because in a completely full universe one body can move outward only if another moves inward, the outward pressure of the planet is resisted by the pressure of the vortex particles which it is displacing. At some point, claimed Descartes, each planet will reach an equilibrium distance from the center where its outward tendency is just balanced by the outward tendency of the primary and secondary matter that it would have to displace to move outward. Thus the stable orbits of the planets are established. The whole universe is composed of indefinite numbers of similar vortices, each centered on its own sun. The Cartesian theory remained the dominant replacement for Aristotle's crystalline sphere cosmology for nearly fifty years after its full development in Descartes's *Principles of Philosophy* of 1644. It was compatible with the general structure of Copernican mathematical astronomy, and above all it appealed to an increasing sense of European familiarity with mechanisms.

Bernard Fontenelle's *Conversations on the Plurality of Worlds,* published in 1686, presented the vortex theory for a popular audience and was one of the most widely read secular works of the century. Here Fontenelle expressed clearly the sense of comfort that a mechanical theory of the universe seemed to offer. A philosopher strolls through a garden in the evening with a lovely countess; the conversation naturally turns to the moon and stars as part of the romantic setting; but then it veers off into astronomical theory:

> "I perceive," says I, "the world has become so mechanical that I fear we shall quickly become ashamed of it; they will have the world to be in large what a watch is in small, which is very regular, and depends only on the just disposing of the several parts of the movement. But pray tell me, madame, had you not formerly a more sublime idea of the universe? . . . Most people have the less esteem for it since that have pretended to know it."
> "I am not of their opinion," says she, "I value it more since I know it resembles a watch; and the more plain and easy the whole order of nature seems to be, the more to me it appears admirable."[36]

A similar attitude toward mechanical explanations also informed the empiricist tradition. In spite of the fact that in his mature works,[37] Bacon renounced the basic premises of mechanism and insisted upon the presence of "spirits" in bodies, most of his followers understood him to be a mechanical philosopher because in his early years he had supported the mechanistic atomic theories of Lucretius and Democritus—especially in *Cogitationes de Rerum Naturae* (ca. 1603) and *De Principiis atque Originibus* (ca. 1612). When Bacon died, these previously unpublished papers were turned over to Bacon's literary executor, William Boswell, who was particularly interested in the mechanical philosophy. Boswell published Bacon's atomic and mechanical writings separately as *Scripta in*

Natruali et Universali Philosophia in 1653, thus making Bacon seem a full-fledged mechanist.[38]

Whatever Bacon's attitudes might have been, the mechanical-corpuscular philosophy was made a central tenet of the empirical school by the chemical philosopher Robert Boyle. In *The Excellency and Grounds of the Mechanical Hypothesis* (1674), Boyle explained both what he understood the mechanical or corpuscular philosophy to be and why he felt it was particularly valuable. Boyle did not claim, as the ancient materialists had, that the mechanical philosophy could account for mental or spiritual phenomena. "The philosophy I plead for reaches but to things purely corporeal," he wrote, "and distinguishing between the first origin of things and the subsequent course of nature, teaches that God indeed gave motion to matter; but that, in the beginning, he so guided the various motions of the parts of it as to contrive them into the world he designed they should compose. . . ."[39] Once God has created the material universe, all the physical phenomena of the world are "physically produced by the mechanical properties of the parts of matter [operating] upon one another according to mechanical laws."[40]

According to Boyle the single greatest reason for accepting this philosophy is "the intelligentness or clearness of its principles and explanations."[41] All other forms of explanation involve principles that are confusing and about which people disagree, but, "as to the corpuscular philosophy, men do so easily understand one another's meaning, when they talk of local motion, rest, magnitude, shape, order, situation, and contexture of material substances; and these principles afford such clear accounts of those things that are rightly deduced for them alone; that, even such peripatetics [Aristotelians] or chemists [Hermetics] as maintain other principles, acquiesce in the explanations made by these, when they can be had; and seek no further. . . ."[42] That is, the "familiarity" of mechanical devices had rendered them almost totally unproblematic.

Among the exuberant disciples of Boyle, Henry Power best illustrates the tremendous late seventeenth-century enthusiasm for a combined empirical and mechanical philosophy:

> These are the days that must lay a new foundation of a more magnificent philosophy, never to be overthrown, that will empirically and sensibly canvass the phenomena of nature, deducing the causes of things from such originals in nature as we observe are producible by art, and the infallible demonstrations of mechanics.[43]

MECHANICAL BEASTS AND BODIES

It was the extension of the mechanical philosophy beyond cosmology, inanimate physics, and chemistry to problems associated with living beings that precipitated its great popular impact during the seventeenth century. The initial move in this direction came in experimental anatomy at the University of Padua. Here, Fabricius ab Aquapendante, principal teacher of William Harvey, analyzed the functioning of the valves in the veins mechanistically, ca. 1578:

> The Mechanism which nature has devised is strangely like that which artificial means has produced in the machinery of mills. Here mill-wrights put certain hindrances in the water's way so that a large quantity of it may be kept back and accumulated for the use of the milling machinery. These hindrances are called . . . sluices and dams . . . behind them collects in a suitable hollow, a large head of water, and finally all that is required. So nature works in just the same way in veins (which are just like the channels of rivers) by means of floodgates, either singly, or in pairs.[44]

Forty years later in his posthumous *De Motu Locali Animalium,* Fabricius was still exploiting mechanistic metaphors. The problems of animal motion formulated in this work were largely resolved only near the end of the century by Giovanni Borelli (1608–1679) using the new Galilean mechanics. A similar mechanistic approach was employed by Fabricius's most famous pupil, William Harvey, whose investigations of the motion of the heart and the blood depended upon viewing the heart as a mechanical pump, or "water bellows."[45]

Fruitful as the mechanical philosophy was in the form of "iatromechanics" (life-mechanics) in the hands of experimental physiologists like Fabricius, Harvey, John Lower, and Richard Mayow, it was used by these men in a limited way. They remained fully aware that to think that the operations of a human body were *like* the operations of a machine was *not* to think the human body identical with a machine. However, Descartes and his rationalist colleagues insisted upon the complete reduction of human and animal physiology to mechanics; and their presentation of mechanical physiology and mechanical animal psychology had a substantial impact upon general intellectual life, leading to virtually endless religious and political ramifications.

Just as Kepler's fascination with mechanical explanations can be linked to his experiences with Burgi's planetarium clock, so too can Descartes's initial fascination with mechanistic interpretations of animal and human behavior be traced to experiences with a particular set of automata. In 1614, at least five years before his sustained absorption with philosophical issues began, the young Descartes suffered the first of a series of minor nervous breakdowns and was sent for a two-year rest at the village of St. Germain-en-Laye near Paris. Here he visited the royal gardens where the Francini brothers had recently constructed a series of spectacular hydraulic and mechanical statues of men, animals, and birds that danced, sang, and moved in a variety of ways. Much later, in *L'Homme*, Descartes wrote about how he was startled and impressed when he unwittingly started these automata moving by stepping on plates hidden in the walkways. He even named his only child Francine after the creators of these mechanical marvels.[46] But immediately the youthful Descartes began to plan the construction of two animal automata—a flying pigeon and a pheasant hunted by a spaniel.[47] There is no evidence that Descartes actually constructed his planned automata, but there is evidence that sometime in the early 1620s he underwent the same kind of conceptual inversion that Kepler had experienced two decades earlier. His friend Isaak Beekman reported in 1631 that Descartes had been talking for several years about how animals could be understood as nothing but complicated automata.[48] During the early 1630s Descartes apparently developed detailed mechanistic explanations of physiology in works that were published only much later. In the *Discourse on Method* of 1637, the first published hints of Descartes's mechanical interpretation of man and beast appeared.

In Part V of the *Discourse*, Descartes presented a summary of the physiological discoveries he had made following his method. In particular he emphasized that external stimuli "cause the members of an [animal] body to move in as many different ways, and in a manner as suited whether to objects that are presented to its senses or to its internal affections as can take place in our own case apart from the guidance of the will." Descartes continued: "Nor will this appear at all strange to those who are acquainted with the variety of movements performed by the different automata, or moving machines fabricated by human industry, and that with the help of but a few pieces compared with the great multitude of bones, muscles, nerves, arteries, veins, and other parts that are found in the body of each animal."[49] Thus far, Descartes had done no more than appeal to automata as suggestive analogs to animals. In particular,

Pl. 2. Diagram of the gardens at Saint-Germain-en-Laye, showing the vogue
for mathematical regularity in gardens. It was in these gardens that
Descartes became fascinated with automata.
(Royal Library, Stockholm)

he had ruled out the possibility that animal responses to events come about as the result of the operation of some non-material sensitive soul that recognizes situations (though it may not be conscious of this recognition) and directs the animal to respond in certain ways. But Descartes went on to claim that people who have experience of automata "will look upon the body as a machine made by the hands of God, which is incomparably better warranted and adequate to movements more admirable than any machine of human invention."[50] To anyone who might have thought that animal behavior could not be accounted for without some special animal soul, he insisted that he had managed to show that "were there such machines exactly resembling in organs and outward form an ape or any other irrational animal, we could have no means of knowing that they were in any respect of a different nature from these animals."[51] Descartes had not argued so far that animals are automata, but he had insisted that, in so far as we can *know* about animal behavior, it is possible to account for it on the assumption that animals other than man are simply complex, God-created machines.

Man, of course, was different. Descartes argued that only irrational animals can adequately be mimicked by machines and that our own behavior is similar only to the extent that it is not guided by the will. Man alone has a rational soul capable of *thinking* rather than simply responding in a mechanical and unvarying way to stimuli. The first clear illustration of Descartes's distinction between human, rational responses and animal, mechanical responses to stimuli was given by Marin Mersenne in *Harmonie universelle* (1634). For an analysis of animal and human responses to musical sounds, Mersenne drew from conversations with Descartes and probably from the unpublished works which Descartes had summarized in Part V of the *Discourse on Method*. Concluding his section on hearing in man, Mersenne wrote:

> I say that the ear does not know sounds, and that it serves only as an instrument and an organ that serves to make them pass into the mind which considers their nature and their properties. Consequently, animals have no knowledge of these sounds, but only a representation without knowing if what they apprehend is a sound or a color, or some other thing; so one can say that they do not act so much as they are acted upon, and that objects make an impression on their senses such that what follows is necessary, just as it is necessary that the wheels of a clock follow the weight or spring which drives them. But man, having been touched by sounds, considers their nature and their properties, distinguishes them from other objects and forms a very certain knowledge of them.[52]

Like Aristotle, Descartes distinguished between the "lower" vegetative and sensitive souls which humans shared with other animals (vol. I, chap. 5) and the rational soul which was distinctively human. But he broke away from the Aristotelian tradition by insisting that the rational soul was non-extended, and therefore non-material, while both the body and lower souls were completely material—not simply *like* machines, but mechanisms pure and simple. *L'Homme,* published posthumously in 1662, concludes with an unambiguous statement of this position:

> the functions [of a human body] follow in a completely natural way from the disposition of the organs alone neither more nor less than the movements of a clock or other automaton flow from their counterweights and their wheels, in such a way that there is no occasion to conceive in it any other vegetative or sensitive soul, no other principle of movement or of life, than its blood and its fluids agitated by the heat of the fire in its heart, which has no other nature than those fires which are in inanimate bodies.[53]

By insisting that human bodies are mechanisms and that human intelligences or rational souls are totally separate and incorporeal entities, Descartes gave shape to one of the most fundamental problems of modern philosophy and one of the most pervasive features of the Western understanding of human nature into the present. If Descartes is correct, our world is irreducibly dualistic—there is *matter* and there is *mind,* two quite distinct realms of existence—and the great problem for those who want to make sense of our experience is to explain how the two realms can possibly interact. The problem is particularly acute because the body—like the Francini automata and the discarded physical universe of the ancient Gnostics (vol. I, chap. 2) is subject to totally determinate causal chains in all of its movements and actions. How then is it possible to talk of the will as initiating activity? Can the mind really have any effective power over the body? If the realm of intellect and consciousness is distinct from that of material body, how is it that matter can ever act upon the mind so that we can know about the material world? These were central questions for philosophers throughout the seventeenth and eighteenth centuries, and they remain fundamentally unresolved even today.

Some of Descartes's contemporaries, of course, simply refused to accept the notion that animals and human bodies were mere mechanisms without any special vital character. The *Discourse on Method* appeared in June 1637, and by September the professor of philosophy at the University of Louvain had already written to friends attacking it, fearing that "what Descartes says about the so called animal soul, others might say about the rational soul—i.e., that it might be mechanistically explained."[54] This fear soon became a major focus of opposition to Descartes. When the manuscript of Descartes's *Meditations on First Philosophy* was sent out for review in 1640, even though it made no mention of the human body-machine, one of the frequent objections made was that some were "likely to assert that man also is without sensation and understanding, and that all his actions can be effected by means of dynamical mechanisms and do not imply mind at all."[55] Such objections, though irrelevant to the arguments of the *Meditations,* were certainly legitimate results of the impact of Descartes's teachings in general; indeed, one of the most brilliant responses to the vexing problems posed by Descartes's dualism was that of Thomas Hobbes, who claimed to demonstrate that rational thought is nothing but motion in some internal substance of the head.[56] That is, Hobbes did exactly what Descartes's conservative critics had feared: he materialized even the mind.

In spite of Hobbes and the conservative critics, most continental intellectuals adopted positions at least positively influenced by Descartes or consistent with Cartesianism in three basic ways. First, they committed themselves to mechanistic physics and elements of iatro-mechanics. Second, they gave great weight to rationalistically grounded hypothetico-deductive schemes. Third, they agreed that there must be some non-material thinking substance. Physicists like Jacques Rohault and Christian Huyghens, metaphysicians like Spinoza, Nicolas Malebranche, and G. W. Leibniz, religious figures as divergent as the Port Royalist Antoine Arnold, the Cambridge Latitudinarian Henry More, and the Bishop of Meux, Jacques Boussuet, and literary figures like Fontenelle and Louis Racine all acknowledged their debt to Descartes, even though they often disagreed with him and with one another on particular issues.

PREVIEW OF CONCERNS ABOUT THE CULTURAL IMPACT OF THE MECHANICAL PHILOSOPHY

In *The Death of Nature: Women, Ecology, and the Scientific Revolution,* Carolyn Merchant expresses the view that the mechanical philosophy has had powerful and often undesirable implications for human behavior. According to Merchant, "the image of the earth as a living organism . . . had served as a cultural constraint, restricting the actions of human beings . . . . As long as the earth was considered to be alive and sensitive, it could be considered a breach of human ethical behavior to carry out destructive acts against it."[57] After the mechanical philosophers had transformed nature into a "system of dead corpuscles,"[58] however, those traditional constraints were removed. "Because nature was now viewed as a system of dead inert particles moved by external rather than inherent forces, the mechanical framework itself could legitimate the manipulation of nature."[59]

Turning to the implications of the new theory for society, Merchant points out that "organismic theory emphasized interdependence among the parts of the human body, subordination of individual to communal purposes . . . and vital life permeating the cosmos. . . . "[60] The mechanical philosophy, on the contrary, "had associated with it a framework of values based on power, fully compatible with the direction taken by commercial capitalism."[61] People were now perceived as independent and competing entities, fully understandable apart from their roles in the larger social order and interacting through the collision of individual interests. The fact that theory now legitimized self-interested activities and competition encouraged such behavior.

There is no doubt that such implications *could* be read out of the mechanical philosophy, but if we are to make the historical claim that it *did* significantly encourage exploitation of the earth and a more self-centered and competitive society, it is necessary to show not only that a handful of intellectuals recognized such implications but also that the behavior of a substantial group of people was affected by the claims of those intellectuals. It is not enough to show that mechanistic social and psychological theories adequately describe post-Renaissance behaviors; it is necessary to establish the plausibility of the claim that they have in some sense *pre*scribed those behaviors.

Strictly speaking, we may not be able to demonstrate that the switch to a mechanical view of the world *caused* less constrained, less careful, and less sensitive treatment of the natural environment or more impersonal treatment of human beings. What we can do is to show that people were aware of the prescriptive implications of their theories and that their behavior was consistent with them.

Because theories do provide explanations of behavior, they may be used to rationalize or excuse behavior undertaken for reasons quite different from those given by the theories. The very fact that a theoretical framework can effectively excuse behavior and increase its social acceptability may make that theory a positive factor in increasing the frequency of that behavior. The Cartesian animal-machine theory provides a fascinating case study of how the new mechanical philosophy really did seem to influence the behavior of a group of its adherents in a direction like that suggested by Merchant.

If one believes that animals react mechanically to stimuli and do not feel grief and love or suffer pain, then any restraints against harsh treatment of animals grounded in empathy or sympathy are removed. Descartes and adherents to his strict form of animal mechanism clearly did believe that animals had no sensations or emotions and that this "fact" licensed uncaring attitudes toward them. Descartes, for example, wrote to Henry More that the mechanical understanding of animals was not so much cruel to animals as liberating to men, who could now kill and eat animals without feeling guilty.[62] According to Nicolas Malebranche, anyone who believes that dogs really love their masters is guilty of making the dog "a little man with great eyes and four feet." Sainte-Beuve reports being shocked when Malebranche, who was widely admired as among the kindest of men, gave his dog a solid kick, telling Sainte-Beuve not to worry because the dog had no sensations.[63] Similarly, the Cartesian poet Louis Racine wrote of vivisection:

> Though to suffering eyes, the devoted dog
> Seems undeservedly nailed to a platform,
> Anatomical victim to cruel scholars.
> His exposed body proves to be only a piece of machinery,
> Which consoles the vivisectors as they contemplate,
> Delightedly following with their scalpels,
> The movements of a heart from which the passion is gone.[64]

That Cartesian theory was actually used to justify extreme behavior is also suggested by reports of the treatment of dogs at Port-Royal-des-Champs, a hotbed of Cartesian philosophy in the 1660s:

> They administered beatings to dogs with perfect indifference, and made fun of those who pitied the creatures as if they had felt pain. They said that the animals were only clocks; that the cries they emitted when struck were only the noise of a little spring which had

> been touched, but that the whole body was without feeling. They
> nailed poor animals up on boards by their four paws to vivisect
> them and see the circulation of the blood, which was a great subject
> of conversation.[65]

Such examples show that at least some individuals did use mechanistic theory to justify activities traditionally considered abhorrent.
Moreover, though vivisection had been practiced before the rise
of mechanical philosophy, it became a standard feature of iatro-
mechanical studies, whereas it was discouraged by the Jesuit supporters of the sensitive animal soul and by many empiricists—like
Samuel Johnson—who were not mechanists.

Most of those scientists who accepted the mechanical philosophy did not proceed to abuse their pets, but continued to treat
them affectionately as if they had both sensitive and limited rational capacities. It is a mistake to assume that acceptance of any
theory—by itself—*forces* one to act in any particular way. Almost
all of us have available a variety of explicit and implicit and often
contradictory theories which *may* be appealed to in any given situation. Our choice of theory may depend on institutional, psychological, and intellectual factors. Today, for example, in many religious
sects and in professional athletics, theories which imply sexual
equality are seldom if ever used; but in many businesses and educational institutions covered by affirmative action laws, they are
frequently—though not always—appealed to. When one is trying to
buoy up one's hopes about getting a job, one usually presumes that
decisions are made rationally and equitably; but when one is trying
to prepare oneself for failure, it is much easier and more comforting
to suppose that decisions are based on connections, conspiracies,
or irrelevant criteria. A physicist, fully aware of relativistic physics,
may choose classical physics for solving a particular problem because the latter is easier to work with and because the resulting solution will be adequate to his or her ends.

Thus the awareness and acceptance of a theory merely makes
available behavior which may be forbidden or discouraged within
the frameworks offered by common sense (which always involves
an implicit theoretic framework) or by other more formal theories.
When a particular theory is absorbed within the common sense
views of a society or when it is strongly identified with some particular institution, however, that theory will produce a preferred
range of behaviors, just as the seventeenth-century mechanical

philosophy encouraged students of medicine to experiment with vivisection.

THE SPIRIT OF GEOMETRY APPLIED TO POLITICS: THE WORKS OF THOMAS HOBBES

In the dedicatory letter to *De Corpore,* written about 1644, Thomas Hobbes laid out his claim to being the founder of a scientific approach to society:

> Galileus in our time . . . was the first that opened to us the gate of natural philosophy universal, which is the knowledge of the nature of motion. . . . The science of man's body, the most profitable part of natural science, was first discovered with admirable sagacity by our countryman, Doctor Harvey. Natural philosophy is therefore but young; but civil philosophy is yet much younger, as being no older . . . than my own *de Cive.*[66]

The theories of man and society which Hobbes produced were not widely accepted during the seventeenth and eighteenth centuries. They contained religious ideas so odd that they appealed to atheists, and political ideas so unusual that, in spite of Hobbes's conservative intentions, they were adopted only by radicals. Thus because they seemed to inspire both immorality and revolution, Hobbes's theories were generally feared and detested by all respectable persons. Yet these theories were so constructed that almost no political thinker of the next three centuries could ignore them or avoid the issues that they raised. In this sense the works of Hobbes, like those of Plato and Aristotle, were of vastly greater importance over the long run than their contemporary impact would have suggested. Indeed, his ideas are much more central to the political and social perceptions of ordinary people living in the mid-twentieth century than they were to those of the mid-seventeenth century.[67]

Another important sense in which Hobbes's works were like those of the ancient Greeks was that they appeared in a period of unusual social, religious, and political instability during which older understandings of social relations and institutions were becoming inadequate to cope with social conditions. The social and political transformations of the early modern period were remarkably similar to those of Hellenic Greece. Once again a society dominated by an aristocracy closely linked to the land and given its political meaning by its military role was being transformed by the growing importance of commerce. Virtues like obedience, observance of one's proper station, valor, and cooperation—virtues closely associated with military discipline and emphasized in hierarchical social theories—were becoming irrelevant to most men living in a society dominated by highly competitive international trade, by rapid urbanization, and by growing numbers of merchants, tradesmen, and craftsmen who were beginning to view themselves as equal or superior to "gentlemen" of the traditional landed nobility. As in classical Greece, these changes led to civil discord and sometimes to violent conflict. Throughout Europe, the seventeenth century was a time of nearly constant turmoil—of religious and dynastic warfare, of trade wars, and, in England, of protracted civil wars involving religious, social, and political strife. Conflict and instability were thus preeminent facts of seventeenth-century life, and they had a major bearing upon the formulation of Hobbes's social theories and upon their assimilation into Western political consciousness.

Our central concern with Hobbes will be the degree to which his political theory depends upon scientific ideas and methods. It is important to recognize, however, that his ideas eventually emerged as politically important because they served so well to acknowledge the central role of conflict and competition in social and political life and to celebrate the rise of an increasingly commercialized society. They both explained developments that had not been considered central to society and politics before and tended to authorize and legitimize those very developments.

Hobbes was born in 1588, son of a poor and all but illiterate vicar. His father died when he was young, and his wealthy merchant uncle provided for his education, sending him to Oxford in 1603. There, like many of his contemporaries, Hobbes both absorbed a great deal of Aristotelian learning and came to despise it. When he left Oxford in 1608, he became a tutor, companion, and traveling secretary in the family of William Cavendish, in whose service he remained with a few brief interruptions (one spent as amanuensis to Francis Bacon) until his death in 1679.

As a young scholar Hobbes was cut much from the mold of such early Italian civic humanists as Coluccio Salutati. His principal concerns were certainly politics, law, and the state rather than natural philosophy. Moreover, he had picked up enough nominalist and skeptical ideas to be absolutely convinced that a science of natural philosophy like that sought by Aristotle—i.e., a knowledge of the necessary relations among natural phenomena—was impossible, though he continued to believe that wisdom regarding human affairs was attainable. Hobbes would have agreed with Salutati that "[human affairs] have their origin not in external things, but in us. . . . Thus we know them with such a certainty that they cannot escape us and that it is not necessary to seek them among external facts. . . . They contain man's natural reason which every sound intelligence can understand by reflection and discussion."[68] Wisdom regarding human affairs was badly needed in Hobbes's day as King and Parliament headed toward confrontation, and Hobbes responded in typical humanist fashion by translating Thucydides's *History of the Peloponnesian Wars* (1629).

Hobbes's position regarding primacy of politics and society never changed; nor did his nominalist position regarding natural science. But within a few months after his translation of Thucydides appeared, Hobbes's intellectual career took a dramatic turn. He discovered Euclidean geometry, and he was entranced. Coming upon a copy of Euclid's *Elements* lying on a table at a friend's home, he was amazed that the improbable theorem he first saw could be rigorously derived from Euclid's initial definitions and postulates.[69] Immediately he began to view mathematical systems and mathematical reasoning as a guiding model for *all* fields of thought.

In the *Elements of Law*, begun during 1631, Hobbes placed all previous philosophers into one of two camps: the *dogmatici,* who had failed for two thousand years to establish any effective moral or political philosophy, and the *mathematici,* who were able to produce a system of knowledge which was both certain and useful because "they proceed from most low and humble principles, evident even to the meanest capacity; going on slowly, and with most *scrupulous ratiocinations.*"[70] A few years later, in *De Cive* (1642), he continued to insist that

> whatsoever assistance doth accrue to the life of man, . . . whatsoever things they are in which this present age doth differ from the rude simpleness of antiquity, we must acknowledge to be a debt which we owe merely to geometry; . . . if the moral philosophers had as happily discharged their duty, I know not what could have been added by human industry to the completion of that happiness which is consistent with human life.[71]

Within about a decade of his introduction to geometry, Hobbes abandoned his initial belief that the first principles of Euclid's geometry were self-evident; but he did not abandon his belief in the truth of geometrical propositions. His new commitment to geometry fit particularly well with his older nominalist leanings. Now geometrical propositions were seen to be true because they were rigorously deducible from a set of fundamentally *arbitrary,* but self-consistent, initial definitions. Whether such terms as point, line, plane, angle, and triangle signify objects which exist in some external "real" world may be problematic; but the relations between these terms established in geometry are true, regardless, for they are consequences of definitions which are agreed upon by all who use them. This notion of the *conventional* character of truth and of "scientific" systems of thought remained a central feature of most of Hobbes's subsequent philosophizing. As he wrote at the beginning of *Leviathan* (1651), his greatest political work:

> Seeing that *truth* consisteth in the right ordering of names in our affirmations, a man that seeketh *precise* truth, had need to remember what every name he uses stands for; and to place it accordingly; or else he will find himself entangled in words, as a bird in lime-twiggs; the more he struggles, the more belimed. And therefore, in Geometry (which is the only science that it hath pleased God hitherto to bestow on mankind) men begin at settling the significations they call *Definitions*; and place them in the beginning of their reckoning.[72]

The fact that geometry or any other science with hypothetico-deductive form produces statements which are true only by convention does not mean for Hobbes that all arbitrary deductive systems are equally interesting or that they are irrelevant to the phenomenal world. Indeed, only sciences whose terms are ultimately relatable to sensory experience are of value to him. The means of connecting experience and science lay for Hobbes in a revised and reversed method of resolution and composition. According to Hobbes, the compositive stage precedes the resolutive. Initially, "the truth of the first principles is *made* and *constituted* by ourselves, whilst we consent and agree about the appellations of things."[73] Then we find out "by the appearances or effects of nature, which we know by sense, some ways in which they may be, I do not say they are, generated."[74] The relationships discovered in this resolutive stage are not made by us but are placed in natural objects by the Author of Nature and are merely observed by us. If we can show that some range of natural phenomena is resolvable in such a way that all of the particular experienced relationships can be understood as special cases of the universal principles relating terms of a science that we have *composed,* then we can say that the theory explains our experiences— even though we cannot be sure that it is the only theory that could do so—and we have accomplished the most that is possible in attaining knowledge of non-human nature. That is, we have discovered a system of true statements which provide a *possible* explanation of a range of experiences.

Hobbes believed that the sciences of man and society could, like the science of inanimate natural bodies, be constructed on the geometrical or hypothetico-deductive model. During the early 1630s he decided to produce a grand scientific system in three parts. The first part would cover traditional physics; the second, man and his special characteristics; the third, "Civil governments and the duties of subjects."[75] Moreover, the conclusions of the first part would provide the basic definitions and postulates of the second, producing a single unified science encompassing *all* phenomena and grounded in the basic definitions of natural philosophy. This plan was finally realized to Hobbes's satisfaction with the publication of *De Corpore* in 1655; but in the meantime Hobbes produced the third, "political," part before he had completed the first two. England was moving rapidly toward civil war, and as Hobbes later wrote, "my country some few years before the civil war did rage, was boiling hot with questions concerning the rights of dominion and the obedience due from subjects, the true forerunners of an approaching war; and this was the cause which ripened and plucked from me this third part. Therefore, what was last in order, is yet come forth first in time."[76]

Though the decision to write the third part before the first two is easy to understand in terms of practical goals, it raises serious questions about the relationship between the three parts. How could Hobbes know what the basic postulates of the third part would be before the first and second parts were done? Hobbes gave two different kinds of answers to this question—one in *De Cive,* when he was still partial to a self-evident theory regarding first principles, and one in *Leviathan* after he had become recommitted to the radical nominalist position discussed above. Both answers have created doubts that Hobbes's political philosophy really depends upon his natural philosophy and his theory of human psychology at all.[77] Since it is my primary aim to argue that crucial elements of Hobbes's political thought *do* depend upon the methods, attitudes, and concepts of mathematics and natural science, it is important to confront Hobbes's own statements about how the science of politics *could* at once precede and be related to natural philosophy.

In *De Cive* Hobbes wrote that he could justify his immediate movement to a science of society because "grounded on its own principles sufficiently known by experience it would not stand in need of the former sections."[78] Although the assumptions of political philosophy could in principle be derived from the conclusions of a science of man, those conclusions can be known *prior* to their derivation. Like Salutati and other humanists, then, Hobbes still distinguished in 1642 between sensory experience of a supposedly external nature which could provide only contingent and particular knowledge and introspective "experience" which could generate the self-evident and universal principles needed for constructing genuine sciences of man and society. Even if these principles could, in theory, also be derived from natural philosophy, they need not be. From this point of view, though natural science may well be consistent with a science of man and a science of society, it is certainly not a logical prerequisite for them.

In *Leviathan*, Hobbes's response to how a science of politics might come into existence before the sciences of nature and man is more closely connected with his later nominalist and conventionalist notion of geometry. From this perspective "science is allowed to men through . . . *a-priori* demonstration . . . of those things whose generation depends on the will of men themselves."[79] The state, like Euclid's geometry, is an artificial creation of man:

> By *art* is created that great Leviathan called Commonwealth or State (in Latin, *Civitas*) which is but an artificial man . . . the matter thereof and the artificer; both of which is man.[80]

Since the state is a human creation, it can be known and understood by merely establishing those conventions or "pacts" and "covenants" which created it and which, "set together, and united, resemble that *fiat,* or the 'Let us make man,' pronounced by God in the creation."[81] Stated later in a slightly different form: "The skill of making, and maintaining commonwealths, consisteth in certain Rules, as doth Arithmetic and Geometry; not (as in Tennis-play) in practice. . . . "[82] Under these circumstances not only is a science of politics logically independent of natural philosophy, but is also logically independent of any science of man, the conclusions of which need not be presumed in order to *construct* the state or a science of the state. This *logical* independence does not, of course, mean that Hobbes's political or psychological assumptions did not owe a great deal to his understanding of natural science. As a matter of historical fact, they did.

Whether considered independently or as part of a single deductive system, Hobbes's theories of nature, man, and society clearly derived their form from a Hobbesian version of the Cartesian *esprit géométrique.* But Hobbes demonstrated his enthusiasm for mathematics in ways that directly influenced the content as well as the form of his theories of man and society. Perhaps the most important example of such an influence is related to Hobbes's definition of "reason," which is crucial for both psychological and political concerns. In *Leviathan* Hobbes wrote:

> When a man reasoneth, he does nothing else but conceive a sum total, from addition of parcels; or conceive a remainder, from subtraction of one sum from another . . . . These operations are not incident to numbers only, but to all manner of things that can be added together, and taken out of one another. For as arithmeticians teach to add and subtract in numbers; so the geometricians teach the same in lines, figures (solid and superficial), angles, proportions, times, degrees of swiftness, force, power, and the like. . . . Writers of politics add together pactions to find men's duties; and lawyers, laws and facts, to find out what is right and wrong in the actions of private men. In sum, in what matter soever there is place for addition and subtraction, there also is place for reason; and where these have no place, there reason has nothing to do.[83]

This computational definition of reason, which emerges as a key by-product of the Hobbesian geometrical spirit, is a far cry from Plato's notion of reason as ruler over man's passion, or from the reason of Clement of Alexandria which could somehow grasp God's hidden messages to man, or from the reason of the Hermetic magi which was understood to be capable of discovering the occult virtues of things through signatures or sympathies. It is, rather, a very important ancestor of our modern notion of reason as purely instrumental—a kind of computer which can only serve, not rule, nonrational ends.

Recent work in cognitive psychology strongly suggests that human brains do *not* function to organize experience solely through processes like the computational activities of digital computers.[84] But the extremely limited concept of reason introduced into our culture by Hobbes has frequently been used by those who would establish criteria for the "rationality" of our beliefs and actions. Moreover, this notion of rationality has become normative as well as descriptive. What does not conform to the canons of mathematical reason is often seen as both irrational and valueless.

Once one assigns unique value to calculative rationality, there is a tendency to assign special significance and value to quantification. Hobbes, for example, is among the earliest theorists to assign value to a human being solely in terms of the price his labor might bring: "The value, or worth of a man, is, as of all other *things,* his price, that is to say, so much as would be given for the use of his power. . . . "[85] It may well be the case that the treatment of men as things whose value can be set in terms of their productivity owes much more to the early modern institution of mercenary standing armies and to the growth of wage labor in the service of commercial capital than to the musings of theorists like Hobbes or the political economists who followed him in the seventeenth century. Nonetheless, there can be little question that the mental sets of such mathematician-moralists was such that they actively welcomed the chance to quantify and objectify human behavior. In the words of Alexandre Koyre, they sought to expunge "all considerations based on [traditional notions of] value, perfection, harmony, meaning, and aim, because these concepts, from now on *merely subjective,* cannot have a place in the new ontology."[86] Furthermore, as we shall see in detail later, these authors fought to get their ideas used in formulating public policies; so they legitimized the increasing tendency to treat men as mere objects. Whether or not they initiated the trend, they certainly encouraged it.

Computational reason and the mathematical structure of thought which it implies result both in a devaluation of *poetic* expression as a legitimate guide to human understanding and in a more generalized attack upon "imagination" and the worth of analogy and metaphor. Those who believe that "truth" can only be a property of propositions arrived at by the logical manipulation of terms whose meaning derives solely from their unambiguous definitions must deny validity of thought processes that give credence to arguments dependent on analogical or metaphoric relationships. Hobbes was by no means the first author to point this out, nor did his explicit analysis and rejection of metaphoric language mean that he never used analogical insights. Indeed, the introduction to *Leviathan* constitutes a powerful expression of the macrocosm-microcosm analogy modified to fit the mechanical philosophy.

In spite of the fact that Hobbes used analogies and metaphors extensively, his insistence that the use of "metaphors, tropes, and other rhetorical figures in the search for *truth* can lead only to absurdity"[87] expressed an important theme associated with the new seventeenth-century notion of reason. And it provided one important source of seventeenth-century criticism of traditional literary forms. In his discussion of intellectual virtues and defects in the *Leviathan,* while he acknowledged the potential entertainment value of poetry, Hobbes insisted that Fancy, which involves the recognition of similitude and which is often expressed in metaphor, can very easily become a form of madness when it is not carefully controlled[88] by judgment. John Locke, borrowing heavily from Hobbes on this issue, also argued that while poetry offers "pleasant pictures and agreeable visions," these are often "not perfectly conformable" to truth.[89] Consequently in his *Thoughts on Education* Locke insisted that parents of a child with poetic tendency "should labor to have it stifled and suppressed. . . ."[90] Even poets of the later seventeenth century tended to accept the Hobbesian judgment. Thus John Dryden writes that "imagination in a poet is a faculty so wild and lawless, that, like a high-ranging spaniel, it must have clogs tied to it, lest it outrun the judgment."[91] Though such attitudes did not kill poetry in the seventeenth century, the literary historian Basil Willey insists that with few exceptions poetry lost its traditional sense of deep seriousness and became a mere ornament.[92]

By the eighteenth century, even though his political and religious thoughts were still highly suspect, the Hobbesian notion of reason had become widespread among European intellectuals. David Hume illustrates how tyrannical the concept of calculative reason could become. He concludes his *Essay Concerning Human Understanding* (1749) with the assertion that we should ask of any book whether it contains either a discussion of facts or "any abstract reasoning concerning quantity or number," and tells us that if it does not, we should "commit it then to the flames: for it can contain nothing but sophistry and illusion."[93]

THE MECHANICAL PHILOSOPHY APPLIED TO MAN AND STATE

We have seen that Hobbes's encounter with Euclidean geometry led him to cast his political thought in a special hypothetico-deductive form and to redefine the concept of reason in a peculiarly restricted way that has had a major role in shaping subsequent Western thought. In addition, during the early 1630s Hobbes hoped to develop a unified science of nature, man, and society in which the "higher" sciences would depend for their principles on the conclusions of the "lower" ones. Political circumstances led Hobbes to produce his political theory first and forced him to claim that while his political philosophy would be consistent with his natural philosophy, it was not logically dependent upon it. In spite of their supposed logical independence, I claimed that Hobbesian assumptions about man and society *did* owe a large debt to his assumptions derived from natural philosophy. In this section we return to assess the historical relation between Hobbes's physics, psychology, and politics.

In 1630 Hobbes left England for the continent with the youngest son of William Cavendish. During this trip, he added not only geometry but also the mechanical philosophy to his central interests. Subsequently he sought out and spoke with Galileo, whose studies of motion Hobbes counted as the true beginning of natural science. At Paris he began a lasting friendship with Marin Mersenne and acquaintances with Mersenne's distinguished circle of scientific friends, including René Descartes and the French reviver of Epicurean atomism, Pierre Gassendi. On returning to England, Hobbes formulated his plan to create his three-stage philosophical system and began to draft the first two stages—a physics based on the mechanistic assumption that all natural phenomena are produced by matter in motion, and a theory of human behavior based on a similarly complete mechanism without appeal to any spiritual entity distinguishing humans from animals. Though these first and second stages reached publishable form only in 1650 and 1655, three manuscript versions dating from between 1631 and 1637 are known;[94] so the basic elements of his mechanistic theories of nature and man were available to Hobbes around 1640 when he switched to consider the third section on society and the state.

We will basically ignore Hobbes's physics. With minor exceptions it was so similar to that of Descartes that the latter accused Hobbes of plagiarism in connection with his doctrine of colors,[95] leading to a coolness between the two great mathematical-mechanical philosophers until a partial reconciliation around 1648. On *human* nature, however, Hobbes went his own distinctively materialist direction along the lines more of Gassendi's renewal of ancient atomism than of Descartes's dualism.

Some consideration of Hobbes's theory of human nature is crucial as a prelude to understanding his politics because Hobbes, like Spinoza, insists that society be understood in terms of its constituent parts—namely, humans. Like Spinoza he justifies this demand in terms of the traditional mechanical analogy:

> For as in a watch, or some such engine, the matter, figure, and motion of the wheels cannot be well known, except it be taken in sunder, and viewed in parts; so to make a more curious search into the rights of states, and duties of subjects, it is necessary (I say not to take them in sunder, but yet that) they be so considered, as if they were dissolved, that is, that we rightly understand what the quality of human nature is, in what matter it is, in what not, fit to make up a civil government.[96]

In *De Cive* Hobbes laid out a bare-bones presentation of human nature in the first chapter because he felt under great pressure to get on with his political analysis, but in the more thorough *Leviathan* he devoted nearly as much space to Part I, "Of Man," as he did to Part II, "Of Commonwealth."

The starting point of Hobbes's theory of man lies in the idea of a "vital motion" that constitutes life itself. Though the theory that life consists of vital motions has a long history in atomist philosophy, Hobbes makes it clear that his use of the notion has a more immediate source in seventeenth-century science. This vital motion, he writes, "is the motion of the blood, perpetually circulating (as hath been shown from many infallible signs and marks by Dr. Harvey, the first observer to it) in the veins and arteries."[97] When the vital motion of the heart ceases, life ceases, so *every man seeks with all of his resources to maintain his own vital motions,* or life. All human behavior derives from this fundamental principle. To see how political behavior does so, we must follow Hobbes through his theory of sensation and into his ethical theory.

Men and women interact with their environment and come to know and interpret events only mechanically. External objects in motion strike or press on our organs of sense. In some senses (e.g., taste and touch) the impacts are direct, but in others (e.g., vision and hearing) the motions of distant bodies may be transmitted through some elastic medium. The motions imparted to each sense organ are then transmitted to the brain, where they produce appearances, ideas, conceptions, or thoughts—we call these sensations—which can be nothing but "some internal notion in the sentient, generated by some internal motion of the parts of the [external] object, and propagated through all the media to the innermost part of the organ [of sensation]."⁹⁸ The important thing to note here is that sensations are nothing but motions and that sounds, colors, and heat exist only as ideas—i.e., motions—in us and are not in the bodies we commonly speak of as having them. Such so-called secondary qualities must be distinguished from the primary qualities which bodies really do possess—i.e., extension, shape, hardness, and motion.

Some of the motions that constitute sensations remain slowly decaying in the brain. We call these memories; and their recollection allows for those comparisons and combinations with incoming sensations or with other memories that constitute reasoning and imagination. But some of the sensation motions proceed from the brain to the heart where they play an even more important role. They act either to reinforce or to diminish those vital motions which constitute life. The centrality of these motions to all human behavior is most vividly stated in Chapter 7 of the *Elements of Law*, though the same basic argument is repeated and elaborated in *Leviathan*, Chapter 6. Hobbes writes:

> [The motions which proceed to the heart] must either help or hinder
> that motion which is called vital; when it helpeth it is called *DE-
> LIGHT,* contentment, or pleasure, which is nothing really but mo-
> tion about the heart, as conception is nothing but motion about the
> head; and the objects which cause it are called pleasant or delight-
> ful . . . and the same delight with reference to the object is called
> *LOVE*; but such motion as weakeneth or hindereth the vital motion,
> then it is called *PAIN*; and in relation to that which causeth it,
> *HATRED.* . . . This motion, in which consisteth pleasure or pain is
> also a solicitation either to draw near to the thing that pleaseth, or
> to retire from the thing that displeaseth. And this solicitation is the
> endeavour or internal beginning of animal motion, which, when the
> object delighteth is called *APPETITE: and when it displeaseth is
> called AVERSION,* in respect of displeasure expected, *FEAR.* . . .
> Every man, for his own part calleth that which pleaseth and is de-
> lightful to himself, *GOOD,* and that *EVIL* which displeaseth him.⁹⁹

Pleasure and Pain (associated with an increase or decrease of vital motion) here become the source and the *only* source of human *passions.*¹⁰⁰ Humans act only out of passion; and traditional moral terms like good and evil take on meaning only in connection with an individual's desires and fears relating to certain stimuli. Reason does not, as it once presumably did, provide guidance regarding the ends or goals of human activity. It serves only to assess the most effective way to maximize pleasure and minimize pain.

Finally, Hobbes incorporates the ancient Epicurean assumption that pain avoidance, or fear, is a more powerful goad to action than desire. This means that human striving can never end. Each human must constantly and ceaselessly strive to increase security against the loss of vital motions and must compete with all other humans for the wherewithal to protect self-interests. Thus, writes Hobbes:

> I put for a general inclination of all mankind, a perpetual and restless desire of power, that ceaseth only in death. And the cause of this, is not always that a man hopes for a more intensive delight than he has already attained to; or that he cannot be content with a moderate power: but because he cannot assure the power and means to live well, which he hath present, without the acquisition of more. . . .[101]

The implications of the ideas expressed in the last few pages were revolutionary, extensive, and frightening to many. Hobbes's *Leviathan* and *De Homine* were the immediate sources of an ever flowing tradition of pleasure-pain psychology and ethics from the fountainhead of ancient atomism. This tradition, in turn, transformed educational theory and practice, giving rise to what we now call "liberal" political theory (see chap. 6). The central premise of Hobbesian politics lies in the claim that human beings are incessantly preoccupied with their own self-preservation. This claim follows logically from the psychological theory we have been discussing, but it does not necessarily presuppose it. Indeed, in *De Cive* Hobbes bypassed the psychology and simply stated:

> In the first place I set down for a principle by experience known to all men, and denied by none, to wit, that the dispositions of men are naturally such, that except they be restrained through fear of some coercive power, every man will distrust and dread each other, and as by natural right he may, so by necessity he will be forced, to make use of the strength he hath, toward the preservation of himself.[102]

All we can say with absolute certainty is that in Hobbes's unpublished writings the psychology preceded its egoistic political implications, though in the published works the focus on self-preservation precedes the full expression of the mechanistic psychology. Hobbesian egoistic political behavior is fully consistent with his mechanistic psychology, but it has a potentially independent source in introspective experience.

No one read Hobbes during the seventeenth and eighteenth centuries solely in connection with the *logic* of his argument. Especially in Part I of *Leviathan,* Hobbes offered not simply a new argument but also a new system of metaphors for the interpretation of humanity and its social institutions—a system of metaphors and symbols replacing the decaying microcosm-macrocosm analogies and exploiting the increasingly familiar and appealing images associated with the new mechanical philosophy. Hobbes's new central image was the mechanical body-in-motion whose meaning was totally independent of its context and which changed its course upon collision with other like bodies. Michael Walzer cleverly illuminates the implications of Hobbes's theory:

> The body-in-motion upon which he builds his system is a symbolic figure. It represents the individual human being but in a very special way: no longer is he a member of the body politic; no longer does he have a place in a hierarchical system of deference and authority; no longer do his movements conduce to universal harmony. Instead, the individual is alone, separated from his fellows, without a master or secure social place; his movements, determined by no one but himself clash with the movements of other, identical individuals; he acts out chaos.[103]

Once Hobbes has created his isolated, purely self-centered individuals, the fact that they are essentially identical becomes of paramount importance. In the absence of civil society, Hobbes insists that there can be no natural leader; for no individual is strong enough to protect himself and the goods he desires from the rapacity of others who might band together to rob him. A cruel balance of powerlessness and of unregulated competition, or a war of each against every other exists in the state of nature and makes life under such conditions intolerable.

Hobbes had no desire to foster an egalitarian or democratic society. His sympathies lay with monarchical government and with a hierarchically arranged social order. Yet his mechanistic system did make his theory appeal to lower-class radical egalitarians, and it forced those opposed both to Hobbes's conservative use of his symbols and to the radicals' acceptance of them to acknowledge the claim that, outside of the rules established to govern society, all men are equal. John Locke appropriated the claim for political liberalism in his *Second Treatise on Civil Government* (1688). He tried to justify it by arguing that the natural state of man is one of equality, "there being nothing more evident than creatures of the same species . . . promiscuously born to all the same advantages of nature, and the use of the same faculties, should also be equal, one amongst another without subordination or subjection."[104]

Even the most cursory knowledge of organic species demonstrates the falsity of Locke's justification, for social hierarchy is the norm among animals. Among atoms in motion, however, there can be no such hierarchy, and the *mechanical* notion of species' equality became enshrined in liberal phrases like that from the American *Declaration of Independence*: "We hold these truths to be self-evident, that all men are created equal. . . ." That men are born equally endowed with physical and mental abilities was ultimately abandoned in liberal political theory; but it left behind the crucial claim that they are born, nonetheless, "equal in respect of their rights."[105] Whatever other Christian sources it may have had, the egalitarian doctrine of natural rights was certainly encouraged by the mechanistic imagery through which it was incorporated into liberal political theory. Since Hobbes's mechanistic system was adopted not only by some of his conservative colleagues but also by radical theorists and by every member of the increasingly important liberal tradition, Walzer seems justified in making an additional claim on behalf of Hobbes: whether they agree with his conclusions or not, "For two hundred years there is hardly an English writer, hardly a coffee house conversationalist, who is not a successor to Hobbes."[106]

Given Hobbes's picture of individuals in a state of nature—one in which every human is engaged in a "war of all against all" and in which the life of each must be "solitary, poor, nasty, brutish, and short"—the basic need for a civil society becomes obvious. Humans must somehow arrange for their own security, and the only way to do so is to agree to create some entity capable of establishing and enforcing peaceful co-existence. For Hobbes, there seemed only one way to ensure that the state could be made strong enough to fulfill its basic peace-keeping obligation. All individuals had to renounce all of their natural claims, or "rights," except that of immediate physical self-defense, investing them in an absolute sovereign which was to have complete authority to create and embrace those social rules it deemed necessary. Just as Euclid defined the mathematical terms and fundamental laws to which all mathematicians must assent, the sovereign was to create the laws of civil society to which all citizens must assent.

To those who would object that placing unlimited power in the hands of a sovereign who might misuse it was unwise, Hobbes retorted,

> . . . the estate of man can never be without some incommodity or other; and the greatest, that in any form of Government can possibly happen to the people in general is scarce sensible, in respect of the miseries, and horrible calamities, that accompany a civil war: or that dissolute condition of masterless men, without subjection to Laws and a coercive power to tie their hands from rapine, and revenge . . . all men are by nature provided of notable multiplying glasses [their passions and self-love] through which, every little payment appeareth a great grievance; but are destitute of those perspective glasses [moral and civil science] to see afar off the miseries that hang over them, and [that] cannot be avoided without such payments.[107]

Subsequent liberal theorists, from Locke to the present, have sought to limit the power and authority of the state in order to reclaim as many of men's "natural rights" to do what they can get away with as is consistent with mutual self-preservation. But all have accepted the Hobbesian premise that the state comes into existence as a necessary evil which must be endured solely in order to ensure protection of themselves and their private acquisitions. This premise is closely tied to the view of human beings as naturally undifferentiated and disassociated bodies-in-motion—a view totally at odds with the hierarchical cosmologies of the pre-modern Western world.

The retention of natural rights within civil society is usually associated with the terms *Liberty* or *Freedom*—political notions that Hobbes defines explicitly in terms of his mechanical worldview:

> Liberty or freedom signifieth (properly) the absence of Opposition (by opposition, I mean external impediments of motion) . . . whatsoever is so tied, or environed, as it cannot move, but within a certain space, which space is determined by the opposition of some external body, we say it hath not liberty to go further. And so of all living creatures, whilst they are imprisoned, or restrained, with walls or chains; and of water, whilst it is kept in banks, or vessels . . . we say they are not at liberty. . . . *But when the words free and liberty are applied to anything but bodies, they are abused; for that which is not subject to motion, is not subject to impediment.*[108]

When we use the term *freedom* in connection with human beings, we can thus mean only that there is no external *physical* impediment that blocks men and women from what they have a will to do.

Contrary to older political and religious usages that defined liberty and freedom as being unregulated by human or divine law, Hobbes suggests that outside of the system created by civil law people are essentially not free because they can be blocked from their desire by the arbitrary and capricious behavior of their competitors. Within civil society, however, people are doubly free or at liberty. In one sense, they are free even to disobey the law:

> So a man sometimes pays his debt, only for fear of imprisonment, which because no body hindered him from detaining; was the action of a man at liberty. . . . And generally all actions which men do in commonwealths for fear of the law, [are] actions which the doers had liberty to omit.[109]

From this point of view, freedom exists under law because laws normally punish only past transgressions. Hobbes has no tolerance for those in civil society who clamor for greater liberties because "if we take liberty, in the proper sense, for corporal liberty; that is to say freedom from chains and prisons,"[110] all men living under any system of laws manifestly have it. Hobbes presumes that they will be incarcerated only for willfully and "freely" choosing to disobey the law.

For Hobbes there is a second and perhaps even more important sense in which civil society creates a special and secure domain of liberty and freedom: "In cases where the sovereign has prescribed no rule, there the subject hath liberty to do, or forbear according to his own discretion."[111] Having been made safe, humans will be allowed by the sovereign liberty, "to buy, sell, and otherwise contract with one another, to choose their own abode, their own diet, their own trade of life, and instruct their children as they see fit; etc."[112] Much of the post-Hobbesian liberal tradition focuses its attention on assuring that this largely commercial domain of unrestricted—and therefore free—human activity remain as large as possible. The Hobbesian and liberal notion of liberty forces abandonment of any attempt to view civil society as anything but a vehicle for creating a secure setting for private—egoistic—activity. The body-in-motion view of man offers only this negative interpretation of the state. It does not direct us positively toward the good life as Classical and Christian political theories had; it merely saves us from the presumed disastrous consequences of anarchy, leaving us free to do nothing but continue our competitive interactions with others.

The Hobbesian emphases on mechanistic imagery, quantitative analysis, and rigorously reasoned arguments not based on the authority of Scripture or classical precedent gradually became the norm in discussions of man and society. They certainly produced a changed understanding of society. Now society was to be analyzed in terms of individualistic rather than communal aims and competitive rather than cooperative social interactions.

The change of focus from communal goals and cooperative interactions grounded in custom—i.e., from what sociologists call *Gemeinschaft*, or community—to individual goals and competitive interactions governed by contracts—i.e., to what sociologists call *Gesellschaft*, or society—characterizes the move from pre-modern to modern culture. More accurately, it characterizes the move from pre-modern to modern *understandings* of social and political institutions.

The shift from *Gemeinschaft* to *Gesellschaft* is usually associated with the rise of commercial capitalism and with a "market economy." As early as 1787, John Millar, the liberal social theorist and student of Adam Smith, expressed his dismay at the transformation of life associated with commercial societies and the "mercantile spirit":

> In a rude age, where there is little industry, or desire of accumulation, neighboring independent societies are apt to rob and plunder each other; but members of the same society are attracted by common interest, and are often strongly united in the bands of friendship and affection, by mutual exertions of benevolence, or by accidental habits of sympathy. But in a country where nobody is idle, and where every person is eager to augment his fortune, or to improve his circumstances, there occur innumerable competitions and rivalships which contract the heart, and set mankind at variance. In proportion as every man is attentive to his own advancement, he is vexed and tormented by every obstacle to his prosperity, and prompted to regard his competitors with envy, resentment, and other malignant passions.
>
> The pursuit of riches becomes a scramble in which the hand of every man is against every other. That there is no friendship in trade is an established maxim among traders. Every man for himself, and God Almighty for us all, is their fundamental doctrine.[113]

It is but a short step from this picture, with its obvious debt to the Hobbesian notion of the warfare of all against all, to Marx's famous thumbnail sketch of change in society brought about by the rise of the bourgeoisie:

> The bourgeoisie, wherever it has got the upper hand, has put an end to all feudal, patriarchal, idyllic relations. It has pitilessly torn asunder the motley feudal ties that bind man to his "natural superiors," and has left remaining no other nexus between man and man than naked self interest, callous "cash payment." It has drowned the mostly heavenly ecstasies of religious fervor, of chivalrous enthusiasms, of philistine sentimentalism, in the icy water of egotistical calculation. It has resolved personal worth into exchange value, and in place of the numberless indefeasible chartered freedoms, has set up that single unconscionable freedom—Free Trade.[114]

Within the Millar-Marxian perspective, social reality was transformed in the sixteenth century, and then liberal political ideology was rapidly created to mirror and rationalize the transformation. Hobbesian political theory is thus nothing but an immediate and almost automatic product of bourgeois self-consciousness which in turn is an immediate and automatic response to the existence of market-dominated society and wage labor. This simplistic view has been challenged by Alan Macfarlane and others, who have demonstrated that English life had, in fact, seen the economic dominance of private property rights, wage labor, production for markets rather than subsistence, and contractual rather than customary relations from the late thirteenth century.[115] Yet for nearly three hundred years, no supportive bourgeois consciousness and political ideology had emerged. Only after Hobbes had introduced a *new* conceptual apparatus for interpreting social actions as grounded solely in self-interest and for valuing men solely in terms of their price—a conceptual apparatus drawn from the new mathematical and mechanical natural philosophy of the seventeenth century—was bourgeois consciousness *or* liberal political ideology possible. There can be no doubt that the Hobbesian elements in subsequent liberal political thought would have been stillborn if they had not served the needs of a growing commercial middle class that sought personal profit and that detested economic regulation through government monopolies. We might wonder, however, if the pace at which extreme self-interest and private acquisitiveness shifted from being viewed as a sign of social deviance to a sign of social normality during the seventeenth and eighteenth centuries was not accelerated by its emergence as a fundamental premise authorized by the materialist mechanical philosophy.

It was certainly the case that many of his contemporaries did view Hobbes as an active agent of social corruption that threatened social stability rather than as a neutral describer of a new social order. Though most early anti-Hobbesian polemics focused on the "atheistical" implications of his writings, social and political implications were almost inevitably linked to Hobbesian irreligion. Thus from Alexander Ross's broadside, first published in 1653, through the proclamation in 1663, that prohibited the reading of *Leviathan* at Oxford, the work was almost uniformly characterized as "a piece dangerous both to government and religion" and as "destructive of all government in church and state."[116] Even according to many proponents for the rising commercial classes, Hobbes encouraged blatant and rapacious greed, self-interest, and uncontrolled pursuit of profit, threatening to destroy all vestiges of Christian morality and social orderliness, and giving unfair advantage to the ungodly and amoral.[117] That is, even liberal thinkers felt that Hobbes encouraged unacceptable excesses in the operation of the emerging market society which they sought at once to defend and control.

Even more important for our purposes than the widespread seventeenth-century perception that Hobbes's argument was a real and active threat to social order as well as a goad to unbounded egoistic behavior was the almost equally pervasive belief that his social philosophy was a direct consequence of his mathematical mechanistic-materialist natural philosophy. Accordingly, the most effective way to controvert his political views would be to demonstrate the operation of immaterial, spiritual forces in the natural world[118] and to establish that the universe was the product of a providential plan rather than the result of the mere chance encounters of atoms. We shall return again to this theme in Chapter 3.

Empiricist Political Science in the Seventeenth Century   2

Hobbes produced what was—in the long run—the most powerful seventeenth-century attempt to provide an understanding of society and the state in terms of the new scientific attitudes and concepts. His reputation as a scientific social theorist, however, was eclipsed for at least half a century by two men who built upon his ideas, but rejected his monarchical tendencies and his almost exclusive emphasis on deductive systems. Both of these men showed much stronger empiricist tendencies than Hobbes. One became the outstanding republican theorist of the age, the other turned political theorizing away from the traditional concern with governmental forms toward an emphasis on policy analysis.

THE POLITICAL ANATOMY
OF JAMES HARRINGTON

Among the most fascinating and at least temporarily influential responses to Hobbes's *Leviathan* and the political crises that it addressed was James Harrington's *The Commonwealth of Oceana,* published in 1656. David Hume wrote in 1777, "[Oceana] is the only valuable model of a Commonwealth that has yet been offered to the public,"[1] comparing it favorably to Plato's *Republic* and More's *Utopia.* Moreover, after Montesquieu's *Esprit des lois* and Locke's *Second Treatise of Civil Government,* Harrington's *Oceana* was cited more frequently by the framers of the American Constitution than any other source.[2]

Harrington had read *De Cive* and *Leviathan* closely, and he had great admiration for their author. Though he violently disagreed with Hobbes on numerous political and methodological issues, he wrote:

> Nevertheless in most things I firmly believe that Mr. Hobbes is, and will in future ages be accounted the best writer, at this day, in the world. And as for his Treatise of human nature and of Liberty and necessity, they are the greatest of new lights and those which I have followed and will follow.[3]

Harrington's admiration for Hobbes is hard to understand, for as we shall see, Harrington disliked the Hobbesian hypothetico-deductive method. He was neither a mechanist nor a materialist; and he violently opposed the Hobbesian emphasis on egoism as the exclusive motive for human behavior. Nonetheless it is almost certainly true that Harrington was sincere in his comments about Hobbes and that the starting place for many of his arguments lay in Hobbesian insights.

Like Hobbes, Harrington wrote out of an immediate concern about political events associated with the English Civil War and simultaneously hoped to establish knowledge of society on a lasting and scientific footing. Thus he insisted that there are laws or principles *to be discovered* which regulate social phenomena as surely as there are laws which describe physical phenomena.[4] And he is certain that the institution of lasting and effective political structures demands that men take cognizance of such principles:

> To make principles or fundamentals, belongs not to men, to nations, nor to human laws. To build upon such principles or fundamentals as are apparently laid by God in the inevitable necessity or law of nature, is that which truly appertains to men, to nations, and to human laws. To make any other fundamentals, and then build on them, is to build castles in the air.[5]

Like Hobbes, who was obsessed with the implications of a single great idea—that self-interest is the foundation of all social relations—Harrington was also committed to a single guiding principle—that political arrangements (in his word, the "superstructure" of the nation) follow from the structure of ownership of economic resources, especially land. In fact, Harrington's followers linked his scientific claims with the discovery of this principle. John Toland wrote in his introduction to Harrington's *Works*:

> That empire follows the balance of property, whether lodged in one, in a few, or in many hands [Harrington] was the first that

ever made out; and [it] is a noble discovery, whereof the honor
solely belongs to him, as much as those of the circulation of the
blood, of printing, of guns, of the compass, or of optic glasses,
to the several authors. . . . This plain truth . . . is the foundation
of all government.[6]

Similarly, John Adams wrote that Harrington had discovered in his
principle that political power reflects economic power "as infallible
a maxim in politics as that action and reaction are equal in
mechanics."[7] Finally, Harrington was, like Hobbes, a fundamentally secular theorist, treating the Bible predominantly as an historical document rather than as a source of revelation: "To hold that
the wisdom of man in the formation of a house or of a government,
may go upon supernatural principles, is inconsistent with a commonwealth, and as if one should say 'God ordained the temple,
therefore it was not built by Masons,'"[8] wrote Harrington. On the
contrary, he insisted, "Government is of human prudence, and
human prudence is adequate to man's nature."[9]

The divergences between Hobbes and Harrington in political
theory can be most easily understood in terms of methodological
disagreements. Though both sought scientific understanding of
political phenomena, their commitments to scientific method were
radically different. Hobbes's understanding of science was vastly
more sophisticated than that of Harrington and focused on the
certainty of hypothetico-deductive systems and the contingency of
sensory experience. To the extent that Hobbesian theory can be said
to be *grounded* in experience, that experience was predominantly
introspective. Though vastly less sophisticated about his methodological commitments, Harrington followed Baconian ideas and insisted that a science of social phenomena be grounded in generali-
zations about social life induced from the observation of men in
societies.

The crux of his methodological differences with Hobbes is
expressed by Harrington in three places, all of which appeal to
William Harvey's anatomical dissections as the ideal model of
inductive empirical science. In *The Commonwealth of Oceana*,
Harrington first attacked Hobbesian method in connection with
Hobbes's criticism of Aristotle's and Cicero's analyses of the rights
of citizens in republics. Hobbes claimed that these ancient authors
"derived those rights not from the principles of nature, but transcribed them into their books out of the practice of their own commonwealths, as grammarians describe the rules of language out of

poets."[10] To this Hobbesian criticism Harrington responded that "[it] is as if a man should tell famous Harvey that he transcribed his circulation of the blood not out of the principles of nature, but out of the anatomy of this or that body."[11] Of course, political principles must be derived from the observation of particular political organizations, Harrington implied. Where else could they come from? When Matthew Wren attacked Harrington on the grounds that Harvey's success in deriving general results from dissecting particular bodies did not imply that the same method would be as successful in generating general results about political bodies, Harrington admitted that the analogy might be inexact; but he extended the medical analogy in his essay, *A System of Politics*. In this instance Harrington defended Machiavelli's empirical approach to corruption in government, writing that "neither Hippocrates nor Machiavelli introduced diseases into man's body, nor corruption into Government, which were before their times; and seeing they do but *discover* them, it must be confessed that so much as they have done tends not to the increase, but to the cure of them, which is the truth of these two authors."[12]

The great virtue of the empirical, or "anatomical," method of seeking knowledge of political behavior as opposed to the Hobbesian "geometrical" method is that while the latter may appeal by its simplicity and clarity, only the former can do justice to the subtlety and complexity of social reality:

> Certain it is, that the delivery of a model of government (which must be of no effect, or embrace all those muscles, nerves, arteries, and bones, which are necessary to any function of a well ordered Commonwealth) is no less than political anatomy. If you come short of this, your discourse is altogether ineffectual; . . . you may perhaps, be called a learned author; but you are obscure and your doctrine is impracticable.[13]

Harrington chided Hobbes for trying to make a king "by geometry" and claimed that such a king could be safe only if he had a "parliament of mathematicians."[14]

Harrington sought a detailed knowledge of political principles so that he could offer a "practicable" model for a commonwealth—one that could be used immediately in building a constitution for humans as they really exist. It was this feature of his work that appealed to David Hume, to the drafters of the constitutions for the colonies of New Jersey and Pennsylvania, and to the framers of the American Constitution.

The authors of most prior works on politics, including Hobbes, sought primarily to rationalize the status quo or to authorize the wide acceptance of political institutions which were, in practice, imposed from above. Even the writers of most utopian tracts shared Thomas More's sense that their works involved a great deal of wishful thinking.[15] Harrington, however, explicitly sought to influence political policy during the Protectorate through the publication of *Oceana*; and though he failed to move Cromwell, many of his ideas were adapted by the colonial proprietors and by Whig politicians like Henry Nevile, Andrew Fletcher, Walter Moyle, John Toland, Robert Molesworth, John Trenchard, and Thomas Gordon.[16]

In Harrington's mind, Hobbes's greatest oversimplification involved the claim that self-interest lay at the foundation of all social interactions. No doubt it is true that *interest* is the driving force of society; but careful observation shows that men are driven not by *self*-interests alone. Quoting from Hugo Grotius, Harrington observed:

> . . . tho it may be truly said that creatures are naturally carried forth to their proper utility or profit, that ought not to be taken in too general a sense; seeing that diverse of them abstain from their own profit, either in regard to those of the same kind, or at least of their young.[17]

That is, creatures of all kinds demonstrate altruistic as well as purely selfish motives. Indeed, Harrington insisted that historical study shows that however men tend to behave, most acknowledge that there is a common or public interest that transcends their private interests. Harrington, in fact, argued that there are *three* basic levels of interests which motivate men—self-interest or "private" interest; factional, or party interests, often associated with what we would now call "class" interest; and the public interest, associated with the greater good of the whole society.

Even though men may agree in some vague way that there is a public good that is greater than their own private good, Harrington acknowledged that Hobbes was correct in saying that in *most* instances "a man does not look upon reason as it is right or wrong in itself, but as it makes for or against him."[18] What is needed, then, for the construction of the best kind of commonwealth is some way of making certain either that people will see that the common interest is ultimately best for them or that people are somehow blocked from actions that would place private and party interests above public ones. Harrington opted for the latter course, arguing that "such orders may be established as may, nay must, give the upper hand in all cases to the common weight or interest, notwithstanding the nearness of that which sticks to every man in private."[19] A marvelously apt story illustrated what he had in mind:

> At Rome I saw [a pageant], which represented a kitchen, with all the proper utensils in use and action. The cooks were all cats and kitlings, set in such frames, so tried and so ordered that the poor creatures could make no motion to get loose, but the same caused one to turn the spit, another to baste the meat, a third to scim the pot, and a fourth to make green-sauce. If the frame or your common-wealth be not such, as causeth every one to perform his certain function as necessary as this did the cat to make green-sauce, it is not right.[20]

To understand how this framing can be accomplished, Harrington returned to an historical, empirical analysis of *power* in society. For as long as men have the power to serve themselves or their factions in society, they will do so. Once again Harrington began his analysis of power where Hobbes left off:

> As [Hobbes] said of the law: that without the sword it is but paper, so might he have thought of the sword, that without a hand it is but cold iron. The hand which holds this sword is the militia of a nation, and the militia of a nation is either an army in the field or ready for the field upon occasion. But an army is a beast that has a great belly and must be fed; wherefore this will come unto what pastures you have, and what pastures you have will come unto the balance of property, without which the public sword is but a name or mere spit frog.[21]

According to Harrington, empirical analysis of the recent history of England as well as the ancient history of the Israelites, Greeks, and Romans shows that whenever a single individual controls the bulk of the land in a nation—as did the Persian kings—he will rule in his own self-interest as an absolute monarch. Whenever the land is controlled by a relatively small noble class—as was true throughout medieval Europe, including "Gothic" England—the king will rule in collusion with the aristocracy in such a way that the interests of the aristocracy are served at the expense of those of the people. But whenever the wealth (usually land) of a nation is controlled by a relatively large number of men, then conditions are ripe for the rise of an "equal commonwealth," capable of serving the public interest. No stable government can exist which does not conform to its economic base; but empirical study shows that change can be initiated either in the political superstructure or in the economic infrastructure. That is, in some cases a king may appropriate land to shore up his position, while in other cases, once a king has lost land to the aristocracy or to the people, he will be forced to accept a diminution of political power.

Harrington argued that by the seventeenth century the control of land in England had moved into a large enough number of hands that an equal commonwealth best be established; and since he was certain that the public interest could be served only in such a commonwealth, he argued for a change in political superstructure rather than reversion to a more centralized pattern of land ownership. In order to ensure continuance of widespread landowning, he urged the passage of an "agrarian" law that would forbid the inheritance of estates larger than a certain maximum size, forcing abandonment of the English tradition of primogeniture, the passage of entire estates to the eldest male heir.[22]

Harrington's commonwealth is said to be "equal" because every independent citizen, regardless of his extent of wealth, is presumed to have an equal stake in the community. The term "independent" here is crucial, for unlike more radical democrats in the seventeenth century, Harrington followed classical republican sources in presuming economic independence to be a prerequisite for political existence. One who is a servant, a wage laborer, or a tenant farmer must be subordinate to his employer and may be supposed to be subject to coercion that will force him to express the interest of his employer rather than his own interests in public activity. If such persons were given a political role to play, they would tend to exert their power in their employer's interest, and those with great wealth could appropriate political power to serve their own ends. The goals of an equal commonwealth—the commitment to public rather than factional interest—would be subverted. Gerard Winstanley and other Levellers, facing this same argument, used it to justify the abandonment of private ownership of economic resources; but for Harrington such a radical step was unthinkable. The only reasonable response was to exclude those without property from citizenship.

So far, I have suggested that Harrington's political ideas are concerned predominantly with notions of *interest* and *power*. But there is a third critical factor in Harrington's political theory which also derives from classical republican ideals rather than from either the Hobbesian or empiricist scientistic traditions. This factor is associated with the term *authority* and has to do with the question of who is capable of best serving the public interest in an equal commonwealth. Economic power does not in itself legitimize or "authorize" political power, for political authority should be granted to those who possess what Harrington calls "goods of the mind"—e.g., wisdom, prudence, courage. Economic power however, depends primarily on goods of fortune—e.g., accidents of circumstance. Harrington was convinced that there is a "natural aristocracy"[23] of the mind and that virtually all men, "if they be not idiots, perhaps [even] if they be,"[24] are capable of recognizing their natural leaders. Thus in an equal commonwealth the best government will be realized if all citizens elect—through some form of secret ballot—[25] a body of the most able to lead them.

In order to be certain that this elected body not constitute a new faction, governing in its own interest, Harrington insisted upon rotation in office. No one should serve in any political office, appointive or elective, for more than three years at a time, and no one should be eligible for immediate re-election or appointment. In this way, every person must live as a private citizen under the laws he was involved in establishing or enforcing while serving in a public capacity. Even if the most able men are chosen to legislate, the legislative body will be subject to factional strife and prone to make decisions which serve particular interests rather than public ones. To control this problem Harrington divided the legislature into two houses, one of which proposes and debates legislation, and one of which accepts or rejects it without debate. In this way any interest group that dominates the debating House or Senate will be overcome by the choosing house, or assembly, just as when two girls divide a cake, equity is established when one divides and one chooses.[26] This scheme for a bicameral legislature was widely adopted or adapted in the American colonies and is enshrined in the Senate and House established under the U.S. Constitution.

Since it is impractical for the legislative body to enforce the laws that it creates—both because of its size and its diversity—there must also be another branch of government, the *executive,* to execute the laws.[27] Once more, executive officers must be controlled both by being subject to the laws while they serve and by being subject to rotation in office, which precludes the establishment of a permanent civil service interest.

Finally, Harrington faced the problem of how to elect officials in a commonwealth so extensive that potential elective officials could be known only to small portions of the total population. The commonwealth was to be divided (by any competent surveyor) into a number of equal electoral districts (fifty for *Oceana*). In each district popularly elected local legislative assemblies proposed and elected a selection of their own membership to serve in the national Senate and House. Once again Harrington pointed out the virtues of such an arrangement—it ensured not only that voters had direct knowledge of those they vote for, but also that no regional interest could dominate the national legislature. There is good reason to believe that this model guided the discussion of the division of France into geographically equal "departments" during the French Revolution,[28] the discussions of representation at the American constitutional conventions, and the establishment of the now awkward electoral college system for electing the President and Vice President of the United States.

There is one great claim made by Harrington regarding government of the kind which he recommended in *Oceana.* Though the government is one established through human prudence, Harrington's oft-repeated aim was to establish an "empire of laws and not of men," an aim which he took from the Aristotelian claim that "in a well ordered commonwealth not men should govern, but laws."[29] This goal must be understood both as an explicit opposition to the Hobbesian notion that law is merely the command of the sovereign power—and therefore fundamentally arbitrary—and as a consequence of Harrington's confidence that when the institutions of government are properly ordered, it will be literally impossible for any individual or faction to impose its will upon the public in opposition to the public interest. Laws will be the product of collective, institutional considerations rather than the reflection of powerful interests. Though perhaps overly optimistic, this remains an explicit goal of democratic/republican governmental structures which seek to protect the public against even tyrannies of the majority.

Harrington's utopian work *Oceana* met one of its first sustained criticisms in the *Considerations upon Mr. Harrington's Commonwealth of Oceana, Restrained to the First Part of the Preliminaries,* by the monarchist and later fellow of the Royal Society, Matthew Wren, which appeared in 1657. This work is particularly interesting from our perspective because it—and Harrington's responses—illustrate the degree to which scientific imagery was beginning to infiltrate political polemics following Hobbes's lead. Wren was not an admitted Hobbesian, but he outdid even the master in drawing support for absolute monarchy from mechanical analogs. Against Harrington's claim that power should be disbursed in government by balancing different interests against one another, Wren argued:

> The liberation he speaks of in government is of the same nature with a perpetual motion in mechanics; for as the one is not attainable as long as there is in matter a resistance to motion, or a propensity in it by which it is determined to some other motion, so will the other be impossible as long as men have private interests and passions by which they are biased from the public ends of government. And as all motion will grow faint and expire unless renewed by the hand of a perpetual mover, so there is no form of laws or constitution of government which will not decay and come to ruin, unless repaired by the prudence and dexterity of those who govern.[30]

According to Wren, the driving force of social activity is private or factional interest. If, on the one hand, the government is set up to allow the clash of interests, the result will be mutual paralysis rather than government in the public interest. This is so because all of people's energies will be used up blocking one another, and there will be no force left over to drive the state in any positive direction. On the other hand, if men give up their freedom to act to an undivided sovereign power, that uncontested power will be available to be used by those who govern in the interest of the governed.

Harrington responded to Wren's criticism in *The Prerogative of Popular Government* (1658) with yet another blast at those who would turn politics into a mechanical and mathematical subject rather than one guided by empirical studies:

> Let me tell him that in politics there is nothing mechanic, or like it. This is but an idiotism of some mathematician . . . the motion of [men] is from the hand of a perpetual mover, even God himself, in whom we live and move and have our being; and to this current the politician addeth nothing but the banks . . . [which] may stand as long as the river runneth.[31]

To historians of politics and political theory, Harrington is particularly interesting for two principal reasons. First, many of the institutional suggestions embedded in *Oceana* were incorporated in colonial constitutions and in the United States Constitution. Charles Blitzer, one of Harrington's recent biographers, even goes so far as to claim:

> If one wishes to discover the original *rationale* of many of the characteristic features of the Government of the United States—

such, for instance, as the written constitution, the secret ballot, the rotation of membership in the Senate—one must look not to *The Federalist* or to the other writings of our founding fathers, but rather to Harrington's *Commonwealth of Oceana*.[32]

Though none of these suggestions was uniquely Harrington's— most come in one way or another from classical political theorists like Aristotle and Cicero—there is evidence from such founding fathers as James Madison, James Otis, and John Adams, that their immediate source for these ideas was Harrington's writings. Thus Harrington figures in the story of the founding of the United States political system. Second, Harrington's fundamental insight—that political forms are intimately related to economic considerations in a society—gives him a special interest to economic determinists and to students of Marxism, to whom "he [can] be seen as a primitive forerunner of Marx, or, alternatively, as the misguided apologist/ prophet of the Bourgeois revolution."[33]

My concern with Harrington has been somewhat different—to assert his role as a polemicist in favor of an empirical and scientific approach to politics and to see in his works and the responses of his critics an illustration of how Hobbesian notions of a science of politics began to permeate seventeenth-century political discourse. The natural question that arises is whether Harrington's empirical scientism had any important bearing on his basic political ideas. As J. G. A. Pocock, among others, has convincingly argued, Harrington was above all "a classical republican, and England's premier civic humanist and Machiavellian."[34] He was an admirer of *ancient* prudence in opposition to the moderns, and his emotional attachment to republicanism was derived in large part from his love of the Venetian Republic, which he visited as a young man. From this perspective Harrington's scientistic rhetoric had little to do with his fundamental political ideas. It is true, however, that Machiavelli's classical republicanism was itself scientistic to the extent that it insisted upon a realistic, naturalistic, or "objective" investigation of political behavior. In Machiavelli's famous words:

> Since . . . it has been my intention to write something which may be of use to the understanding reader, it has seemed wiser to me to follow the real truth of the matter rather than what we imagine it to be. Imagination has created many principalities and republics that have never been seen or known to have any real existence; and how we live is so different from how we ought to live that he who studies what ought to be done rather than what is done will learn the way to his downfall rather than his preservation.[35]

Harrington's internalizing this dimension of Machiavelli's works accounts for the fact that no institutional proposal is brought forth in *Oceana* without appeal to specific historical precedent. If *The Commonwealth of Oceana* blends political institutions in a new way aimed at producing an eternally stable polity, at least each institution is one which has been demonstrably workable in some human society. Thus it would seem that Machiavelli's methodological attitudes were at least in part responsible for the particular way in which Harrington chose to formulate the political institutions of *Oceana.*

When we come to consider Harrington's formulation of the relationship between economic circumstances and political superstructure, there can be no question that he understood himself to be acting as a political scientist, discovering a general principle which underlies immediate experience. Harrington saw his principle as the natural law that alone links classical republics, medieval feudatories, and modern nations. They appear superficially different, but they can all be recognized as specific cases governed by the law of "the balance or dominion of property."[36] He argued that Aristotle and Machiavelli had both recognized special cases of the general law in discussions of why republics cannot long survive the rise of great inequalities of wealth between the many and a few.[37] And he shows that Bacon nearly recognized a second kind of special case in praising Henry VII's wide distribution of land to undermine the power of the wealthy nobility.[38] He rightly claims, however, that he alone has been able to recognize that such insights are merely special cases which can account for the stability or instability of popular republics (democracies), aristocracies, and monarchies alike in both the ancient and the modern world; and he draws from many specific examples a support for his contention. Because Harrington's general principle of the balance of property distinguishes him from earlier theorists and because this principle was the product of a search for general laws underlying social phenomena, there seems little question that his scientistic bent was germane to this aspect of his theorizing.

THE POLITICAL ARITHMETIC
OF WILLIAM PETTY

Harrington's political writings represent one of the first stages in the popular and relatively unsophisticated attempt to appropriate

the methods and authority of empiricist natural science for practically oriented political ends. In contrast, William Petty's works
represent the early attempts of a well-trained mathematician and
natural scientist to extend scientific methods to issues concerning
human behavior and public policy. Moreover, Petty's works represent more the mentality of the Baconian "projector" than that of the
classical political theorist. Though he had definite opinions about
the large questions of who should govern and what form government should take, Petty served Charles I, Cromwell, Charles II, and
James II with almost equal enthusiasm; at the same time, he was
an active member in radical political clubs. He first devoted his
technical expertise to building a massive fortune of his own; then
he focused his attention on policies that he felt *any* government
should follow to encourage trade and commerce and to promote national strength and the efficient exploitation of human and natural
resources. He was, in fact, among the first to identify politics with
the management of an economy and to suppose that an objective
science of society could replace the exercise of power by special interests as the foundation for political policy making.

Petty's background incorporated the best training available to
any scientist during the early seventeenth century, when specialization was the exception rather than the rule. Born into the family of
a relatively poor clothier at Romsey in 1623, Petty reported to his
earliest biographer that his greatest enjoyment as a young boy involved "looking on the artificers, e.g., smyths, the watchmakers, car-
penters, joiners, etc., until he could have worked at any of their
trades."[39] At age thirteen he went to work as a cabin boy on an English merchant ship; but after about a year, he broke his leg and
was set ashore in France. There he managed to get admitted as
a student to the Jesuit University of Caen where he focused on
mathematical studies and supported himself by trading in costume
jewelry and by tutoring a French naval officer in navigation. Next,
he served briefly in the Royal Navy, where he developed a life-long
fascination with naval architecture; but in 1643 he left England to
escape the Civil War and began studying medicine at Leyden, in
Holland. Late in 1645 he moved to Paris where he continued his
anatomical studies and became closely associated with Thomas
Hobbes, reading Vessalius with him and becoming acquainted with
the scientific circles connected with Marin Mersenne and Pierre
Gassendi. At this point he seems to have developed an interest in
the mechanical philosophy which later led, among other things, to

the publication of "A New Hypothesis of Springing or Elastique Motions."[40]

Soon Petty was back in England, trying to sell a new invention for "double writing," and making friends with Samuel Hartlib, the London advocate of Andraean and Baconian ideas, and with Robert Boyle, the great English chemist. Both Hartlib and Boyle were deeply concerned with leading a reform of society that would be grounded in the reform of knowledge, and probably in their circle Petty developed his intense interest in using scientific knowledge for social purposes. In 1648 Petty went to Oxford where he finished his medical degree and became assistant, then successor, to Thomas Clayton, Professor of Anatomy at Oxford. While at Oxford, Petty associated with Robert Boyle, John Wilkins, John Wallis, and the rest of the so-called "invisible college" which was one of the groups that formed the Royal Society of London after the Restoration. During his Oxford period, the invisible college usually met at Petty's lodgings;[41] later, when the Royal Society was created, Petty became a founding member, serving for a time as its vice president. In 1649 Petty's friends also procured him the position of Reader in Music at Gresham College, and in 1650 he was admitted to the London College of Physicians.[42]

Thus, at age 27, Petty stood as a secure and central figure of the British scientific community. He had received the best mathematical and medical educations available; he was well connected with prominent scientists in both France and Great Britain; he had feet in both the "rationalist-mathematical" camp and the "empiricist-anatomical" camp; and he held prestigious—and relatively lucrative—positions at Oxford and in London. The poor boy from Romsey had a very respectable annual income of 480 pounds and an apparently unlimited future as a London physician and natural scientist. Yet for reasons that remain unclear, Petty threw all of this over late in 1651, changing the focus of both his academic and his economic interests. He took a leave of absence from Oxford and left London to become physician to the Parliamentary army in Ireland. Though he retained this position until 1660, he made an even more radical change of direction which was formalized in 1654.

In order to account for this change of direction, we must understand the circumstances of the English army in Ireland. Ever since the Middle Ages, Ireland had nominally been subordinate to the English crown; except for a small region near Dublin, Ireland's was essentially an independent feudal society well into the sixteenth century. Under Elizabeth, a virtual reconquest of Ireland took place, with large plantations being established by English projectors like Sir Walter Raleigh, Richard Boyle, father of Robert Boyle, and the City of London. These wealthy individual and corporate projectors were almost uniformly Anglican, whereas the indigenous population was almost uniformly Catholic. During the early seventeenth century Ireland was in constant turmoil, the new English Protestant landlords vying with the older Irish Catholic nobility for control, with a population of poor renting farmers being squeezed to enrich both groups. As the civil wars in England approached, Charles I's governor, the Earl of Strafford, stepped in to establish Ireland as a base of support for the king, thus clashing with both the older Irish nobility and the new English aristocracy, who had been running things in their own interest. Finally, in October 1641, there was a rebellion led primarily by the older Irish Catholic nobility. In order to finance a campaign against the Irish rebels, the king agreed to set aside land forfeited by the rebels to repay advances made by English financiers and eventually to serve in lieu of cash payments to those who served in the army. From 1642 to 1652, nearly constant warfare and plague decimated much of Ireland, just as the Thirty Years' War decimated much of the Germanies. During this period the Irish population probably declined from about 1,466,000 to about 850,000. The later phase of the war, of course, was carried out by Parliamentary armies under Cromwell and Ireton, after the execution of Charles I.

When Petty arrived in Ireland as physician to the army of occupation, one of the chief problems facing the government was the division of land to satisfy war debts and to pay the increasingly restless soldiery; but before a division of lands could be made, the boundaries of forfeited lands had to be established. The Irish Surveyor General, Benjamin Worsley, proposed a two-stage survey of the relevant area (about two-thirds of Ireland) which would be completed within thirteen years—a hopelessly protracted project, given the demands of the unpaid army. At this point Petty offered to organize and supervise a detailed survey to be accomplished within thirteen *months* at a lower cost.

Carrying out this survey (the Down Survey) required an intricate organizational plan, and it is hard to figure just why Petty undertook the job. Many of his former friends neither understood nor sympathized with his decision. On one level the survey of Ireland offered an immense economic opportunity. Its successful completion—coupled with some associated shrewd land speculation—made Petty one of the richest men in Ireland within just a few years. Perhaps equally important it offered him an opportunity to exploit his scientific interests in a truly spectacular way. The technical side of the problem was trivial for a man with as much practical mathematical training as Petty had acquired in France and in the navy; but the efficient organization of human activity provided a tremendous challenge. Seven years later, looking back on his decision to conduct the Down Survey, he wrote in his "Reflections on Persons and Things in Ireland":

> I hoped thereby to enlarge my trade of experiments from bodies
> to minds, from the motions of the one to the manners of the
> other, thereby to have understood passions as well as fermenta-
> tions, and consequently to have been as pleasant a companion to
> my ingenious friends, as if such an intermission from physics had
> never been.[43]

The key to Petty's proposal was the application of the method of resolution and composition from natural science to the practical organization of a massive human enterprise. As Petty later formulated the method of resolution in drafting rules for the Dublin Philosophical Society, one must "analyse and divide complicated matters into their integral parts."[44] In accordance with this principle, Petty wrote that he "divided the whole art of surveying into its several parts, viz. 1, Field work; 2, protracting; 3, casting; 4, reducing; 5, ornaments of the maps; 6, writing fair books; 7, examination of all and every premises."[45] For each job Petty employed people with appropriately chosen skills. In particular, he speeded up work by training literally thousands of unemployed foot soldiers to do the basic chain work that involved little skill but great stamina.

One brilliant example of Petty's use of the division of labor appears in connection with his organization for providing instruments needed for the survey:

> ". . . scales, protractors, and compass cards, being matters of ac-
> curate division," he had prepared "by the ablest artists in London."
> For less complicated instruments, Petty "thought of dividing both
> the art of making instruments as also of using them into many
> parts," and set up his own workshop, furnished with wiremakers for
> the making of measuring chains, watchmakers for the production of
> magnetic needles, complete "with their pins," turners for the wooden
> parts, founders for the brass parts, etc. Particularly shrewd and far
> sighted was the employment of another workman, "of a more versa-

tile head and hand," whose task it was to touch the needles, adjust the sights and cards, and generally adapt every piece to each other.[46]

At each stage every task was reduced to its simple elements, and workers were trained to do just that part of the job which they had to do. In order to monitor the operation and reduce collusion and error, Petty even trained highly skilled auditing teams to do spot checks on the work at every level. The contract for the Down Survey was signed 11 December 1654; and in spite of horrendously wet weather, guerrilla attacks on his surveyors, and changing requirements on the part of the government, Petty completed the survey on 1 March 1656, at a net personal profit of 9,000 pounds. With much of this money he bought up lands at a small fraction of their value from soldiers who had acquired them; so by late 1657 he had acquired nearly 70,000 acres of Irish farmland worth about 7,000 pounds per annum. Even though title to some of the land was subject to dispute, Petty had become a man of vast property.[47]

From late 1657 to the end of his life, much of Petty's time and energy was devoted to managing and defending his title to his Irish estates. We, however, will be more interested in his growing concern with political policy. In part this concern was self-interested, for England's policy toward Ireland was important in determining his own fortunes; but Petty's analysis of political issues transcended self-interest and constituted an impressive step in the formulation of a science which he variously called political anatomy and political arithmetic—and which we now call economics.

Through the influence of Henry Cromwell, Petty was elected to a seat in Parliament from West Loo in Wales, and he moved to London in 1658. The next year found him one of the most active and argumentative members of James Harrington's political discussion group, the Rota Club, where he sharpened his awareness of the theoretical side of the political issues that he faced as a wealthy landowner and member of Parliament.[48] Then, in 1660, Petty began writing a series of works attempting to provide a scientific foundation for political policy formation. In two places Petty offered short statements regarding his general goals for these works. In an unpublished dialogue supporting establishment of a registry of lands, commodities, and inhabitants for Ireland—an idea which he initially proposed in 1660—Petty wrote, "God send me the use of things, and notions, whose foundations are sense and the superstructures mathematical reasoning; for want of which props so many governments do reel and stagger, and crush the honest subjects that live under them."[49] This passage clearly states Petty's belief that we need both factual information and describable laws that involve social phenomena as the basis for policy formation to ensure the stability of governments and the welfare of inhabitants.

With Petty we are far from the classical notion of political wisdom derived from historical awareness, and we are in the school of mathematical politicians so much detested by Harrington. But Petty's quantitative approach to politics differed from that of Hobbes as well. Like Hobbes, he was fundamentally ahistorical and opposed to the classical tradition of political theory. He even repeated on his own behalf the Hobbesian claim that "had he read much, as some men have, he had not known as much as he did."[50] But unlike Hobbes, Petty assumed nothing about individual human motives or psychology and spurned introspective experience in favor of an empirical approach to large-scale social and economic phenomena:

> The method I take . . . is not yet very usual; for instead of using only comparative and superlative words, and intellectual arguments, I have taken the course (as a specimen of the political arithmetic I have long aimed at) to express myself in terms of *Number, Weight,* or *Measure*; to use only arguments of sense, and to consider only such causes as have visible foundations in nature; leaving those that depend upon the mutable minds, opinions, appetites, and passions of particular men, to the consideration of others: really professing myself as unable to speak satisfactorily upon those grounds (if they may be called grounds), as to foretell the cast of a dye or to play well at tennis, billiards, or bowls, (without long practice) by virtue of the most elaborate conceptions that ever have been written *De Projectilibus et Missilibus,* or of the Angles of Incidence and Reflection.[51]

Petty does not deny the possibility of a science of society grounded in individual psychology any more than he denies that, in principle, the motions of a dye or a tennis ball are regulated by the fundamental laws of mechanics. Instead he argues that there are regularities in aggregate social behavior—we would say that there are statistical regularities—which can be observed, quantified, and much more easily used for forecasting purposes than the presumably more basic science of individual psychology.

Petty himself was most pleased with his *Political Arithmetic,* a work composed late in his life and published posthumously in 1690. Indeed, in a letter to Southwell dated 19 March 1678, on the topic of algebra, he boasted:

Archimedes had algebra 1900 years ago, but concealed it. Diophantus had it in great perfection 1400 years since. Vieta, Descartes, Roberval, Harriot, Pell, Oughtred, van Schoten and Dr. Wallis have done much in this last age. It came out of Arabia by the Moors into Spain, and from thence hither, and W[illiam] P[etty] hath applied it to other than purely mathematical matters, viz: to policy by the name of *Political Arithmetic,* by reducing many terms of matter to terms of number, weight, and measure, in order to be handled mathematically.[52]

In spite of his pride in *Political Arithmetic,* which was largely devoted to proving the economic superiority of England over France and Holland, Petty's most penetrating and interesting application of scientific method to policy issues appears in two of his earliest works, *A Treatise of Taxes,* first printed in 1662, and *Verbum Sapienti and on the Value of People,* written about 1665. His subsequent writings offered little more than detailed applications and minor extensions of the principles established in these two works.

The most fundamental principle underlying all of Petty's economic arguments is that it is a terrible mistake to think that "the greatness and glory of a Prince lyeth rather in the extent of his territory than in the number, art, and industry of his people, well united and well governed."[53] So Petty spent much effort trying to get the government to encourage immigration to England. Though the notion that the wealth of a nation lies principally in products of *labor* rather than products of land was contrary to the arguments of Harrington, it was almost a commonplace in the seventeenth century. What Petty managed to do was to *quantify* the claim and show (not very precisely, it is true) just how much more valuable the people of a nation were than its other capital goods, including land. Petty estimated that the population of England and Wales, ca. 1665, was six million and that they expended, for all purposes, about forty million pounds sterling each year.[54] Petty identified this forty million pounds with what we would call the Gross National Product (GNP) of England. Through ingenious indirect calculations he established that the total capital assets of England were:

144 million in land
 30 million in housing
  3 million in shopping
 36 milion in livestock
  6 million in money
 <u>31</u> million in durable goods
250 million total capital.[55]

Since Petty knew that rents on land returned, on the average, somewhat under six percent per annum, for a total of 8 million pounds, and that money and other movable assets returned slightly more than six percent per annum, the annual return on the 250 million pounds of capital assets could be set at very nearly 15 million pounds.[56] Then the difference between the 40 million GNP and the 15 million return on capital could come only from the labor of the people. So the people account for 25/40, or 62.5% of the national income, whereas the land accounts for only 8/40 or 20%. Assuming that humans return about six percent per annum—i.e., that the average working life is roughly sixteen years, which was probably a reasonable estimate—then the human "capital" of England was worth 416.66 million pounds and, according to Petty, "each head is worth 69£, or each of the 3 million workers is worth 138."[57] Thomas Hobbes may have been one of the first to declare that the value of a man can, in principle, be measured in monetary terms, but to Petty we owe the first theoretical calculation of a precise figure. It is interesting to note—as Petty did—that slaves at that time were selling for roughly their expected return over eight years, which is what Petty's workers were calculated to be worth.[58]

Petty's demonstration of the relatively high value of people vis à vis land had innumerable policy implications—at least as he saw it. On the most simplistic level, it justified Petty's opposition to all wars of territorial expansion, since it was much surer and more efficient simply to encourage population growth—both through procreation and immigration.[59] On another interesting subject, Petty's argument implied that crimes should be punished not by death, mutilation, or imprisonment, but by suitable fines or service to the nation. As Petty wrote regarding punishment for theft by the indigent, "Why should not insolvent thieves be rather punished with slavery than death? So as being slaves they may be forced to as much labor, and as cheap fare, as nature will endure, and thereby become as two men added to the commonwealth, and not as one taken away from it. . . . "[60] For those who could afford to pay fines, a rather careful calculation was necessary to figure out the appropriate level of fine. In cases of theft, for example, the probability of being caught in any given type of crime had to be calculated and its inverse multiplied times the profit for any given crime to determine the break-even point for thievery. Fines then had to be calculated to be slightly higher than this value in order to discourage the practice.[61] The high value of people also formed the basis for Petty's special argument in favor of religious toleration. Not only will toleration invite the immigration of valuable artisans—as it did in Holland—but if the state charges a fee for being a dissenter, it also gains revenue in a relatively painless way.[62]

More important than these side issues was the implication of Petty's version of the labor theory of value for general tax policy. Any tax tending to reduce levels of economic activity should be avoided.[63] In the first place this principle is true with respect to the total amount of taxes demanded, if those taxes are demanded in money rather than in kind:

> If the largeness of a public exhibition [tax] should leave less money than is necessary to drive a nation's trade, then the mischief thereof would be the doing of less work, which is the same as lessening the people, or their art and industry; for a hundred pounds passing a hundred hands for wages, causes ten thousand pounds worth of commodities to be produced, which hands would have been idle and useless, had there not been this continual motive to their employment.[64]

By the same token, Petty recognized that if properly spent, taxes can be a desirable device for both stimulating the economy and producing income redistribution:

> Taxes, if they be presently [i.e., immediately] expended upon our own domestic commodities, seem to me to do little harm to the whole body of the people, only they work a change in the richness and fortunes of particular men; and particularly by transferring the same from the landed and lazy, to the crafty and industrious.[65]

Petty analyzed the virtues and the vices of a huge variety of traditional and innovative taxes in order to optimize income with minimal negative social impact. Traditional land taxes became problematic because they failed to tax most of the nation's wealth and fell most heavily on those landowners whose rents could not be changed.[66] Customs duties were complicated to enforce, encouraged smuggling, and because they raised the price of an exported commodity beyond what foreigners were willing to pay, they often discouraged productivity.[67] Poll taxes or head taxes were unfair and inefficient because they failed to extract what the wealthy might pay and they forced undue burdens on the poor. And the sale of monopolies, while they might occasionally encourage invention and entrepreneurial activity, more often served to inhibit trade.[68] Only two types of taxes generally received Petty's blessing. The first, tithing or the donation of a small portion of all goods and services to the state, was reasonably fair except that manufactured articles which went through several stages of production might be multiple taxed. Moreover, tithes were inconvenient to collect because of the seasonal nature of agricultural production in particular. Much better was the imposition of what Petty called an accumulative excise tax or what we now call a retail sales tax. For Petty this had the great virtues that it (1) made the tax most nearly proportional to the enjoyment of wealth—i.e., to consumption; (2) avoided multiple taxation of manufactured goods, and therefore encouraged industry more than a tithing or "income" tax; (3) encouraged thrift and capital accumulation by *not* taxing raw materials and productive capacity, but rather finished goods; and (4) offered a relatively even and constant source of revenue.[69] Note that two of the major reasons for the excise or retail sales tax derive from Petty's insistence that national wealth comes primarily from putting people to work.

For the most part, Petty's recommendations on taxes were shrewd and impeccably argued; yet his advice was ignored, as was the advice contained in almost all of his other works on policy for Ireland, immigration policy, naval affairs, and policy regarding shipping and trade. Petty was deeply disappointed about his failure to have any immediate impact on policy; and we should consider whether this failure was in any way tied to the "objective" and scientific character of his arguments. Petty was not ignored because he either lacked access to policy makers or was personally out of favor with government leadership. During the Protectorate Petty served as close advisor and friend to Richard Cromwell's brother Henry, in Ireland, remaining a supporter of Henry even after the Restoration. This personal connection might have antagonized Charles II after the Restoration, but it seems that Petty's wit and interest in naval architecture charmed the king and that Petty was on close personal terms with both Charles II and James II, serving as Commissioner of the Navy for Charles.

In the face of his close personal relations with the monarchy after the Restoration, we have to look at the character of Petty's advice itself to understand why it was ignored. In the first place, totally aside from the way in which they were supported, far too many of Petty's recommendations smacked of the recently rejected leveling tendencies of the radical groups connected with the revolutionary armies. Among the many such causes supported by Petty were care for the helpless and poor, public employment for those who could not get jobs, prison reform, more equitable taxation, liberty of conscience and religion, a liberalized franchise tending toward universal manhood suffrage, equalized representation, as well as the reduction and simplification of law, medicine, education, the Church, and government.

This political radicalism and Petty's insistence on sticking to his principles in the face of a strong political reaction are at least indirectly related to a slightly different kind of "radicalism" associated with Petty's apparently sincere but naive belief that his scientific method allowed him to argue "without passion or interest, faction or party; but as I think, according to the Eternal Laws and Measures of Truth."[70] Like other scientists, Petty felt committed to truth and capable of deriving his conclusions from well-founded first principles. He was thus inclined to commit himself to the best possible "rational" solution to any problem he was considering, with relatively little regard for existing institutions, personal feelings, or other political exigencies. Charles II expressed his frustration with this aspect of Petty's advice during the late 1670s, remarking that Petty was one of the best and most able commissioners of the navy that he had ever appointed, but that his effectiveness as a policy advisor was limited because "the man will not be contented to be excellent, but is still aiming at impossible things."[71] Thus with Petty, the scientistic approach to policy formulation was already faced with a paradox that it continues to face in the latter part of the twentieth century—courses of action which are highly "rational" may nonetheless not be reasonable or accomplishable within their complex cultural and political settings. When this happens, the scientistic policy advisor will almost certainly be seen as trying not to *describe* social conditions, but to "*prescribe* social and political institutions to bring them into conformity with the dictates of mathematical reasoning."[72]

This problem with Petty's arguments shows up in a particularly poignant way in connection with an essay, "On the Growth of the City of London" which Petty forwarded to his close friend, cousin, and correspondent Robert Southwell in 1686. In this essay Petty proved on purely economic grounds that London should grow to over five million people by 1800, at which time, less than 4.5 million people would inhabit the rest of England. Southwell wrote back to Petty with a kind of patronizing and exasperated good nature:

> Now as to the grandeur of London, would not England be easier and
> perhaps stronger if these vitals were more equally dispersed? Is
> there not a tumor in that place, and too much matter for muting and
> terror to the government if it burst? Is there not too much of our
> capital in one stake, liable to the ravage of plague and fire? . . . Will
> not the resort of the wealthy and the emulation of luxury, melt down
> the order of superiors among us and bring all toward levelling and
> Republican? I know these things do not concern your matters of
> fact which are so strangely ascertained, but [they] are my political
> reflections upon your arithmetic.[73]

What Petty dearly needed was to leaven his often brilliant rational
analyses of situations with common sense. By a purely economic
analysis he could never convince his county cousin that London
should dominate England, nor could he convince Charles II and
James II, both of whom found sanctuary in France at critical
junctures, that England's economic interests should be forwarded
by developing anti-French foreign policy, no matter how correct he
might have been in narrowly economic terms.

Petty's exclusive focus on quantitative and methodological
arguments produced an additional negative reaction because they
probably made him seem even more callous, insensitive, and un-
feeling than he really was. Though he was almost always an advocate of support for the poor, his "objectivity" could lead him into
rather brutal claims. Thus, in explaining how ten percent more
taxes could be raised across the board Petty wrote:

> Laboring men work, 10 hours *per diem*, and make 20 meals *per*
> week, *viz.* 3 a day for working-days and two on Sundays; whereby it
> is plain, that if they could fast on Friday nights, and dine in one
> hour and a half, whereas they take two, and spending 1/20 less, the
> 1/10 above-mentioned might be raised.[74]

Of course Petty is correct (well, nearly correct, since he assumes that
all of the workers' wages go for meals), but he seems totally insensitive both to physiological needs and to established cultural
traditions involving the number of meals in a day and the length of
rests from labor. Similarly, in his preface to *The Political Anatomy
of Ireland* (ca. 1672) Petty seems oblivious to the feelings of the
Irish people—with whom he greatly sympathized—in writing what
must be one of the most insulting sentences in all of serious
literature:

> . . . as students in medicine practice their inquiries upon cheap and
> common animals . . . I have chosen Ireland as such a political ani-

mal . . . in which, if I have gone amiss, the fault may be easily mended by another.[75]

It can hardly be surprising that Jonathan Swift found Petty's *Political Anatomy of Ireland* an ideal target for bitter satire in *A Modest Proposal for Preventing the Children of Poor People in Ireland from Being a Burden to Their Parents or Country and for Making Them Beneficial to the Public* (1729). In the *Modest Proposal*, Swift perfectly apes Petty's passionless, analytic style:

> I calculate there may be about two hundred thousand couples whose wives are breeders, from which number I subtract thirty thousand couples who are able to maintain their own children . . . this being granted, there will remain a hundred and seventy thousand breeders. I again subtract fifty thousand, for those women who miscarry, or whose children die by accident or disease within the year. There only remain a hundred and twenty thousand children of poor parents annually born. The question therefore is, how this number shall be reared and provided for . . . ?[76]

Swift cites traditional ploys, dismissing them all, and then offers the "rational" but totally outrageous suggestion that, since they contain all necessary nutrients and are undoubtedly tender, 100,000 children should be sold yearly for food, the proceeds going to support not only the remaining 20,000 children, but all of the parents as well.[77] In a shattering conclusion he again insists upon his total objectivity and disinterestedness in making his proposal:

> I have not the least personal interest in endeavouring to promote this necessary work, having no other motive than the public good of my country, by advancing our trade, providing for infants, enriching the poor, and giving some pleasure to the rich. I have no children by which I can propose to get a single penny, the youngest being nine years old, and my wife past childbearing.[78]

Swift was far from alone in finding objectionable the early modern attempts to reduce traditional topics of moral and political discourse to a science. He and his conservative colleagues were concerned that two activities equivalent in the cash value of their economic products seemed to count as scientifically equivalent even though they might be manifestly different in the pleasure that they give to the participants or in their relationships to traditional mores and customs. In social activities the relationship between "rationality" in the Hobbesian-Pettyesque computational sense and "reasonableness" in a given cultural context is extremely problematic. The failure to distinguish between the two erected substantial barriers to taking scientific approaches to social issues. These barriers remained in place until the scientific notion of reason became more widely adopted among intellectuals and the power of traditional morality associated with Christian revelation became weakened. This simultaneous spread of scientific authority and the decay of traditional religious influences in European culture will be one of the central themes explored in the next chapter.

The Religious Implications of   3
Newtonian Science

According to one of the most widely accepted interpretations of the relationship between science and religion, seventeenth-century scientific developments—dominated by the mechanical philosophy in a special form developed by Sir Isaac Newton—were almost single-handedly responsible for transforming European culture to one in which secular replaced sacred authority as the foundation for ethics, morality, and politics.[1] Science proceeded simultaneously (1) to rob traditional Christianity of its emotional power by forcing the attention of its leaders away from the central issue of salvation through Christ into an increasingly arid defense of theological and social doctrine, (2) to seduce able intellectuals away from Christianity by undermining traditional claims about religious knowledge, and (3) to offer an intellectually satisfying non-religious justification of ethical, moral, and political imperatives. In this way, science replaced religion as the dominant institution of modern Western culture. In this chapter we focus on the claim that seventeenth-century science managed somehow to erode both the emotional appeal and the intellectual support of Christianity.

Most of those who argue that early modern scientific developments forced a decline in the emotional power of religion begin by claiming that Christianity depends for its special emotional power on the belief in God's immanence and direct providential activity in the world. Like several other mystery religions that emerged and spread in Hellenistic culture, Christianity was successful because it offered a God that listened—that is, a God that could respond to human needs or pleas by intervening in the normal flow of events to modify them. Seventeenth-century science supposedly under-mined belief in a providentially active God because the scientists were successful in portraying nature "as a machine running by itself without external aids."[2] Such a mechanical universe was totally in-sensitive to thinking beings, and it contradicted the traditional no-tions of divine providential activity and miraculous intervention in the world. The very survival of religion depended upon some kind of accommodation between science and religion, and that accom-modation "came more and more to mean the adjustment of Chris-tian beliefs to conform to the conclusions of science."[3] By driving Christianity away from its focus on providence and an actively in-volved God, it removed the emotional appeal of Christianity.

Such an interpretation is seductively simple, but it is both un-satisfactory and misleading. It completely fails to explain how the scientists, working in a relatively new and powerless cultural spe-cialty, managed to gain enough influence to force members of the well-established religious specialty into paying attention to their claims. And it fundamentally misstates both the aims and beliefs of the major seventeenth-century scientists. Not only did most of them—including Isaac Newton—*not* depict the universe as running without external aids, but they also continued to believe in the pos-sibility, even the necessity, of both God's ordinary providential ac-tivity and his miraculous intervention in the universe. Moreover, when they thought of accommodation between religion and science, they were much more likely to demand that scientific concepts be modified to conform to their religious beliefs than vice versa.[4]

It is true that one of the *long-term* consequences of early mod-ern interactions between science and religion was the modification of some doctrines among major segments of the religious com-munity to establish their compatibility with an increasingly pres-tigious scientific tradition, but this long-term transformation of religion was preceded by a period in which key scientific elements were enthusiastically welcomed into dominant religious institu-tions without calling for any obvious religious accommodations and in which many scientists viewed the support of Christianity as a central goal of scientific activity. Changes in science during the eighteenth and nineteenth centuries *were* threatening to main-stream Christianity primarily because the religious community had come to depend so deeply upon scientific theories and habits of mind associated with the mechanical philosophy and with Isaac

Newton to defend traditional notions of God's providential and miraculous activity.

THE NATURAL THEOLOGY TRADITION

From the time of the early Church Fathers there has been an important and continuous Christian tradition that knowledge of God, understanding of Scripture, and recognition of one's Christian duties can all be enriched through study of the "book" of nature.[5] In general, it has been agreed that scriptural revelation provides sufficient guidance for a successful Christian life and that the study of nature merely allows one to look more deeply into "the secrets of God's wisdom," so that "the mind may rise *somewhat* higher thereby to behold his glory."[6]

Within Continental Catholicism the natural theology tradition received a substantial new emphasis during the 1560s, when the Council of Trent established St. Thomas Aquinas's *Summa Theologica* as a central element of Church doctrine. According to St. Thomas, some theological propositions are capable of being demonstrated from the nature of the universe. God's existence, for example, can be established through Aristotelian arguments for the necessity of a first cause or unmoved mover (vol. I, chap. 4). Such natural theological arguments are not sufficient or even necessary for salvation; but when available, they can be relied upon without scriptural support. Moreover, according to St. Thomas, we can be completely confident that no proposition drawn from natural theology can possibly conflict with any proposition drawn from scriptural revelation.

With the rise of the Jesuit educational system in the late sixteenth century, Thomistic natural theology and the Aristotelian science from which it drew became increasingly important to Catholic intellectual life. Furthermore, with the emergence of the mechanical philosophy in the early seventeenth century, Father Marin Mersenne, among others, attempted to show that Cartesian science could be as easily applied to natural theological ends as that of Aristotle.[7] Another cleric, Pierre Gassendi, undertook the much more challenging task of trying to bend Epicurean atomism to natural theological ends.[8] And by the late seventeenth century, the Oratorian order in France had undertaken the teaching of modern science as well as the development of a full-blown Cartesian natural theology to replace the Aristotelianism of the Jesuits.[9]

THE SPECIAL ROLE OF NATURAL THEOLOGY WITHIN ANGLICANISM

Though the long tradition of natural theology was continued and even intensified on the continent to parallel increased levels of scientific activity in the seventeenth century, in England it took on a much more central religious role in connection with a series of crises in Church politics which occurred between about 1580 and 1688. The English National Church (the Anglican Church) had been established by Henry VIII in the early sixteenth century. Lutheran in its doctrinal orientation, but with critical residues of Catholic rituals and governance, the Anglican Church was under tremendous pressure to continue its reformation in the direction of Calvinist doctrine, ritual, and governance during the reign of Elizabeth I.

Virtually all Elizabethan Anglicans were vehemently anti-Catholic as a result of persecutions carried out during Mary's brief Catholic reign. But when the Anglican Church was restored under Elizabeth, several factions sought influence and control. At one end of the spectrum, a faction led by clergy who had fled to Geneva and Frankfurt during Mary's reign (the Marian exiles) sought to turn the Church in an increasingly Calvinist direction. These men denounced its hierarchical form of governance and opposed the ornate ritual of the Anglican liturgy. Above all, they insisted that Scripture alone expressed all Christian doctrine and duties, so that anything one does or believes that is not explicitly commanded in Scripture is to be considered sinful. Those who agreed with these positions gradually came to be called "Puritans."

At the other end of the spectrum, a faction led primarily by clergy who had remained in England during Mary's reign and whose attachment to the crown and the hierarchical governance of the Church was intense, bitterly opposed any erosion of the authority of the queen and her appointed bishops or any changes in the liturgy or character of the service. This faction viewed the Puritans as a threat to both church and state. They sought to evict Puritan leaders from their church positions and to have them persecuted in a variety of ways.

Elizabeth sought a compromise between these two groups, whose combined support was crucial to her effectiveness as a ruler. To defend this compromise, Elizabeth's moderate Archbishop of Canterbury, John Whitgift, chose a young Oxford-trained clergyman named Richard Hooker. Trained under a Puritan tutor, but with close personal associations with the Archbishop of London, Hooker had friends in both camps and was trusted by all. Through the late 1580s, Hooker engaged in a sermonizing duel with his nominal subordinate and the chief Puritan spokesman, Walter Travers; then in 1591, he resigned his living in London to complete *The Laws of Ecclesiastical Polity*. This work was intended by Hooker to provide the intellectual justification for ameliorative policies, though his sponsors, including Edwin Sandys, who actually paid for the printing of Books 1–4, had hoped for a much more aggressively anti-Puritan stance and for greater support for the autonomy of the Anglican Church with respect to Parliament. In Hooker's mind, Anglican doctrine, ritual, and ecclesiastical authority had to be supported without appeal to either Catholic tradition or scriptural imperative; and a reformed church had to be justified without accepting the Puritans' extreme scriptural emphasis. Finally, an attempt had to be made to reject Puritan claims without driving the would-be reformers out of the National Church and into opposition to Elizabeth.

Hooker's work is extremely important for our purposes because in order simultaneously to reject Catholic tradition and Puritan appeal on Scripture as adequate foundations for Christianity, he focused attention on the religious importance of natural reason and natural theology. He did not deny that scriptural revelation was necessary to salvation; but unlike earlier proponents of natural theology, he did explicitly claim that Scripture was not *sufficient* in itself without the support of natural reason and its religious manifestation, natural theology. Given both scriptural revelation *and* natural theology, *no* additional source of religious knowledge— neither church councils, patristic tradition, nor the special authority of a pope—was necessary. In Hooker's words:

> It sufficeth therefore that Nature and Scripture do serve in such full sort that they both jointly and not severally either of them be so complete that unto everlasting felicity we need not the knowledge of anything more than these two may easily furnish.[10]

In the long run, Elizabeth's and Hooker's hopes for avoiding a schism between Calvinist reformers and the Anglican Church failed. But Hooker's *Laws of Ecclesiastical Polity* provided guiding principles for the development of Anglican doctrine and governance into the present. More important for our purposes, it stimulated a rash of natural theologizing among religious moderates, or "Latitudinarians," throughout the seventeenth century as they sought to head off the proliferation of nonconformist sects prior to the Civil War, to restore a comprehensive (inclusive) Anglican Church with the restoration of Charles II, and finally to justify their own acceptance of the Glorious Revolution of 1688 when James II was deposed.

Many of the major British scientists of the seventeenth century, including the physician Walter Charleton, the chemist Robert Boyle, the naturalist John Ray, the polymath John Wilkins, and Isaac Newton himself, became involved in the tradition of Anglican natural theologizing initiated by Hooker. Although they did not all agree on every issue, at least two key elements were accepted by the vast majority. First, Anglican natural theology emphasized God's continuing providential role in the universe. Scientific arguments were brought to bear that God was *not* merely the creator of the universe; he was also the continuing efficient cause of all that happens. In Hooker's words, he is "both the creator and the *worker* of all in all."[11] Second, Anglican natural theology deviated from most traditional forms in emphasizing the probable rather than the certain nature of both natural scientific and religious knowledge. Ultimately only God can fully know himself or his creation. Through rational analysis of the evidence accessible to our senses, we can develop a high degree of confidence in our probable knowledge; but "insomuch as the true idea of nature is proper only to that eternal intellect which first conceived it, it cannot be but one of the highest degrees of madness for dull and unequal man to pretend to an exact, or adequate comprehension thereof."[12] Since nature and Scripture are equally the creations of an infinite intellect and equally confronted by finite humans, even the interpretation of Scripture is a merely probable activity. One should not only adopt the interpretation best supported by available evidence but also keep in mind the possibility of other legitimate interpretations and be very cautious about insisting that everyone conform to a single opinion on minor details.

THE SPECIAL CHARACTER
OF NEWTONIAN SCIENCE

Like many of his scientific colleagues, Isaac Newton was deeply concerned with religious issues throughout his career. In fact, his religious manuscripts exceed his writings on natural philosophy in extent. Moreover, because of his tremendous scientific reputation, Newton's religious opinions were sought and respected by many. Initially, as we shall see, Newton's scientific and religious ideas seemed to support the Latitudinarian emphases on God's continuing providence and the necessity for evaluating the evidence for any interpretation of Scripture. But over time, as both Newtonian science and Newtonian religious arguments were extended in the eighteenth century, they created some of the most severe challenges to the religious tradition from which they emerged.

In the century prior to Newton's publication of the *Principia* (1687), British scientists and Latitudinarian theologians had emphasized the limits of human knowledge and the humility and toleration engendered through scientific study. But Newtonian science seemed to transcend the presumed limits of human knowledge.[13] Even John Locke, who had been peculiarly concerned to emphasize the limits to human knowledge, viewed Newtonian science with a special kind of awe: "Though the systems of physics that I have met with afford little encouragement to look for certainty . . . yet the incomparable Mr. Newton has shown, how far mathematics, applied to some parts of nature, may upon principles that matter of fact can justify, carry us in the knowledge of some, as I may so call them, particular provinces of the incomprehensible universe."[14] Moreover, Locke suggested that if other scientists modeled their work on that of Newton, "we might in time hope to be furnished with more true and certain knowledge . . . than hitherto we could have expected."[15] On this score there is little question that Locke mirrored a widely held understanding of Newton's work and one that was shared by Newton himself.

The question we must consider first, then, is what was it about the nature and content of Newton's scientific work that seemed to legitimize a new optimism regarding the human capacity to achieve true and certain knowledge; for it was at least in part the presumed greater certainty of Newtonian science that made its extension to religious and social topics so appealing.

Newton's first scientific paper was submitted to the Royal Society of London in 1672, when he was already thirty years old. He had been offered the Lucasian Professorship in Mathematics at Cambridge three years earlier and in 1664 and 1665 had experienced a period of scientific activity that provided the foundation for virtually all of his subsequent accomplishments. Newton provided a brief and remarkable account of this period—often called his *annus mirabilis,* or miracle year—in which he was forced by the plague in Cambridge to retire to his mother's house at Woolsthorpe for a time:

> I found the Method [the calculus] by degrees in the years 1665 and 1666. In the beginning of the year 1665 I found the method of approximating series and the rule for reducing any dignity of any Binomial into such a series [i.e., that for all *x*, *a*, and *n*,
>
> $$(x + a)^n = x^n + n \cdot ax^{(n-1)} + \frac{n\ (n-1)}{2}\ a^2x^{(n-2)} + \frac{n\ (n-1)\ (n-2)}{2 \cdot 3}$$
>
> $$\cdot\ a^3x^{(n-3)} + \frac{n\ (n-1)\ (n-2)\ (n-3)}{2 \cdot 3 \cdot 4}\ a^4x^{(n-4)} \ldots + a^n]$$
>
> The same year in May I found the method of tangents of Gregory and Slusius, and in November had the direct method of fluxions [differential calculus], and the next year had the Theory of colors, and in May following I had entrance into ye inverse method of fluxions [integral calculus]. And the same year I began to think of gravity extending to ye orb of the Moon, and having found out how to estimate the force with which a globe revolving within a sphere presses the surface of the sphere, from Kepler's Rule of the periodical times of the Planets . . . I deduced that the forces which keep the Planets in their orbs must be reciprocally as the squares of the distances from the centers about which they revolve: and thereby compared the force requisite to keep the Moon in her orb with the force of gravity at the surface of the earth, and found them to answer pretty nearly. All this was done in the two plague years of 1665 and 1666, for in those days I was in the prime of my age for invention, and minded Mathematics and Philosophy more than at any time since.[16]

If this statement can be taken seriously, in the course of less than two years Newton had made discoveries in three different fields, each of which was as important and dramatic as that for which any Nobel Prize has been subsequently granted. In effect, he invented calculus, made one of the major discoveries in modern optics, and formulated the major principles of his law of universal gravitation.

Though it was nearly forty years before he finished working

out the details and publishing his results, by the end of 1666, Newton had the fundamental insights which made him the foremost mathematician, experimental natural philosopher, and astronomical theorist—indeed the most celebrated scientist—of his or any other age.

The first clue the world at large had of Newton's mastery of scientific theory came when he sent his first major communication, a "New Theory about Light and Colors," to the Royal Society of London. An analysis of this paper and the controversy that it created provides important insights into Newton and his science. At the heart of Newton's first paper are a series of experiments performed by allowing a narrow beam of light to pass through a prism or series of glass prisms with triangular cross section. Very similar experiments, which produced a rainbow effect or "spectrum," had been done since the thirteenth century and by René Descartes, Robert Hooke, Barrow, and a number of other seventeenth-century students of optics. Speculating about why some rays of light seemed to be bent more after traveling through a prism than others, he wrote:

> I began to suspect whether the rays, after their trajection through the prism, did not move in curved lines, and according to their more or less curvity tend toward different parts of the wall. And it increased my suspicion when I remembered that I had often seen a tennis ball, struck with an oblique racket, describe such a curved line.[17]

If the rays of light were composed of "globular" bodies, and if they attain a rotation in their oblique passage from one medium into another, he then suggested that they should move in curved paths just like tennis balls.[18] New experiments showed this not to be the case, for once the rays had passed through the prism, they traveled in straight lines rather than curved ones. So Newton immediately dropped his speculation regarding the possibility that light is like tennis balls and moved on to an *experimentum crucis* (crucial experiment), which his tests of this failed hypothesis suggested. Now he placed a screen with a small hole roughly midway between the initial prism and the wall. The initial prism was slightly rotated to allow only a small portion of the spectrum to fall on the hole in the screen, and a second prism, similar to the first was placed behind the hole to refract the light once more.

The results were startling. A second refraction did not spread the light out into a full spectrum. Indeed, red remained red, and it was bent again as much as it had been in the first refraction. Blue remained blue and was refracted again through an angle much greater than that through which red was bent (see fig. 3). From these results, Newton concluded that "light itself is a heterogeneous mixture of differently refrangible rays."[19] Colors in turn "are not qualifications of light, but original and connate properties, which are in divers rays divers . . . "[20] for experiments showed that no attempts to change a particular spectral color were effective.

Newton insisted that colors are to be defined quantitatively by their differing indices of refraction. Finally Newton showed that white light, which could be dispersed into the spectral colors, could also be reconstituted by bringing the spectral colors back to a focus using reversed refractions:

> Hence, therefore it comes to pass that *whiteness* is the usual color of light; for light is a confused aggregate of Rays induced with all sorts of colors as they are promiscuously darted from the various parts of luminous bodies.[21]

In his concluding remarks, Newton offered one additional critical inference from his experiments and one methodological claim, both of which signaled important tendencies in his thought. Reflecting briefly on "what light is," he argued that

> since colors are the qualities of light, having its Rays for their entire and immediate subject, how can we think of those Rays as qualities also, unless one quality may be the subject of and sustain another; which is in effect to call it a *substance*. We should not know bodies for substances, were it not for their sensible qualities, and the principle of those being now found due to something else, we have as good reason to believe that to be a substance also. . . . But, to determine more absolutely what light is, after what manner refracted, and by what modes or actions it produceth in our minds the Phantasms of Colors, is not so easy. *And I shall not mix conjectures with certainties.*[22]

Several features of this first paper are worth special consideration in connection with the reputation that Newton and Newtonian science achieved both for its emphasis on experiment, quantification, and logical rigor, and for its explicit renunciation of "mechanical" causes and of "hypothetical" reasoning in general. In the last sentence Newton absolutely insisted that conjectures and certain-

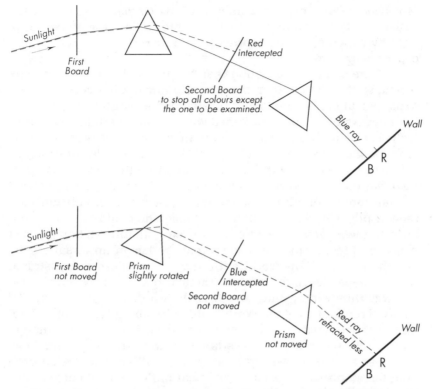

Fig. 3. The experimental configuration for Newton's analysis of the composition of white light.

ties be rigorously distinguished in his scientific writing. Later, in a famous comment added to the second edition of the *Principia*, he reiterated this attitude in speaking of gravity and of his unwillingness to speculate about its causes:

> Hitherto I have not been able to discover the cause of [the] properties of gravity from phenomena, and I feign no hypothesis; and hypotheses, whether metaphysical or physical, whether of occult qualities or mechanical, have no place in experimental philosophy. In this philosophy particular propositions are inferred from the phenomena and afterward rendered general by induction. . . . And to us it is enough that gravity does really exist and act according to the laws which we have explained, and abundantly serves to account for all the options of the celestial bodies and of our sea.[23]

Moreover, in the very first sentence of the *Opticks* (1704), Newton insisted, "My design in this Book is not to explain the properties of light by hypothesis, but to propose and *prove* them by reason and experiments."[24]

There can be no question that Newton and his followers took seriously the notion that there is a radical difference between hypothetical explanations and full-fledged scientific accounts that were understood to be established with as much certainty as empirically based knowledge could attain. In the early optics paper, Newton thought, for example, that he had been able to *demonstrate,* as a consequence of his experiments, that white light was a composite of various colored rays, that colors were uniquely characterized by their indices of refraction, and that light rays were substances—though precisely what kind of substances he could not say. Similarly, in the *Principia,* Newton convinced both himself and the vast majority of his competent readers that the law of universal gravity and the system of the world built upon it had been demonstrated without dependence on any hypothetical reasoning. Newton's system was thus widely understood, in the words of Jean d'Alembert, as "the True System of the World" which had been "recognized," not merely constructed or supposed.[25] Moreover, it was widely, though improperly, understood to be certain and not merely probable. Newtonian science thus seemed of a substantially different kind both from the empirically based mechanical philosophy of persons like Walter Charleton and Robert Boyle, who made no claims to "certainty," and from the mathematically based philosophy of Descartes, who purported to be able to *demonstrate* the necessity of certain features of the cosmos from basic metaphysical or theological assumptions about the goodness and power of God.

Yet there is something odd about Newton's claim that he does not "mix conjectures with certainty" or "feign hypotheses." His writings are filled with conjectures—such as the conjecture that light is globular and resembles tennis balls, which is clearly presented in the first optics paper. Moreover it is undeniable that he *did* speculate about the cause of gravity—not only privately, but also in print. It has even been argued very convincingly that, so far as the study of experimental natural philosophy in the eighteenth century is concerned, Newton's conjectures and hypotheses—presented both in the General Scholium to the third edition of the *Principia* and in the Queries presented in various editions of the *Opticks*—

were *more* important than the antihypothetical tradition of the *Principia*.[26]

How, then, are we to understand Newton's outspoken hostility to hypotheses and conjectures?[27] How are we to understand the response to this hostility on the part of subsequent Newtonians? First, it seems to me that scholars like Lord Keynes, Thomas Kuhn, and Frank Manuel are correct in seeing a fundamental source of Newton's *obsession* with rigor and certainty in his psychological makeup. According to Keynes:

> . . . in vulgar modern terms, Newton was profoundly neurotic of a not unfamiliar type, but—I should say from the records—a most extreme example. His deepest instincts were occult, esoteric, semantic—with profound shrinking from the world, a paralyzing fear of exposing his thoughts, his beliefs, his discoveries in all nakedness to the inspection and criticism of the world.[28]

According to Kuhn, "Newton's fear of exposure and the correlated compulsion to be invariably and entirely immune to criticism show throughout the controversial writings."[29] In the face of competent criticism, Newton usually responded with hostility, denials that he did as critics claimed, and threats to withhold future publications. Only when faced with obviously incompetent and nonthreatening criticisms was he able to respond openly and without antagonism.[30]

At one level Newton's unwillingness to admit his own fallibility led to fairly straightforward distortions. When Hooke criticized Newton for "supposing" in his first paper that light was a body, for example, Newton insisted:

> Had I intended any such hypothesis, I should somewhere have explained it. But I knew that the properties, which I declared of Light, were in some measure capable of being explicated not only by that but many other mechanical *hypotheses*. And therefore I chose to decline them all, speaking of Light in general terms, considering it abstractly . . . without determining what that thing is.[31]

In this instance one gets the distinct impression that Newton protested too much. He probably realized that Hooke had caught him in a failure of rigor and was trying to bluff his way out.

As time went on, two different things happened. On the one hand, Newton increasingly lost his ability to recognize some of the assumptions implicit in his own work—or more accurately, certain assumptions, such as the existence of perfectly void absolute space, of absolute and uniform time, of atoms, and of nonmechanical forces, became so obviously true to Newton that he could not see them as *suppositions.* Thus, as Alexandre Koyre has pointed out, "The expression 'hypothesis' seems to have become . . . one of those curious terms, such as 'heresy,' that we never apply to ourselves, but only to others. As for us, *we* do not feign hypotheses, *we* are not the heretics. It is *they,* the Baconians, the Cartesians, Leibniz, Hooke and others—they feign hypotheses and they are the heretics."[32] To a much larger extent than we might like to admit, the success of Newton's explanatory schemes blinded many of his followers—even into the twentieth century—to the *assumed* character of such Newtonian constructs as space and time. Indeed, not until the work of Albert Einstein did the "obvious" character of these Newtonian notions face a serious and widely convincing challenge within the physics community. On the other hand, Newton's unwillingness to be wrong produced something more important than mere assertions that he did not mix conjectures with certainties. It forced him, in the face of criticism, to develop a theory of scientific method that really did seem to ensure something approaching certainty—at the same time that it constricted the aims of scientific theorizing. Again he began to formulate these ideas in response to a criticism of his first optics paper, but this time he was responding to the generally favorable comments of the Frenchman, F. Pardies.

Newton willingly acknowledged that a number of hypotheses might well explain his experimental results, but now he insisted that "the doctrine which I explained concerning refraction and colors, consists only in certain properties of light, without regarding any hypothesis, by which those properties might be explained."[33] The difference between this response and the one to Hooke's critique is that Newton here makes the lesser claim that no hypothesis had been used to *establish* the properties of light. Strictly speaking, it was irrelevant whether he took light to be a substance because no special assumption regarding the nature of light had been employed to determine its properties. He added that

> the best and safest method of philosophizing seems to be first to enquire diligently into the properties of things, and establishing these properties by experiments and then to proceed more slowly to hypotheses for the explanation of them. For hypotheses should be *subservient only* in explaining the properties of things, but not assumed in determining them; unless so far as they may furnish experiments. For if the possibility of hypotheses is to be the test of the

truth and reality of things, I see not how certainty can be obtained in any science; since numerous hypotheses may be devised which shall seem to overcome new difficulties.[34]

In this early and unguarded statement Newton presents an attitude toward hypotheses which seems consistent with both his concern for rigor and his subsequent use of imaginative conjectures. Strictly speaking, scientific knowledge is inferred directly from experimental results and not from hypotheses; but hypotheses may well be of use in suggesting experiments—as the "tennis ball" hypothesis of his first optics paper suggested that refracted light travels in curved lines in the air. Later Newtonians would express this notion by saying that hypotheses are like the temporary scaffolding used by masons in constructing a building: the stability of the main structure is independent of the scaffolding, which serves merely to make the job easier.[35]

The brilliant French popularizer of Newton's work, Voltaire, emphasized the master's abandonment of hypothetical systems— and his distinctiveness among seventeenth-century philosophers *on this point*—in a famous preface to the first French translation of the *Principia,* a translation done mostly by his sometime mistress Madame du Chatelet:

> If there were still somebody absurd enough to defend subtle and twisted (screw-formed) matter, to assert that the Earth is an encrusted Sun, that the Moon has been drawn into the vortex of the Earth, that subtle matter produces gravity, and all those other romantic opinions that replaced the ignorance of the Ancients, one would say: this man is a Cartesian; if he should believe in Monads, one would say he is a Leibnizian. But there are no Newtonians as there are no Euclideans. It is the privilege of error to give its name to a sect.[36]

Newton did not believe in principle that the necessary and sufficient causes of such phenomena as the refraction of light or the gravitational attraction between two bodies were ultimately undiscoverable. Indeed, he persisted throughout his career in seeking to discover them, as he persisted in claiming—e.g., in the General Scholium to the *Principia*—that until such causes have been discovered, we should rest comfortable in our knowledge of descriptive laws established on the basis of reason and experiment. Subsequent Newtonians, less confident of our ability to discover the causes of phenomena, insisted that we must "*acquiesce in a law of nature* according to which the effect is produced, as the utmost that natural philosophy can reach, leaving what can be known of the agent, or efficient cause, to metaphysics or natural theology."[37]

Newton's natural philosophy was grounded in a firm belief in the existence of a "real," physical world independent of its human observers. As we shall see, he was insistent that natural philosophy could establish not only the possibility, but even the necessity of the divine creation of the universe. At the same time, there can be no question that the methodological position which Newton developed in connection with his early optical work and which he expressed in the *Principia* was extremely important in spreading the idea that far and away the most important function of the natural philosopher was to establish unimpeachable and hypothesis-independent descriptive laws expressed in mathematical form. On this issue Newton was a key figure in shifting the focus of physical science away from the concern with causal explanations, a concern equally central to rationalist and empiricist traditions of mechanical philosophy before him.

When Newton wrote his major scientific masterpiece, he chose to title it *Philosophiae Naturalis Principia Mathematica* (The Mathematical Principles of Natural Philosophy). This title was almost certainly chosen to suggest both that his natural philosophy had the same scope as that of Descartes—which was given its fullest expression in a work entitled *Principia Philosophiae* (1644)—and that it was distinguished by the special role played by mathematics. To see how this special role for mathematics was developed and to provide a final illustration of the character of Newton's scientific work, we will look briefly at the strategy and a few of the special tactics used by Newton in the *Principia*.

The *Principia* is presented in three books. Book I has the structure of a mathematical treatise, laying out a series of definitions of terms like quantity of matter, quantity of motion, acceleration, and innate, impressed, and centripetal forces. Then three axioms, or laws of motion, are stated; and finally a series of propositions, lemmas, and corollaries are derived regarding masses moving under the influence of central forces—i.e., forces directed toward some point. Two key scholia explain (1) how *relative* spaces, motions, and times, which we ordinarily measure, are related to *absolute* space and time, which Newton presumes to exist, and (2) how the traditional science of mechanics (the explanation of how simple machines work) and the science of ballistics developed by Galileo can be derived from Newton's definitions and axioms.

The axioms and definitions of Book I were obviously "principles received by mathematicians" and "confirmed by abundance of experiments."[38] An extensive literature exists analyzing the sources of Newton's laws and definitions and the subsequent history of their impact. For our present purposes, why Newton chose the particular definitions he did and how they subsequently stimulated such figures as Immanuel Kant, Ernst Mach, and even Albert Einstein to formulate new and interesting thoughts and theories is not important. Strictly speaking, the propositions of Book I, like the propositions of Euclid's *Geometry*, were true whether any entities in the natural world existed at all or had those characteristics that Newton assigned to "matter" and "forces." As long as one used the terms as Newton defined them and the axioms as he stated them, then the propositions followed logically.

Newton's style of thought is beautifully illuminated in the choice of the first few propositions and corollaries of Book I of the *Principia*. Proposition I states: "The Areas which revolving bodies describe about an unmovable center of force do lie in the same immovable planes and are proportional to the times in which they are described." Though he does not tell us why this proposition should be interesting, any of his alert contemporary readers would have recognized that it provided a general statement very similar to a rule of planetary motion recently stated by Johannes Kepler—i.e., that the planets move in elliptical curves such that the radius drawn to one focus describes areas proportional to the times in which they are described (see fig. 4). Newton also demonstrates in the corollaries that Proposition I can be generalized so that it is true for curves defined with respect not only to immovable points but also to points that move uniformly in a straight line. His passion for the greatest possible generality and inclusiveness is one of the most characteristic and impressive features of Newton's work.

If any single proposition of the *Principia* can be said to be strategically most important, it is Proposition II: "Every body that moves in any curved line described in a plane, and by a radius, drawn to a point either immovable, or moving forward with an uniform rectilinear motion, describes about that point areas proportional to the times, is urged by a centripetal force directed to that point."[39] Proposition I had shown that Kepler's observation of the

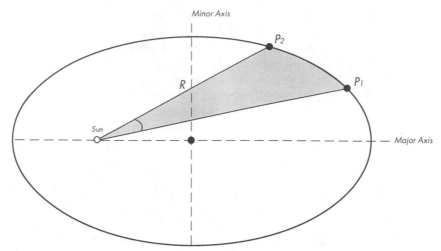

Fig. 4. A Keplerian planetary orbit with exaggerated eccentricity. The planet, *P*, moves so that the shaded area is proportional to the time it takes to move from P_1 to P_2.

motion of planets could be accounted for if one *assumed* the existence of a central force acting toward one focus of an ellipse, but it did not exclude the possibility that some other force might account for the same observed motion. Proposition II now guaranteed that Keplerian motion could exist *only* under the influence of a central force. Once again, in the next proposition, Newton generalized his results to show that even when the center of force was itself moving in some accelerated fashion, a body undergoing Keplerian motion about some other body had to be under the influence of a compound force including a central force toward that body and the accelerative force by which that body was itself moved.

Proposition IV and its corollaries deal with bodies moving in circles and the relationships between the radii of the circles, the periods of motion, and the magnitudes of the forces (see fig. 5). Corollary 6 to Proposition II is of special interest, for it asserts that when bodies move in concentric circles such that the periods of motion (i.e., the times taken to complete one circuit) are to one another as the radii raised to the 3/2 power ($T = r^{3/2}$), then the central forces are to one another as the inverse squares of the radii ($F \propto 1/r^2$). Once again the discerning contemporary reader would have recognized that Kepler and many others had observed that the planets move in nearly circular orbits about the sun such that their periods of revolution were proportional to the 3/2 power of the radii of their orbits.

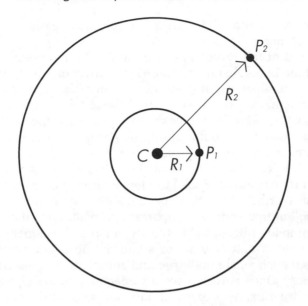

where T is the period of revolution
  F is the force keeping body P moving in circle of radius R,

$$\frac{T_1}{T_2} = \frac{r_1^{\frac{3}{2}}}{r_2^{\frac{3}{2}}}$$

then $$\frac{F_1}{F_2} = \frac{r_2^{\,2}}{r_1^{\,2}}$$

or $F \propto \dfrac{1}{r^2}$

Fig. 5. Illustration of Newton's *Principia*, Book 1, Proposition IV, Corollary 6.

Subsequent propositions in Book I generalized the results of Proposition IV and its corollaries to ellipses and other conic sections as well as circles. They showed how bodies moving under the influence of more than one central force would move. They showed that uniformly dense spherical aggregates of central force-producing bodies acted as if they had all their mass concentrated at their centers. Moreover, they provided a number of additional useful techniques for finding or approximating the orbits of bodies subject to known forces and for inferring the forces necessary to produce known motions.

Book II of the *Principia* can be ignored for present purposes, though it too had substantial long-term impact on the development of the natural sciences. It presented techniques for analyzing motion in resisting fluids. Then it used these techniques to demonstrate that Cartesian vortex cosmology was insupportable and to analyze wave motions; but it played no positive role in establishing Newtonian celestial mechanics.

We now turn to Book III of the *Principia*, in which Newton used the propositions of Book I to "demonstrate the frame of the system of the World."[40] This book begins with a set of four innocuous appearing but tremendously important propositions labeled "Rules of Reasoning in Philosophy" in the second and subsequent editions. The first rule is that "We are to admit no more causes of natural things than such as are both true and sufficient to explain their appearances."[41] Once we discover an adequate cause of some phenomenon, we must not expect redundancy; and we cannot allow an event to be overdetermined. But what did Newton mean by insisting that a cause be "true" if he is not simply begging the question? Rule II helps to clarify this issue; for it begins oddly: "*Therefore* to the same natural effects, we must, as far as possible, assign the same causes." In connection with this rule Newton cited several examples, including the reflection of light in the earth and in the planets. Borrowing from an argument that Galileo had used in the first of his *Dialogues Concerning the Two Great World Systems,* Newton asked his readers to suppose that what we seek to explain is how sunlight is reflected by the moon, creating the illusion that the moon is a light-source. Since we can already be presumed to know the causes of reflections from the objects that we can manipulate on the surface of the earth, those causes are "true." If we can now show that lunar reflections are adequately explained by the laws of terrestrial optics, we will have found a cause which is *both* true *and* sufficient. A true cause is thus one whose operation is known from effects not currently under consideration.

Rule III states that "the qualities of bodies, which admit neither intention nor remission of degrees, and which are found to belong to all bodies within the reach of our experiments, are to be esteemed the universal qualities of all bodies whatsoever." Two things are being claimed in this rule. First, by using the phrase "which admit neither intention nor remission of degrees" to identify the qualities of bodies to be considered here, Newton implicitly accepted the distinction made by Galileo, Boyle, and Locke, among others, between *primary* qualities, which are the properties of bodies themselves and which cannot be increased or decreased without changing the body, and *secondary* qualities, like temperature or sound, which may be varied and which are produced by the powers of the bodies to produce sensations in observers. Second, he insisted much more explicitly that qualities empirically established in that limited class of bodies accessible to direct experience must be assumed to exist even in bodies to which we might not have direct access. Since, for example, all bodies that we can touch are hard or impenetrable, we may—no, must—suppose that even the stars, which we cannot touch, are also impenetrable.

Newton's final rule is "In experimental philosophy we are to look upon propositions collected by general induction from phenomena as accurately or very nearly true, notwithstanding any contrary hypothesis that may be imagined, till such time as other phenomena occur, by which they may either be made more accurate or liable to exceptions."[42] Again, Newton stated two different ideals in one rule. First, he argued that empirical evidence must take priority over *all* theoretical considerations in natural science; so that no piece of well-founded empirical evidence can be rejected because it is inconsistent with someone's theoretical structure—no matter how compelling that theory may be. Second, he allowed that all scientific knowledge is provisional and subject to revision in the face of newly acquired evidence.

Following the Rules of Reasoning, there is a set of six "phenomena" listed, all taken from the observations of contemporary astronomers:

1. Jupiter's satellites move such that the radii drawn from them to Jupiter's center describe areas proportional to their times of motion; moreover their periods of motion are as their radii to the 3/2 power.
2. Saturn's satellites move in the same manner as Jupiter's.
3. The planets orbit around the sun.
4. The periods of motion of the planets about the sun are proportional to their mean distances from the sun raised to the 3/2 power.
5. Each planet moves such that the radius drawn from it to the sun describes equal areas in equal times.
6. The moon moves such that the radius drawn from it to the center of the earth describes equal areas in equal times.[43]

Now Newton readily combined the mathematical propositions of Book I, The Rules of Reasoning, and the Phenomena to form his system of the world. In Proposition I he took Phenomenon 1 and combined it with Propositions III and IV from Book I to establish that Jupiter's moons are moved by forces directed toward the center of Jupiter and proportional to $1/r^2$, where r is the distance from the moon to the planet. The same is true for Saturn's moons.

In Proposition II, he combined Phenomena 4 and 5 with Propositions II and IV of Book I to demonstrate that the planets are moved by central inverse square law ($F \propto 1/r^2$) forces directed to the sun.

The Rules of Reasoning come into play next to demonstrate Proposition III—i.e., that the moon moves under the influence of an inverse square law force directed to the center of the earth. In this case the central character of the force follows directly from the Phenomena and Proposition III of Book I; but since there is no other natural satellite circling the earth, we do not have an observed set of period relations to use in establishing the $1/r^2$ law. At this point, we draw upon Rules 1 and 2. From Proposition I of Book III we know that inverse square forces are *true* causes of lunar motions for Jupiter and Saturn; furthermore, a $1/r^2$ force is certainly *sufficient* to explain the motion of the earth's moon; so by Rule 2 we must assign the same cause to the motion of the earth's moon as to the motions of those of Jupiter and Saturn.

In Proposition IV another critical use of Rules 1 and 2 allowed Newton to identify the $1/r^2$ force causing the moon's motion with the force of gravity previously known to cause the fall of bodies at the earth's surface. Huyghens, Galileo, and others had measured the distance that a body falls in one second at the surface of the earth to be about 15 1/12 Paris feet ($\sim$ 16 English feet). If the gravitational force were an inverse square law force, one could ask how far an object would fall in a unit of time at the lunar distance (60 earth radii) under the influence of gravity. The answer was that it would fall 15 1/12 feet per minute. An independent calculation of the centripetal motion of the moon in its orbit showed that it was falling toward the center of the earth at the rate of $\sim$15 1/12 feet/ minute. Consequently, Newton wrote that

> [by Rules 1 and 2] the force by which the moon is retained in its orbit is that very same force which we commonly call gravity; for were gravity another force different from that, then bodies descending to the earth with the joint impulse of both forces would fall with a double velocity [which is] altogether against experience.[44]

By Rule 2 Newton showed in Proposition V that gravitational forces, being both true and sufficient, must be the sole cause of the motion of the planets about the sun and of the satellites moving about the other planets. In Proposition IV, Newton demonstrated that gravitational forces acting between any two astronomical bodies are proportional to the product of the masses of the two bodies.[45] Finally, in Proposition VII, Newton used Rule 3 to argue that, since we discover gravitational forces between astronomical bodies and between ordinary gross matter on the earth and the earth itself, *we must infer that all bodies in the universe* attract one another according to a law of the form

$$F_{(1,2)} = \frac{gM_1M_2}{r_{(1,2)}{}^2}$$

where $F_{(1,2)}$ is the force drawing masses M_1 and M_2 together at a distance of $r_{(1,2)}$. Moreover, since the inverse square law gravitational force is induced from experience by Rule 4, it could not be challenged on any merely hypothetical or theoretical grounds.

Here is Newton's law of universal gravitation in all of the elegant simplicity, certainty, and glory that seemed to make it the most marvelous of achievements and a sign to everyone that humans might after all triumph over their skeptical doubts and discover the great mysteries of nature. Alexander Pope caught a sentiment that was widespread—though not universal—in writing his famous couplet:

> Nature and nature's laws lay hid in night:
> God said, "Let Newton Be" and all was light.[46]

Jean d'Alembert, the great French mathematician and philosopher, was almost equally enthusiastic, but a good deal more insightful in his assessment of the impact of Newton's work:

The true system of the world has been recognized . . . from the earth to Saturn, from the history of the heavens to that of insects, natural philosophy has been revolutionized; and nearly all other fields of knowledge have assumed new forms—the discovery and application of a new method of philosophizing, the kind of enthusiasm which accompanies discoveries, a certain kind of exaltation of ideas which the spectacle of the universe produces in us—have brought about a lively fermentation of minds. Spreading—in all directions, this fermentation has swept with a sort of violence everything before it which stood in its way, like a river which has burst its dams. . . . Thus, from the principles of the secular sciences, to the foundations of religious revelation, from metaphysics to matters of taste, from music to morals . . . everything has been discussed, analyzed, or at least mentioned. The fruit or sequel of this general effervescence . . . has been to cast new light on some matters and *new shadows on others*; just as the effect of the ebb and flow of the tide leaves some things on the shore and washes others away.[47]

The law of universal gravitation was such a marvelous symbol of hope because it was able to explain a range of detailed phenomena through the use of a single, amazingly simple law. Not only did it unify celestial and terrestrial phenomena, but as its detailed applications to celestial mechanics were developed throughout the eighteenth and early nineteenth centuries, it was able to account for both the gross features of heavenly motions and the most initially puzzling apparent anomalies—phenomena like the precession of the equinoxes; nutation (the wobble of the earth's axis caused by the nonspherical shape of the earth coupled with the tilt of the lunar orbit); and the deviation from closed paths about the sun which the mutual interaction among planets produces. When the planet Uranus was discovered and its observed path seemed distinctly odd, the law of gravitation even led to the prediction and discovery of a new planet, Neptune.[48]

It is not true that Newton's works were *directly* influential in the same way that, say, Aristotle's had been, or that Galileo's or Bacon's or even John Locke's were. They did not form the basis of a kind of Newtonian scholasticism in which every aspiring intellectual had to confront and comment on the *Opticks* and *Principia* line by line; nor were they presented in the kind of popular prose that made Galileo and Locke accessible to huge audiences. The *Opticks* probably was widely read among European scientists, but my guess is that perhaps only fifty men and one woman—Madame du Chatelet—were able to follow the detailed arguments of the *Principia* during the first half century after its publication. Newton's ideas and attitudes entered European intellectual life not directly, but through a series of able popularizations in the eighteenth century.

In England, popular overviews of Newtonian natural philosophy were available in Henry Pemberton's *A View of Sir Isaac Newton's Philosophy* (London, 1728) and Colin McLaurin's *An Account of Sir Isaac Newton's Philosophical Discoveries* (London, 1746), while Newtonian teaching entered the textbook tradition in natural philosophy primarily through Samuel Clarke's extensive notes to his translation of the Cartesian textbook, Jacques Rohault's *Physics* (first ed., 1697; fourth ed., 1718).

Perhaps the most popular of all Newtonian accounts was the Italian *Il Newtonianismo per le Dame*. This work, published by Jean Algorotti at Milan in 1737, went through five Italian editions in three years and was soon translated into French and English. Voltaire produced his superb *Eléments de la Philosophie Newtonienne* with the help of his mathematician/mistress Madame du Chatelet in 1738; and German lay audiences were introduced to Newtonian ideas primarily in Leonhard Euler's *Lettres à une Princess d'Allemagne* (1768–1772). By 1749 fascination with Newtonianism was so pervasive that the French literary dilettante, Jean Baptiste de Boyer d'Agens, could write, "One may see Lawyers forsake the Bar to busy themselves in the study of Attraction, and Divines neglect their Theological exercises for its sake."[49] In England Newtonian natural philosophy even entered children's literature through such marvelous books as John Newbery's *The Newtonian System of the World, Adapted to the Capacities of Young Gentlemen and Ladies* (London, 1761). English pride in Newton's accomplishments led the popular writer Benjamin Martin to proclaim: "We may truly affirm, That it is more Honor to be King of the Learned English Nation, than Emperor of all the world besides."[50]

NEWTONIAN NATURAL THEOLOGY

The first major and influential public attempt to apply Newtonian science to theological ends was undertaken by a Cambridge cleric and classicist, Richard Bentley. When Robert Boyle died in 1690, he endowed a series of annual lectures aimed at setting forth "the Truth of the Christian religion in General, without descending to the Subdivisions among Christians." Tapped to give the first series of lectures of 1691, Bentley decided that since Newton thought that some of the arguments of the *Principia* might be applied to theological ends, he would devote at least one of his eight lectures to the religious implications of Newton's work.

First, Bentley asked the mathematician John Craig to suggest some mathematical texts that he might consult to prepare for reading the *Principia*; but Craig's suggestions were so daunting that Bentley wrote directly to Newton in hopes of finding some easier way. Newton was much more encouraging. "When I wrote my Treatise about our system," he replied, "I had an eye upon such principles as might work with considering men, for the belief of a Deity, and nothing can rejoice me more than to find it useful for that purpose."[51] Moreover, Newton suggested that it would be enough for Bentley to read the early Propositions of Books I and III in order to recognize the major theological consequences of the *Principia.* Eventually, Newton wrote a series of four letters to Bentley that were closely followed in the latter's *A Confutation of Atheism from the Origin and Frame of the World* (London, 1692 and 1693). According to Perry Miller, the Bentley sermons were a marvelous success, setting a pattern for the early Enlightenment, giving believers "the assurance (or perhaps one should say the illusion) that the Newtonian physics, by conclusively showing that the order of the universe could not have been produced mechanically, was now the chief support of faith."[52]

To the traditional arguments for design which had been discussed for several decades in the Anglican tradition, Bentley added three variants, appropriating Newton's theories. Earlier natural theologians had purportedly demonstrated the falsity of ancient atomist claims that the ordered universe we know could have emerged out of chaos through purely chance, mechanical interactions among particles moving in a void. Bentley used Newton's natural philosophy to show additionally that no Cartesian theory which demanded a world without void could account for the universe as we know it.[53] Next, he demonstrated that no initially chaotic or uniformly distributed aggregate of mutually attracting bodies could produce the experienced world. If the universe were finite and gravitating particles had initially been randomly or uniformly distributed, they would in time have fallen into one great clump in the middle;[54] in an infinite universe, however, they could not have created large bodies like the planets and stars which we observe. Even if one could imagine some chance circumstances leading to the formulation of large celestial bodies, there could be no way for mutually attracting bodies to fall together to generate a system like the solar system which has a huge angular momentum.[55] Finally, even if the planetary system could somehow have acquired angular momentum, the planets could not have ended up in nearly circular orbits. The production of any transverse motion of the planets around the sun "could in no wise be attained without the power of the Divine Arm";[56] and that transverse motion should have been adjusted to create nearly circular orbits clearly indicated not just the *power* of that arm but also the *wisdom* with which it was wielded.[57]

A second set of arguments Bentley claimed to be completely original—but they at least paralleled arguments that had long been current among anti-Hobbesian natural theologians for decades. Such corpuscular philosophers as Charleton and Boyle had argued that matter's passivity ensured the need for God both to initiate and to conserve its motions. Now, since Newton had demonstrated that all matter attracts other matter toward itself, Bentley proposed to demonstrate the need for some immaterial, active, divine agent in the production of gravity, just as others had demonstrated the need for a divine initiator and conservator of motions.

Bentley began by offering his own formulation of a basic premise which had been accepted by natural philosophers since the time of Aristotle, who had said that no body can act where it is not: "'Tis utterly inconceivable," Bentley argued, "that inanimate brute matter (without the mediation of some Immaterial Being) should operate upon and affect other matter without mutual contact."[58] Since every particle of matter attracts every other, no matter how distant, according to Newton, either some impulse is conveyed through the air, ether, or some other medium; some effluvium is emitted to the one from the other; or there is a spiritual, immaterial cause of gravity. Gravity cannot be produced by impulse; for "to do this mechanically, the same physical point of matter must move all manner of ways *equally* and *constantly* in the same instant, which is flatly impossible."[59] By the same token, Bentley was convinced that no small particle could conceivably emit "such multitudes of Effluvia as to lay hold of every atom of the universe without missing one."[60] Gravitation is thus "above all mechanism and material causes, and proceeds from a higher principle, a Divine energy and impression."[61]

The later Boyle lecturer and Newtonian, William Whiston, reiterated this point:

> 'Tis now evident, that *Gravity*, the most mechanical affection of Bodies, and which seems most natural, depends entirely on the constant and efficacious, and if you will, the supernatural and miraculous Influence of Almighty God.[62]

Although subsequent Newtonian natural theologians like Samuel Clarke were inclined to admit that the immaterial agent producing gravity might not be the "immediate Finger of God" insisted upon by Bentley,[63] but rather some subordinate instrument, they all agreed that gravity provided conclusive evidence in favor of non-material causes and thus undermined atheistic materialism.[64] Even from the dissenting community came congratulations on this "excellent use of . . . the law of Gravitation to demonstrate the continual Providence and Energy of the Almighty."[65] In this case, as in many others, Newton was personally more cautious than his admirers. He too agreed that gravity implied some agent, "but," he wrote to Bentley, "whether this agent be material or immaterial" he was not absolutely certain.[66]

Bentley's final modification of traditional arguments also addressed the issue of God's continuing providential activity rather than his initial creative activity; but it raised an interesting set of problems and opportunities for Newtonian natural theology. In Proposition X, Book III of the *Principia,* Newton had demonstrated that the planets moved in orbits that were *very nearly* stable. If there were absolutely no resistance to motion and no net gravitational effect on a planet other than that directed toward the sun, then the absolute stability of the solar system would be assured; but the total absence of resistance and of net gravitational forces on planets in the solar system was unlikely. Even more critical—at least in Bentley's argument—in a world in which gravitation was the only universal force, there must be a long-term tendency toward collapse: "For though the universe was infinite, the fixed stars could not be fixed, but would naturally convene together, and confound system with system; for all mutually attracting, every one would move whither it was most powerfully drawn."[67] Though Newton had explained to him that stability might have been possible in an infinite universe in which "the very mathematical center of Gravity of every system be placed and fixed in the mathematical center of all the rest,"[68] both men agreed that no evidence supported the notion that God had so arranged all celestial masses.

Under the circumstances, Bentley insisted, the direct providential activity of God was necessary to conserve the stars in their stable system or to readjust the system occasionally to stave off collapse.[69] In the first Latin addition to the *Opticks* in 1706, Newton publicly lent his name to the notion that even if the universe had not yet required God's special intervention, it would eventually, because of "some inconsiderable irregularities . . . which may have risen from the mutual Actions of Comets and Planets upon one another, and which will be apt to increase, till this system wants a Reformation."[70] And this position was reaffirmed by Newton's able scientific expositor, Colin McLaurin, who wrote that "the Deity has formed the Universe dependent upon himself, so as to require to be altered by him, tho at very distant periods of time. . . ."[71]

Like Newton, McLaurin was not certain whether the "alterations" necessarily would be produced by ordinary or special providence—i.e., by secondary instruments or by the immediate interference of God. Empirical evidence did not yet offer a way to resolve such a question, and, frankly, McLaurin felt that "it does not appear to be a very important question."[72] But if the instability of the Universe and its need for reformation by God was reassuring but unimportant to a few Newtonians, who neither could nor particularly cared to resolve the question of precisely how it might get reformed, that was not true for several very important groups. It was not true for continental opponents of Newtonianism who saw in the imperfection of the Newtonian universe a serious chink in the great man's armor, through which they might attack his whole system of philosophy. It was not true for many Latitudinarians who had found a way to use the imperfection argument for critical ideological purposes. Moreover, it was not true for many High Church opponents of the Latitudinarians, who found themselves having to battle Newtonian science in order to reject the Latitudinarian ideology it supported.

To Leibniz, it was a travesty to think that the omniscient and omnipotent creator could have bungled the job of creation so badly that it needed patching up occasionally. In 1715 he wrote to Princess Caroline of Wales (soon to be Queen Caroline) to express his fears that Newtonianism, which he opposed for both philosophical and personal reasons, was one of the root causes of the deterioration of morals in England. Samuel Clarke—with private help from Newton—undertook a defense of Newtonian natural theology; and the Leibniz-Clarke correspondence, which was published in 1717 after the German's death, elevated the argument regarding the proof for God based on the instability of the universe into a philosophical and theological *cause célèbre,* with long-term consequences unintended by either side.

Leibniz's attacks pushed Newtonian natural theology into an all but exclusive emphasis on what we might call "the argument from imperfection." Not only did it ensure the need for God's ongoing activity in the world but, by analogy, provided a strong argument against the atheistic assertion of an externally existing universe. "A more convincing proof cannot be devised against the

Some of the Principal Inhabitants of y^e MOON, as they Were Perfectly Discover'd by a Telescope brought to y^e Greatest Perfection since y^e last Eclipse; Exactly Engraved from the Objects, whereby y^e Curious may Guess at their Religion, Manners, &c.

Pl. 3. William Hogarth's engraving satirizing the linkages between the mechanical philosophy and Anglican natural theology.

present constituents having existed form eternity than this, that a certain number of years will bring it to an end," wrote Henry Pemberton.[73] Clarke turned the issue directly against Leibniz: "Whether my inference from this learned author's affirming that the universe cannot diminish in perfection . . . [that] the world must needs have been . . . eternal, be a just inference or no, I am willing to leave the learned to judge."[74]

As always, however, for the Newtonians, it was God's present dominion, rather than his past creative act, that was most important; and Clarke found a peculiarly apt way to express the Newtonian view to Caroline, a member of the royal family:

> As those men, who pretend that in an earthly government things may go on perfectly well without the King himself ordering or disposing of anything, may reasonably be suspected that they would very well like to set the King aside . . . , so too those who think that the universe does not constantly need "God's actual government" but that the laws of mechanism alone would allow phenomena to continue, in effect tend to exclude God out of the World.[75]

It was the imperfection of the physical universe which thus served to undermine both atheistic and deistic appropriations of natural philosophy to their own ends.

Clarke's causal linkage of Newtonian natural theology with monarchical political ideas in the passage above raises an important issue regarding the reception of Newtonianism in early eighteenth-century England. Margaret Jacob has argued that the Newtonian emphasis on God's dominion—which combined an emphasis on the regular orderly government of nature through God's ordinary providence with the assertion that special interventions to "reform" the natural world were not merely possible, but necessary—played a special *political* role in England after the Glorious Revolution of 1688. It offered those who benefited most from the revolution—the liberal Anglican clergy and the growing class of capitalists—a way to rationalize their support of a revolutionary change without acceding to any form of revolutionary ideology.

Latitudinarian clergy found themselves in a peculiarly embarrassing position after the Restoration. As members of a state church, they viewed themselves as responsible for the morality and social cohesion of the nation; and as inheritors of a long-standing ameliorative position which had opposed all of the forces for fragmentation that had produced the great Civil War, they were committed to a stable monarchy and social order. Thus they constantly preached the duty of passive obedience to the Crown. At the same time, they were opposed to and appalled by James II's pro-Catholic tendencies. During the period from 1685 to 1688, the liberal clergy found themselves caught between their equally deep-seated anti-Catholicism and their political and social conservatism. As a consequence, most of the Latitudinarian clergy played almost no role in the events leading to the replacement of James, who was sympathetic to the Catholics, with the Protestant line of William and Mary.

After the Revolution, the Latitudinarians found themselves raised to dominance in the Church, largely because High-Church clerics were so committed to divine-right monarchical beliefs that their conservatism vastly outweighed their anti-Catholicism. Thus High Churchmen for the most part opposed William and Mary's accession to the throne; and they found themselves excluded from preferment in the Church, even though they probably represented a substantial majority among the clergy. Under these circumstances, the Latitudinarians had to find a way to defend and justify their position of power, not just to reassure themselves, but, more importantly, to stave off the attacks of the majority party of High Churchmen. Moreover, they had to accomplish this feat without acknowledging that disobedience to the king was justifiable under any foreseeable circumstances. No doubt most of the Latitudinarians really believed in obedience to the monarch; but even if they had not, it was clear that unless they argued that way, they too would lose their positions, which were dependent on the pleasure of the monarch.

No traditional divine-right theory of monarchical legitimacy could have justified the appropriation of power to replace the reigning monarch by Parliament. At the same time, no secular contract theory based on Machiavellian or Hobbesian principles could be accepted, both because they denied the fundamentally religious source of authority and social cohesion, thus contributing to the increase in the alarming "atheistic" trends, and because they threatened stability by specifying conditions under which revolutionary action became legitimized.

In this situation, Newtonian cosmology and natural theology offered a clear and obvious way out. Just as God had created a natural world governed by laws representing his ordinary providence but very rarely demanding his special "reformation," so too had he created a "world politick" in which lawful obedience to the monarch was every Christian's duty. Occasionally that "world politick" came so near collapse that only immediate divine intervention could reform it and put it back on track. Thus, John Tillotson, the Latitudinarian bishop, wrote in his private commonplace book, "I look at the King and Queen as two angels in human shape sent down to us to pluck a whole nation out of Sodom, that we may not be destroyed."[76] Similarly, William Lloyd preached in 1691, "The marks of God's Hand were so visible in it [The Glorious Revolution], at first, and are so daily more and more, that he is blind that doth not see them. There is, enough, one would think, to convince even the Atheist to the belief of a Providence. . . . It is plainly the design of God . . . to establish the Protestant religion in these Kingdoms."[77] Just as *only* God can make a tree, *only* God can make a revolution, and conversely, just as God *can* make a tree, God *can* make a revolution.

From a platform built on revolutionary foundations, one could be perfectly justified in preaching a doctrine of obedience to secular authority. William Whiston, one of Newton's closest associates, made this point most blatantly, insisting that obligations to a monarch may not be abrogated: "Even where great inconveniences arise by the observation of them," the people must maintain their oaths, "and bear the inconveniences of that person's government and oppression till his death; or till God by some other means of his Providence, deprive him of that power; *but no longer.*"[78]

While the political uses of Newtonian natural theology—with its emphasis on special providential activity—probably played a marginal role in spreading Newtonian science within the communities of mathematicians and experimental natural philosophers, it almost certainly enhanced the broader appeal of popular Newtonianism—especially his natural theology—in liberal quarters.

Throughout the eighteenth century, increasingly detailed mathematical analyses of the orbital motions of the planets based on Newtonian assumptions and methods showed that many of the secular (nonperiodic) motions feared by Newton and his early followers disappeared, turning into long-term periodic motions.[79] As a consequence, Leibniz' attempt to discredit Newtonian *science,* to the extent that it depended on ridiculing its purportedly unacceptable theological implications, became untenable. At the same time, a Newtonian support for natural theology, which had been raised to great importance by the controversy with Leibniz and by political circumstances, was so completely undermined that when Pierre Laplace was asked by Napoleon why God did not appear in his *Système de la monde*—the most detailed eighteenth-century exposition of Newtonian celestial mechanics—he is purported to have responded, "Sir, I have no need of that hypothesis."[80]

In fairness to Newton and his more subtle followers such as Colin McLaurin, mathematical proofs that the solar system was vastly more stable than they had expected held no threatening implications for them; for they had acknowledged all along the possibility that ordinary laws might eventually be found to account for any necessary adjustments. But most Europeans knew of Newton's argument only from Leibniz' caricature, from Bentley's version, or from the queries to the *Opticks* (Query 23, 1706 ed. or Query 30, 4th ed., 1730), none of which expressed the qualifications that we recognize. As a consequence, a real sense of disillusionment seemed to set in regarding the support that the system of the universe could give to natural theology. Just over a century after Bentley's impassioned *Confutation of Atheism from the Origin and Frame of the World* (1692), and William Derham's equally enthusiastic and even more popular *Astro-Theology* (1711), William Whewell, the devoted churchman and able scientist commissioned to write the Third Bridgewater Treatise, *Astronomy and General Physics Considered with Reference to Natural Theology* (London, 1833), was forced to admit, "I feel most deeply . . . [that] all the speculator concerning natural theology can do is utterly insufficient for the great ends of religion. . . ."[81] Newtonian science now seemed to show that the universe might well continue on forever without divine intervention; and it became a comfort and support to Deists—those who acknowledged the necessity of a supernatural creator of the universe, but who doubted the existence of a providentially active God. They would have been startled to learn of the master's focus on God's continuing dominion.

NEWTON AND SCRIPTURAL PROPHECY

When Newton wrote his first letter to Richard Bentley in 1692, encouraging Bentley to use the propositions of the *Principia* for supporting natural theology, he ended the letter with a very curious passage: "There is yet another argument for a Deity, which I take to be a very strong one, but till the principles on which it is grounded are better received, I think it more advisable to let it sleep."[82] Very recently, James Force has argued that Newton's "Sleeping Argument" was, in fact, the demonstration of God's dominion through properly interpreted biblical prophecy.[83]

Part of Newton's fascination with the prophetic tradition almost certainly derived from arguments made by both Locke and Boyle that prophecies were quite simply the most easily verified miracles, or "signs," of Divine authority. Boyle had made this point with particular clarity in his *Christian Virtuoso*:

> True prophecies of unlikely events, fulfilled by unlikely means, are supernatural things; and as such, (especially their author and design considered) may properly enough be reckoned among miracles. And, I may add, that these have a peculiar advantage above most other miracles, on the score of their duration: since the manifest proofs of the prediction continue still.[84]

Unlike Boyle and Locke, Newton was interested in prophecies not merely as signs of God's authority, but also as instructions regarding our duty to God. He wrote:

> Giving ear to the Prophets is a fundamental character of the true church . . . the authority of Emperors, Kings, and Princess is human. The authority of Councils, Synods, Bishops, and Presbyters is human. The authority of the Prophets is divine, and comprehends the sum of religion. . . . Their writings contain the covenant between God and his people, *with instructions for keeping this covenant. . . .*[85]

Newton hated the vulgar and fanatical enthusiasts who claimed to know precisely what Scripture meant, but he was convinced that if one approached the mysteries of Scriptural prophecy with the same intellectual commitment and care that one used in approaching the secrets of the natural world, one might gain an absolutely crucial insight into God's dominion. According to Newton, the study of prophecy was particularly important in the late seventeenth century, for it was nearly the end of the time of the [Papal] Antichrist:

> Wherefore it concerns thee to look about thee narrowly lest thou shouldst in so degenerate an age be dangerously seduced and not know it. Antichrist was to seduce the whole Christian World and therefore he may easily seduce thee if thou beest not well prepared to discern him.[86]

Like Locke, with whom he corresponded on the topic, Newton argued that only probable, rather than certain arguments could be found for grounding our faith in Scripture; for if *certain* arguments were available, then all men would be forced to believe. No choice would be demanded; no distinction would exist between the good and the wicked:

> I could wish they would consider how contrary it is to God's pur-
> pose that the truth of this religion should be as obvious and
> perspicuous to all men as a mathematical demonstration. 'Tis
> enough that it is able to move the assent of those which he hath
> chosen; and for the rest who are so incredulous, it is just that they
> should be permitted to die in their sins.[87]

Even though faith in and knowledge of scriptural prophecy would never be granted to all, Newton seems to have been convinced that he had a duty to study and communicate his findings in some fashion in order to prepare for the coming of the millen-
nium. To this end he worked throughout his life to develop a system of scriptural interpretation very much modeled on the same lines as his System of the World—a system that could at least help to guide those for whom God intended his prophecies. The parallels between this system of scriptural interpretation and Newton's System of the World show through in many places, but perhaps most clearly in the introduction to a 550-page unpublished manuscript now housed in Jerusalem:

> First I shall lay down certain general Rules of Interpretation,
> the consideration of which may prepare the judgment of the Reader
> and inable him to know when an interpretation is genuine and of
> two interpretations, which is the best.
> Secondly, to prepare the reader also for understanding the
> prophetique language I shall lay down a short description there-
> of. . . . By which means the Language of the Prophets will become
> certain and the liberty of wresting it to private imagination be cut
> of[f]. The heads to which I reduce these words I call Definitions.
> Thirdly, these things being promised, I compare the points to
> the Apocalypse one with another and digest them in order by those
> internal characters which the Holy Ghost hath for this end im-
> pressed upon them. And this I do by drawing up the substance of
> the Prophecy into Propositions, and subjoyning the reasons for the
> Truth of every proposition.[88]

Newton's was far from the sole seventeenth-century attempt to bring intellectual respectability and a scientific approach to prophecy interpretation. Newton was, in fact, a self-professed follower of Joseph Mede, whose *Clavis Apocalyptica* (1627) serves as a cornerstone for almost all early modern interpretations. Mede had been tutor to Henry More, with whom Newton was in constant communication, and to Isaac Barrow, Newton's mathematics teacher. A tradition of Mede-based interpretations of the prophetic books developed across the intellectual and theological spectrum during the seventeenth century.[89] What distinguished Newton's attempts at interpreting Scripture was his unwillingness to rest with any subjective assurance that God's spirit within was any kind of guarantor of the legitimacy of some particular interpretation, and his consequent emphasis on the possibility of an "objective" interpretation based on nothing but sensory evidence and its rational analysis.

In the first instance this meant to Newton that no extant text of any prophetic book could merely be accepted as unimpeachable. To establish some best or most probable text, all of the techniques of textual criticism being developed by such scholars as Richard Simon and Benedictus Spinoza had to be employed. Masoretic, Aramaic, and Greek texts had to be compared, variant readings analyzed, and choices made. Next, a special dictionary of the historical, political, and religious meanings of the images and symbols of prophetic language had to be established. It was not enough simply to choose meanings for the symbolic language which would make particular prophecies intelligible. That would have been equivalent to *speculating* in natural philosophy by assigning causes that might be "sufficient" to account for a phenomenon without also establishing that those causes were also "true." In biblical prophecy, the equivalent of establishing "true" causes was establishing an unambiguous meaning of prophetic terms that came from sources *outside* the interpretation of prophetic texts themselves. Thus Newton insisted on trying to undertake what Hooker had long before seen as necessary to any adequate interpretation of Scripture—i.e., the development of an historical dictionary that would fix the denotation of words as they were used in the particular historical context in which they were produced, the context of Near Eastern (not merely Hebrew) prophetic writings.

To suggest the significance of this dictionary, we need only look at the first of Newton's "Rules for interpreting the words and language in Scripture" expressed in the Yahuda Manuscript: "To observe diligently the consent of Scriptures and analogy of the prophetique style, and to reject those interpretations where this is not duly observed."[90] Since Newton purportedly discovered a full-blown Near Eastern prophetic tradition in which beasts signified political entities or their kings, he could argue, "Thus if any man interpret a beast to signify some great vice, this is to be rejected as his private imagination because according to the style and tenor [of the prophetic style] . . . a beast signifies a body politique and sometimes a single person that heads that body, and there is no ground in Scripture for any other interpretation."[91] This Newtonian injunction has the same intent as modern "form criticism," which argues that the greatest insight into scriptural meanings comes from comparing passages of Scripture with passages from contemporary literature produced by neighboring peoples in institutional contexts that can be well established. No such activity, according to Newton, could lead to *certain* knowledge of the meaning of any term; it could only establish some probable meaning—in contrast with the idiosyncratic speculations of enthusiasts and vulgar interpreters.

In Rule 4 for "interpreting the words and language in Scripture," Newton provided an admonition which is necessary to eliminate ambiguity and which serves as a critical guide to modern fundamentalist Scripture interpretation: "To choose those interpretations which are most according to the literal meaning of the Scriptures unless where the tenor and circumstance of the place plainly require an allegory."[92] Leaving aside any difficulties there might be in applying such a rule, it is easy to understand why it is necessary. There was no question in any believer's mind that Scripture was partly historical and partly prophetic allegory. But if each interpreter were free to decide which passages to read allegorically and which to read as a plain historical account, there could be no hope of establishing any unique meaning. Newton's rule thus was intended to provide a criterion for deciding which linguistic usages were to be applied to interpreting which passages of Scripture. Ordinary language use was to be assumed for all parts of the Bible for which no evidence demanded the use of the special dictionary of prophetic or "hieroglyphic" language.

Once a highly probable meaning of the symbolic or "hieroglyphic" terms of Prophecy was established, one could then work out possible meanings of the Prophecies themselves. These possible meanings could then be tested against observed political and ecclesiastical events; that is, the symbolic statements of the prophetic texts could be translated into statements about observable events that could in turn be empirically tested. At this point the other "Rules of Interpretation" developed by Newton came into play in order to choose from among possible interpretations the best or most probable. Though Newton never satisfied himself that he had produced an adequate set of interpretative rules, a few are worth noting here both to show how they drew from the same sources as his Rules of Reasoning in Natural Philosophy, and to suggest how they were incorporated into modern fundamentalist Scripture interpretation.

Rule 9 of Newton's "Rules for Methodizing the Apocalypse" is perhaps most interesting as a demonstration of both the continuity between the strategy and tactics of the *Principia* and those of Newton's system of Prophecy interpretation as well as the extent to which both depended on "metaphysical" principles of the kind Newton found so repugnant when they were expressed by Leibniz. Rule 9 reads:

> To choose those constructions which, without straining, reduce things to the greatest simplicity. . . . Truth is ever to be found in simplicity, and not in the multiplicity and confusion of things. As the world, which to the naked eye exhibits the greatest variety of objects, appears very simple in its internal constitution when surveyed by a philosophic understanding, and so much the simpler, the better it is understood, so it is in these visions. It is the perfection of all God's works that they are done with the greatest simplicity. He is the God of order and not confusion. And therefore as they that would understand the frame of the world must endeavor to reduce their knowledge to all possible simplicity, so must it be in seeking to understand these visions.[93]

God's preference for simplicity was *not* simply a proposition derived *from* experience for Newton, although it certainly seemed warranted by the success of the *Principia.* In both the *Principia* Rule 1 and in the Yahuda Manuscript 1.1, Rule 9, Newton's assertion that God acts in the simplest possible way is accepted prior to any interpretation of phenomena or revelation; and it is used to guide and control all interpretations. At one level it is a fundamentally arbitrary assumption that restricts God's freedom to act; and there is no real reason to accept such an assumption except blind faith. At another level, however, some variant of this assumption is at the root of all science; for it states a necessary condition for making phenomena intelligible in the systematic or general sense demanded within the scientific enterprise. Science exists only insofar as men and women seek to express regularities among apparently chaotic events. To admit that God was chaotic or completely arbitrary or that the universe was a chaos rather than a cosmos would be to denounce all hope of achieving the kind of knowledge that scientists seek of the natural world. By the same token, Newton insisted that without the assumption of God's simplicity and constancy, Scripture prophecy would be unintelligible and therefore incapable of acting either as a warrant for faith or as a guide to one's Christian duty.

Rule 11, the last rule of the Yahuda Manuscript, also directly parallels the final rule of the *Principia* in its form and intent. Rule 4 of the *Principia* insists that propositions in natural philosophy established through Newton's method must be accepted, "notwithstanding any contrary hypothesis that may be imagined" because no other *method* could produce such strong support. By the same token, Newton's final rule of Scripture interpretation insisted that we must "acquiesce in that construction of the Apocalypse which results most naturally and freely . . . from the observation of the precedent [*sic*] rules."[94] If we do not, we will be accepting an interpretation "grounded upon weaker reasons, [and] that . . . is demonstration enough that it is false." This assertion, of course, is equivalent to the claim that there is one "objective" interpretation of Scripture, discoverable by anyone capable of applying the scientific method of analysis.

Newton offered just one more critical consideration regarding methods of Prophecy interpretation that had a major long-term impact—especially on nineteenth- and twentieth-century Fundamentalism. Given his admission that his method must produce the most probable interpretation of Scripture, but not a *certain* one, Newton was extremely cautious about predicting the times and circumstances of "future" prophecies. Enough of prophecy had already been fulfilled that we can rest assured of God's providential activity. But not enough had yet been accomplished to allow a full foreknowledge of prophesied events. Thus, wrote Newton, "till then we must content ourselves with what hath already been fulfilled."[95] No passage from Newton has offered greater satisfaction and frustration to the Fundamentalist tradition of prophecy interpreters. On the one hand, to them is entrusted study of the most powerful support for Christianity and the most important guide to Christian life; on the other hand, they are doomed to be *historians* rather than prophets in their own right until the millennium has actually arrived.

Newton did not publish his writings on Scripture prophecy, presumably because he did not think that the world was yet ready to give them a fair hearing. But this does not mean that his ideas regarding scriptural interpretation remained unknown. His *Observations upon the Prophecies of Daniel and the Apocalypse of St. John* was finally published posthumously in a truncated form by his nephew in 1733. More important, almost all of Newton's attitudes toward Prophecy became public through the writings of William Whiston, Newton's deputy and successor in the Lucasian professorship of mathematics at Cambridge. Whiston, according to James Force, was encouraged by Newton to test the waters of public opinion in *The Accomplishment of Scripture Prophecy* (London, 1708), *The Literal Accomplishment of Scripture Prophecy* (London, 1724), and *A Supplement to the Literal Accomplishment of Scripture Prophecy* (London, 1725).

Following Newton, Whiston is certain that bringing a scientific approach to biblical studies is the only way finally to establish the truth of Christianity. On the one hand, he argued that

> till the learned Christians imitate the learned Philosophers and
> Astronomers of the present age; who have almost entirely left of
> hypotheses and metaphysics, for experiments and mathematics; I
> mean till they be content to take all things that naturally depend
> thereon, from real facts, and original records, without the Bypass of
> Hypothesis, or Party, or Inclination; till then I say I verily believe
> that disputes, doubts, Skepticism, and infidelity will increase upon
> us. . . . [96]

On the other hand:

> If once the learned come to be as wise in religious matters, as
> they are now generally become in those that are Philosophical and
> Medical . . . if they will imitate the Royal Society [and] the College
> of Physicians . . . and if they will then proceed in their Enquiries
> about revealed religion by real evidence and Ancient Records, I ver-
> ily believe . . . that the variety of opinion about those matters now in
> the world will greatly diminish; the Objections against the Bible
> will greatly wear off; and genuine Christianity, without either
> *Priestcraft* [i.e., Catholicism] or *Laycraft* [i.e., enthusiastic sec-
> tarianism] will more and more take place among mankind.[97]

To accomplish this state of affairs, one needed only to apply rigorously Newton's methods of interpretation as understood by Whiston. First, in those parts of the Bible that were *not* prophetic, one had to accept that "The Obvious or literal sense of scripture is the True and real one, where no evident reason can be given to the contrary."[98] Second, one had to accept that the language of the prophets, though odd, "is always single and determinate, and not capable of those double intentions . . . which most of our late Christian Expositors are so full of upon all Occasions."[99] Finally, one had to recover the "true" text of the Bible by searching for and comparing biblical texts and by relying on their oldest and least corrupt versions—a task which Whiston undertook in his *Essay towards Restoring the True Text of the Old Testament and for Vindicating the Citations Made Thence in the New Testament* (London, 1722). Whiston acknowledged that complete certainty on these issues was not possible, for new evidence (like the Qumran scrolls excavated in the twentieth century) might always come to light, forcing a reconsideration.

Just as Newton insisted in the *Principia* that his conclusions regarding the natural world had to be accepted "'til such time as other phenomena occur, by which they may either be made more accurate or liable to exceptions," so too did Whiston insist that his interpretation ought to be accepted until new evidence appeared. And he promised "that upon the Appearance of such real Evidence, I will carefully consider it, and determine my judgment on that side of each question, on which the *momentum* shall appear to preponderate."[100]

What reasonable Christian could have asked for anything more?

EIGHTEENTH-CENTURY NEWTONIANISM AND THE DRIFT TOWARD MATERIALISM AND IRRELIGION

From the works of Boyle and Charleton through those of Newton and his early followers, the most odious and unvarying enemy of Anglican natural theologians has been *atheistic materialism*. As we have seen, it was one of the most hard-pressed claims of the early Newtonians that the existence of gravitational forces demonstrates the need for immaterial agency in the world and that the imperfection of the universe demands the special miraculous intervention of God on occasion. The imperfection argument for God's immanence was seriously challenged by technical developments in celestial mechanics during the mid-eighteenth century, as it became clear that the solar system was vastly more stable than Newton had thought. At the same time, a whole series of self-professed "Newtonian" natural philosophers began to turn away from the early Newtonian insistence that matter was intrinsically passive or inert and that such phenomena as gravity demanded an immediate immaterial agent. With rare exceptions these natural philosophers were as concerned with rejecting atheism as Boyle or Newton. When they applied Newton's Rules of Right Reasoning to phenomena, however, they became convinced that since *both* gravitational attraction *and* some repelling force that makes it impossible for two particles to coexist in one place are characteristic of all matter within our experience, we are driven to assign *active* forces to all particles of matter, even though we may not be able to account for or explain the origin of those forces.

For such men, including Roger Boscovich, Joseph Priestley, and James Hutton, matter must be understood as "a substance possessed of . . . *powers* of attraction and repulsion."[101] All of these "Newtonians" continued to believe that the active powers of matter are ultimately derived from God; but they also agreed that because active powers were intrinsic to matter, one did not have to seek the immediate cause of any events outside the material world itself. Like Leibniz, they believed that though God was necessary to sustain those powers in matter which accounted for the laws of nature, he would not intervene directly or specially in the course of nature. Hutton even went so far as to deny the very possibility of miracles in the traditional, special providential, sense: "We must deny the possibility of anything happening preternaturally or contrary to the common course of things."[102] Hutton did *not* see his position as any kind of renunciation of Newtonian natural philosophy. Rather, it seemed to be the consequence of an unwavering commitment to the Rules of Right Reasoning in Natural Philosophy. As such, it had a major impact on the science of geology in the form of Hutton's formulation of the "uniformitarian" notion that all past geological events *must* be attributed to causes that can currently be observed operating in nature. In the early nineteenth century Hutton's uniformitarians produced new problems for biblical studies by forcing a radical reassessment of all accounts of the Noachian flood in Genesis; but for present purposes Hutton's attitudes are more important as an illustration of a trend in Newtonian natural philosophy which had much broader impacts on natural theology and religious doctrines.

The same logic that led many Newtonians to admit attraction and repulsion as intrinsic, active aspects of matter led some into the much more dangerous assertion that the power to *think* might also belong to mere matter.[103] If matter could think, then some of the most central assumptions underlying Christianity—that humans have immortal souls with spritiual attributes that continue to exist after the destruction and decay of the body; so that they can be the recipients of God's rewards and punishments—were in question. Whether they viewed the doctrine of salvation through Christ primarily as a prop for morality or as an end in itself, most eighteenth-century thinkers were convinced that without life after death and the immortality of the soul, religion was both powerless and vacuous. They were also convinced that to make the soul material was to deny that it could be immortal. In the introduction to his *True Intellectual System of the World,* Ralph Cudworth had long ago made clear the implications of giving thought to matter: "[the doctrine of thinking matter] must take away all Guilt and Blame, Punishments and Rewards—rendering a day of Judgement *Ridiculous.*"[104]

Cudworth had addressed his comments to the older Hobbesian understanding of what it meant to extend thought to matter, but eighteenth-century Newtonians operated within a new intellectual context established largely by John Locke. In his analysis of scientific knowledge, Locke argued that for all finite substances, our knowledge extends exclusively to their "nominal" essences rather than their "real" essences; we can know virtually nothing regarding any *necessary* connection among the different characteristics of any substance, and the very notion of substance is extremely vague— consisting of a *supposed* substrate to which a variety of attributes inhere in some unknown and unknowable manner. Consider some substance like gold, for example. From our experience we discover that at ordinary temperatures and pressures, gold is heavy, yellowish, malleable, and tarnish-resistant. Any material which we find to have these attributes we simply call "gold," but we have no way of discovering whether there is some other as yet undiscovered attribute of gold that is necessary to its existence. By the same token, though we know that yellowness and heaviness are coexistent attributes of gold, we have no way of knowing *why.*

In the course of discussing the many limits to knowledge imposed by our inability to discover the necessary connections among attributes of substances, Locke raises the question of whether we can know how matter and thought are related to one another. To this problem he responds:˙

> We have the ideas of *matter* and *thinking*, but possibly shall never be able to know whether any mere material being thinks or no: it being impossible for us . . . without revelation, to discover whether omnipotence has not given to some system of matter, fitly disposed, a power to perceive and think or else joined and fixed to matter, so disposed, a thinking immaterial substance: it being, in respect of our notions, not much more remote from our comprehension to conceive that God can, if he pleases, superadd to matter a faculty of thinking, than that he should add to it another substance with a faculty of thinking.[105]

Locke does not commit himself to the notion that systems of matter can think without aid of an immaterial soul. In fact, he seems to believe that revelation does warrant belief in immaterial souls; but he does insist that there are no grounds in experience or reason to deny that thinking is an attribute compatible with all of the other attributes of matter, and he insists that it is just as difficult for us to imagine how a body might be influenced by an immaterial soul as it is to imagine how matter might think.

The troublesome suggestion that matter might think is made doubly so by the way that Locke handles the notion of personal "identity" in the *Essay*. This issue is important in connection with all analyses of change, for as Aristotle had clearly understood, in all changes there must be some underlying substrate that remains the same to link the initial and final states. The meaning of personal identity becomes essential to any understanding of life after death and the efficacy of rewards and punishments. If damnation or salvation are to be possible, they must be imposed after death on the *same* agent that sinned or repented while alive. The traditional notion that the individual's immaterial soul remained the same before and after death allowed the Christian to resolve this issue.

The assumption that human souls are separable from bodies and constitute the persons who may be rewarded or punished raises its own set of problems, however. It seems to open up the possibility of metempsychosis or reincarnation—i.e., of the transmigration of souls from one body to another. It also raises the critical issue of what it means to say that Christ died for one's sins. If it is really true that the same person may inhabit many bodies or continue unchanged beyond death, then it does not make sense to say that Christ, or any other person, for that matter, *died* at all.

In trying to sort out the notion of human identity, Locke begins by analyzing what we mean by the notion of identity when we apply it to machines and living beings other than humans. If we take a machine, such as a watch, and repair it many times, so that most of the individual springs and gears are different from those that it had when it was first created, we continue to say that it is the same watch. The identity of the watch consists in the "organization, or construction of parts, to a certain end, which, when a sufficient force is added to it, it is capable to [*sic*] attain."[106] By the same token, we can talk about an animal remaining the same, even though it grows and has its individual material particles changed many times over, because it continues as an organized system of particles capable of self-motion. Then Locke takes a critical step:

> He that shall place the *Identity* of man in any thing else, but like that of the other animals in one fitly organized Body taken in any one instant, and from thence under one Organization of Life in several successively fleeting particles of matter, united to it, will find it hard, to make an Embryo, one of years, mad, and sober, the same Man, by any supposition, that will not make it possible for Seth, Ismael, Socrates, Pilate, St. Augustine and Ceasar Borgia to be the same man.[107]

Locke does not seem to want to deny the union of body and soul in humans, but rather to argue that there is something about the organization of matter in each individual that makes it suitable to one and only one particular soul. But Locke could not identify that something; and many readers—enemies and friends alike—read out of the *Essay* a denial of the very existence of immaterial souls. The enemies, in particular, saw in Locke's doctrine of human identity the specter of a purely material and machinelike human being incapable of salvation, free will, or moral responsibility. Richard Bentley launched the public attack on Locke's ideas in his second Boyle lecture of 1692, "Matter and Motion Cannot Think: Or a Confutation of Atheism from the Faculties of the Soul"; and soon a full-scale battle was engaged, with Samuel Clarke as chief spokesman for the immateriality of the soul and freedom of its will, and Anthony Collins as chief defender of the materialist and determinist implications of Locke's arguments.

Of the many debates associated with Latitudinarian natural theology, that over the materiality or immateriality of the soul probably generated the greatest amount of heat. Collins asserted that "human consciousness or thinking is a mode of some general power in matter . . . not necessarily indeed, and at all times . . . but only under a convenient structure and disposition."[108] And he was joined by John Toland and a host of others who found that particular structure in the brain.[109] On the other side, Samuel Clarke insisted that the very idea that we have of thinking is of an act without spatial extension, whereas our idea of matter necessarily contains the notion of extension; thus, thought and matter cannot coexist in a single substance.

Between the time of the Collins–Clarke debates at the beginning of the century and the 1770s, when Priestley wrote his *Disquisitions Relating to Matter and Spirit* (1777), a major transformation had occurred in many natural scientists' ideas of the nature of matter. Though first fully developed in the works of a Serbian Jesuit, Roger Joseph Boscovich, these new ideas were soon widely adopted among British Newtonians.[110] Boscovich denied that extension in its usual meaning was a necessary or even a possible attribute of matter. The fundamental characteristics of matter were rather forces of attraction and repulsion associated with *unextended* points distributed through space:

> The primary elements of matter are . . . perfectly indivisible and nonextended points; they are so scattered in immense vacuum that every two of them are separated from one another by a definite interval; this interval can be indefinitely increased or diminished but can never vanish altogether.[111]

The key features of Boscovich's matter theory are completed by considering the law of forces (see fig. 6) which describes the attractions and repulsions between any two points. The forces are repulsive near the origin, oscillate between repulsive and attractive through an interval small with respect to ordinary distances, and approach an inverse square law of attraction as they reach ordinary macroscopic distances. Thus, there is a continuous and simple law of forces.

Following the careful methods of Newton, his acknowledged intellectual mentor, Boscovich refused to speculate about the causes of the inter-point forces and argued merely that they must be accepted because they can be induced from phenomena:

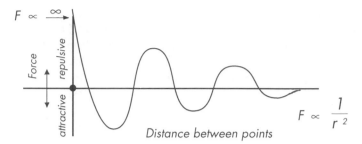

Fig. 6. The Force between two point atoms according to
Roger Joseph Boscovich.

> Whether the law of forces is an intrinsic property of indivisible
> points; whether it is something substantial or accidental added to
> them . . . ; whether it is an arbitrary law made according to his will;
> this I do not seek to find, nor indeed can it be found from the phe-
> nomena, which are the same in all these theories.[112]

With extension removed as a necessary characteristic of matter,
Joseph Priestley, one of Boscovich's chief British champions, immediately pointed out the implications for a materialist interpretation
of thinking. Now Leibniz's argument against thinking matter was
irrelevant, as Priestley claims:

> Since the only reason why the principle of thought, or sensation,
> has been imagined to be incompatible with matter goes upon the
> supposition of impenetrability being the essential property of it, and
> consequently that *solid extent* is the foundation of all the properties
> that it can possibly sustain, the whole argument for an immaterial
> thinking principle in man on this new supposition falls to the
> ground; matter, destitute of what has hitherto been called solidity,
> being no more incompatible with sensation and thought, than that
> substance, which, without knowing anything farther about it, we
> have been used to call *immaterial*.[113]

Priestley insists that thought is as certainly to be found con-
joined to the brains of humans as attraction is to the presence of
bodies or as sound to the presence of moving air. Since there is no
longer any compelling reason to deny thought to matter, we must
allow the general rule derived from phenomena to carry the day and
admit that organized matter can think, even if we cannot imagine
how. Priestley somehow fuses his curious version of materialism
with deep religious commitments; but the most common response
to the spreading acceptance of materialism of any kind within the
scientific community was fear of its moral implications. Joseph Ber-

ington spoke for many in criticizing Priestly for spreading such a dangerous doctrine:

> Materialism is . . . of dangerous tendency, because it contributes to darken the prospects of futurity; because it unbinds the reins to vice, confirming the libertine and unbeliever in their bad opinions and incredulity; . . . finally, it overturns the whole fabric of natural religion, because its injunctions can no longer be enforced, when the professors of it are told, that the same will be the ultimate fate of the virtuous and vicious—utter annihilation.[114]

The major difference between the uneasiness over Locke's suspected materialism at the end of the seventeenth century and the response to the materialism of Priestley and his friends was that in the first case, advanced scientific opinion was overwhelmingly in favor of Bentley's and Clarke's immaterialist arguments, whereas in the latter case the situation was reversed. At least one group of philosophical opponents—the Scottish Common Sense School of Thomas Reid and Dugald Stewart—did try to build a defense of body-soul dualism based upon scientific principles;[115] but for the most part the Scottish physician John Gregory only slightly overstated the case among intellectuals when he wrote to his friend, James Beattle, in 1767:

> Atheism and materialism are the present fashion. If one speaks with warmth of an infinitely wise and good being, who sustains and directs the frame of nature or expresses his steady belief of a future state of existence, he gets hints of his having either a very weak understanding or of being a very great hypocrite.[116]

"Till within these last thirty years," he continued, "the wit was generally on the side of religion." By implication, the wit had moved to irreligion sometime around 1740.

I have no idea why John Gregory placed the turning point among "wits" when he did, but an important symbol of change did occur in 1741 within the Royal Society of London, the leading institution of British science. From its founding in 1662, the Society had been a major center of Anglican Latitudinarian apologetics. Robert Boyle, John Ray, John Wilkins, John Locke, and Isaac Newton had served as its intellectual leaders through its first eighty years of existence. They had carefully blocked Thomas Hobbes from membership, at least in part because of his materialism and presumed atheism, in spite of his undeniable scientific interests and abilities. When Newton died in 1727 after holding its presidency from 1703, another Christian "virtuoso," Sir Hans Sloane, was elected president.

On Sloane's retirement in 1741, however, a radical change took place in this bellwether of the scientific community. After a bitter contest, Martin Folkes, a Deist, was elected president. A telling commentary on Folkes's incumbency comes from the unpublished commonplace book of William Stukey, a continuing fellow who had been closely attuned to the natural-theological commitments of his old friend Isaac Newton:

> He [Folkes] chooses the council and officers of his junto of Sycophants that meet with him every night at Rawthmills coffeehouse, or that dine with him Thursdays at the Miter, Fleet Street. He has a great deal of learning, philosophy, and astronomy. . . . In matters of religion an errant infidel and loud scoffer. Professes himself a good father to all monkeys, believes nothing of a future state, of the scripture, or revelation. He perverted the Duke of Montegue, Richmond, Lord Pembroke, and very many more of the nobility, who had an opinion of his understanding; and this has done an infinite prejudice to religion in general. . . . He thinks there is no difference between us and animals; but what is owing to the different structures of our brain, as between man and man. When I lived in Ormond Street in 1720, he set up an infidel club at his house on Sunday evenings. . . . He invited me earnestly to come thither but I always refused. From that time he has been propagating the infidel system with great assiduity, and made it even fashionable in the Royal Society, so that when any mention is made of Moses, of the deluge, of religion, Scriptures, etc., it is generally received with a loud laugh.[117]

Though some of the most distinguished members of the Royal Society continued to combine scientific and religious commitments—Priestley is perhaps the most illustrious of these—the general trend was undoubtedly toward secularization if not toward an open antipathy toward traditional religious doctrines. When the aggressive anticlericalism of the mid-eighteenth century came under attack throughout British society in response to perceived links between free thinking and the Revolution in France, there was a parallel return to religious orthodoxy among the British scientific elite; but the overwhelming commonality of interests shared by institutionalized science and institutionalized religion which had pervaded seventeenth-century British intellectual life was never to be regained.

SUMMARY

At the end of the sixteenth century the Anglican Church, which had been reestablished when Elizabeth replaced Mary on the throne of England, had faced a set of severe challenges—both external and internal. From the outset it faced challenges both from the continued existence of a substantial Catholic minority and from a perceived expansion of unbelief or atheism which threatened the social order. Internally, it faced fragmentation by the Puritan left which sought radical changes in church governance, doctrine, and ritual, and an increasing emphasis on scriptural religion.

Working at the behest of Elizabeth's Archbishop of Canterbury, John Whitgift, to defend the status quo and to try to pull the splintering Church together, Richard Hooker argued in his *Laws of Ecclesiastical Polity* for an ameliorative position, later called "Latitudinarian," which rejected both Catholic dependence on tradition and the Puritan exclusive emphasis on scriptural religion. Hooker emphasized the role of natural religion as a *necessary* supplement to scriptural religion. He argued not only that much knowledge of God and Christian duty could be learned from nature and with the use of "natural" reason alone, but also that even one's understanding of Scripture was to some extent dependent on a reasonable interpretation of *evidence* which left a probable but not certain knowledge. By focusing on the limits to certainty, Hooker hoped to undermine the pride and assurance which many Puritans seemed to have in those private interpretations of Scripture that seemed to be splitting the Christian community into warring factions.

Throughout the seventeenth century, numerous English scientists sought to respond to the call for support from the most widely read spokesman for English orthodoxy; and this became increasingly true as the sectarian tendencies of the late sixteenth century helped to foment the English Civil War. By the time of the Restoration in 1660, it seemed more important than ever on the one hand to calm religious passions, and on the other to support Christianity against atheistic tendencies, identified now most often with Thomas Hobbes's materialism and destructive egoism.

In the works of the mechanical philosophers Walter Charleton and Robert Boyle, we find sincere, widely read, and widely emulated attempts to demonstrate not only that a mechanical or corpuscular philosophy, properly understood, is consistent with Christian principles, but that it alone can be used to establish the existence, power, wisdom, and ongoing providential activity of God. In particular, the corpuscular philosophy seemed throughout the seventeenth century to undermine both atheistic materialism and the pantheistic or deistic tendencies inherent in systems of natural philosophy that posit the existence of an animate universe governed by some kind of world soul.

With the publication of Newton's *Principia* and *Opticks* in 1687 and 1704 respectively, scientific activities seemed to many to attain a new level of achievement and popularity. Drawn from the *Principia,* natural theological arguments in the popular sermons and writings of men such as Richard Bentley and in the more sophisticated works of Samuel Clarke seemed particularly powerful in supporting the notion of a God who was at once the supreme designer of a magnificently ordered universe and an imminent and providential actor in the affairs of nature and of man. Newton and the early Newtonians even seemed to some to offer a new and conclusive way to interpret and simultaneously establish the Divine character of Scripture by developing new, scientific techniques of Scripture interpretation.

For reasons that must seem fortuitous, perverse, or providential, depending on one's point of view, Newtonian natural philosophy was peculiarly well adapted to provide ideological support for the Latitudinarian Anglicans who had risen to power as a consequence of the Glorious Revolution of 1688, even though they had played virtually no role in encouraging it; for it allowed them to retain their belief in an orderly and stable "world politick" subject to their moral guidance at the same time that they derived their power from the overthrow of a legitimate monarch. God had intervened in the "world politick" in 1688, just as Newtonian science seemed to demonstrate that he must occasionally—though very rarely—intervene to "reform" the natural world.

By the first two decades of the eighteenth century, then, scientific and religious thought had become inextricably intertwined in Great Britain; and this was not just true among the intellectual elite—the so-called "Christian virtuosi" of the Royal Society—but across the spectrum of clergy and literary hacks. Of course not every Christian jumped on the natural theologians' bandwagon. Some simply ignored the new trends, whereas others opposed them. Moreover, there was a much greater divergence of positions and opinions among those who took natural theology seriously than we have been able to consider here. In particular, there were some explicitly anti-Newtonian natural philosophies and natural theologies developed and adapted by both High Church and more radical opponents to the Latitudinarians who appropriated Newtonianism. And there were important groups of both conservative and dissenting religious thinkers—exemplified by Jonathan Swift and William Law—which deeply opposed all attempts to dilute or corrupt religious belief with natural or rational arguments.

As the eighteenth century progressed, the dangers envisioned by opponents of natural theology manifested themselves in the increasingly heterodox positions among those who sought to apply scientific principles to theology and Scripture interpretation. We have briefly considered the materialist thrust of mid-eighteenth-century Newtonianism with its challenges to traditional understandings of the immortality of the soul and the nature of salvation and damnation. We could equally have followed natural theological arguments present challenges to the traditional understanding of God as Three Persons (Unitarianism), to the divinity of Christ (Socinianism), to the continuing providential activity of God (Deism), and to the authority of *any* established clergy (Freethought). In fact, in one of the most disturbing of all extensions of Newtonian techniques of argumentation to religious topics, David Hume managed to turn the Rules of Right Reasoning in Philosophy and the Rules of Methodizing the Apocalypse into tools to discredit natural religion and scriptural revelation alike.[118]

Natural theology persists as a central feature of orthodox Christian apologetics even today, but during the late seventeenth and the eighteenth centuries it helped to establish heterodox positions whose cumulative effect was to encourage precisely that kind of fragmentation of the religious community that its early seventeenth-century promoters had hoped to stop. As one consequence, the moral and ethical force of revealed religion declined throughout Western Europe, leaving scientific rather than religious arguments as the chief source of most beliefs about human nature and the nature of social and political activity. It is to these secular sources of ethical, moral, and political commitments that we turn in the next three chapters.

Liberalisms and Socialisms: The Ideological Implications of Enlightenment Social Science I: The Tradition of Political Economy

4

When Francis Bacon had prophesied early in the seventeenth century that tremendous material advances in society would be grounded in advances in knowledge about nature, he was careful to offer both explicit and implicit denials that innovation in the natural sciences would imply parallel innovations in political and social arrangements. The patriarchal and monarchical society within which Bacon placed Solomon's house in *New Atlantis* assured his readers that while science would raise the standard of living, it would have no bearing on the relative distribution of honor, goods, and power in society.

By the end of the eighteenth century, the vast majority of politically aware men and women—certainly in France, Italy, and America, but probably even in England and Germany—held precisely the opposite view. They held that progress of the mind, led by the growth of the natural sciences, had already begun to force dramatic changes in political and social arrangements; and it was sure to continue to do so without a foreseeable end. Writing to his personal friend and political opponent, John Adams, Thomas Jefferson expressed the widespread optimistic vision of a world constantly progressing in all ways—materially, morally, and politically—as a consequence of scientific advances:

> One of the questions you know, on which our parties took different sides, was on the improvability of the human mind, in science, in ethics, in government, etc. Those who advocated reformation of institutions, *pari passu,* with the progress of science, maintained that no definite limits could be assigned to that progress. The enemies of reform, on the other hand, denied improvement, and advocated

steady adherence to the principles, practices, and institutions of our fathers, which they represented as the consummation of wisdom and *akme'* of excellence beyond which the human mind could never advance. Altho' . . . you expressly disclaim the wish to influence the freedom of enquiry, you predict that that will produce nothing more worthy of transmission to posterity, than the principles, institutions, and systems of education received from their ancestors. I do not consider this your deliberate opinion. You possess, yourself, too much *science,* not to see how much is still ahead of you, unexplained and unexplored. Your own consciousness must place you as far before our ancestors, as in the rear of our posterity.[1]

It will be the aim of this and the following two chapters to explore the central role of science in developments of the seventeenth and eighteenth century which transformed Western political consciousness and which produced the dominant political ideologies of the modern world—liberal, conservative, and socialist.

Some comment on the term *ideology* is necessary, for it is an unavoidable word that carries very different meanings for those who write in the modern liberal tradition than for those who write in the modern Marxist-Socialist tradition. Both liberal and socialist meanings derive from a term that was coined by Destutt de Tracy at the very beginning of the nineteenth century in order to emphasize that humans live and act most immediately in a world of "ideas" rather than in some external, objective, material universe. Our decisions and actions can be based only on what we know, think, or believe about the world; and whatever remains outside our domain of knowledge and beliefs can have no bearing on our actions. The "Ideologues" associated with de Tracy admitted that the set of ideas used by any individual in guiding his or her actions is at least in part socially constructed and that it might contain errors or prejudices; but they were certain that these errors could be avoided through proper education and the rigorous application of critical reasoning.

At this point the two very different modern approaches to ideology diverge. On the one hand, the Marxist tradition focuses on the claim that because our ideas are *socially* constructed, they must reflect the interests and biases of those in power at any given time. "Ideologies" are the means for dominant groups to mask reality and to manipulate and exploit the powerless. Every "ideology" is thus a special form of "false consciousness" about the world, and the admission that we live largely in a world of consciousness rather than in an "objective" reality is a constant reminder to beware of the distortions that exist. On the other hand, while liberals admit that the belief systems we operate in are to a large extent socially determined and reflect the interests of special groups, institutions, or classes, they tend to focus on the constructive role played by those belief systems or ideologies; for unless the people of a given society are bound together by some more or less shared set of beliefs about the nature of the world and desirable aims to be sought in human action, no social cohesion will be possible.

In what follows, unless some explicit qualification is made, I adopt a definition of ideology which is derived from the liberal tradition. An *ideology* is any system of meaning shared by members of society which functions to supply information, give direction, and provide justification for behavior.[2] Such a definition allows us to enter sympathetically into the Marxist concern with how an ideology may reflect some specific and perhaps undesirable social relationships. At the same time it allows us to emphasize both the unavoidability and the constructive functions of ideologies.

In order to structure our discussion of science and ideology in the late seventeenth and eighteenth centuries, we will follow three different approaches to moral and political science which came to be seen as distinctive during this period. Furthermore, we will follow these traditions in two different national contexts—those of Britain and France—because the very different relations between the scientific community and religious, political, and economic elites, as well as the different political structures in the two countries played important parts in shaping the special role of science in guiding political understandings and doctrines in each country. In fact, the major differences between British liberal thought and French liberal thought in the eighteenth century are largely a consequence of the different governmental roles of science established during the half century prior to the French Revolution.

THE PSYCHOLOGICAL, SOCIOLOGICAL, AND ECONOMIC TRADITIONS DEFINED

In Chapters 1 and 2 we discussed the attempts of three seventeenth-century individuals to extend techniques of analysis and conceptual models drawn from the natural sciences into the domains of politics and morality. To the extent that each author sought an immediate impact on events, Hobbes, Harrington, and Petty were largely unsuccessful. Hobbes failed to convince the opponents of monarchical government to submit to the authority of Charles I; Harrington failed to convince Cromwell to establish a republic modeled on *Oceana;* and virtually none of Petty's schemes for revising taxes or governing Ireland was palatable to either Cromwell or Charles II. To the extent, however, that each also sought to legitimize a new way of approaching moral and political issues without appeal to the *authority* of either divine revelation or the classical moralists, they were remarkably successful.

Within little more than fifty years of the publication of *Leviathan,* there were few writers on man and society who could claim serious attention without also claiming that their authority derived from their *scientific* approach to their subject—i.e., to some combination of empirical and rational analyses of humans and their interactions with one another. Moreover, though the scientific approaches of eighteenth-century moral and political philosophers certainly incorporated critically important insights from Newton, Locke, and Hume, as well as new mathematical techniques developed in the course of the eighteenth century, almost all of them derived their major problems and strategies from those articulated by Hobbes, Harrington, and Petty, singly or in some combination.

For present purposes each of these seventeenth-century thinkers embodied a separate scientific tradition with its own special methodological emphasis and its own special aims. Hobbes initiated the tradition that I will label "psychological." The basic aim of this tradition was to ground all analyses of social and political phenomena in a science that began with an investigation of individual human sensations and mental operations, and that sought to use a train of reasoning modeled on Euclidean geometry to deduce moral and political propositions from the nature and needs of individual human beings. Few eighteenth-century authors within the psychological tradition admitted indebtedness to the materialistic and atheistic Hobbes. Instead, they grafted their Hobbesian analyses onto the much more acceptable sensationalist psychology and epistemology derived from John Locke's *Essay Concerning Human Understanding.* But there was a continuous and extremely powerful tradition that was self-conscious in its multiple concerns with individual psychology, systematic analysis of the nature of reason, and the derivation of both moral principles and the structure of social and political institutions from human nature. That the psychological tradition was already viewed as a more or less autonomous tradition of scientific analysis during the eighteenth century is demonstrated by Marie-Jean-Antoine-Nicholas de Condorcet's *Projet de décret sur l'organization générale de l'instruction publique* (ca. 1792) in which he established three lecturers in the moral and political sciences at the institute (i.e., undergraduate college) level. To *one* of these lecturers he assigned (1) analysis of sensations and ideas, (2) scientific method or logic, (3) morality, and (4) the general principles of political constitutions,[3] all of which seemed naturally linked to most eighteenth-century psychological thinkers.

Petty embodied a very different tradition in the social sciences. Concerned with avoiding the thorny problems of individual psychology and of political structure alike, Petty initiated *political economy* by attempting to develop quantitative statistical techniques for arriving at rational policy decisions regarding specific economic and social issues that might be implemented by *any* government interested in augmenting its own power and the wealth of its subjects. Henry Robinson made this perspective particularly clear in *Certain Proposals—for the Advancement of Trade and Navigation* (1652), arguing that "even the worst kind of Government is capable of being so managed, so ordered, that a people may enjoy better days under it, than ever yet did any under the very best of Governments from the beginning of the world."[4]

Political economy was developed in late seventeenth- and early eighteenth-century France and England primarily by pleaders for special interests and by government bureaucrats hoping to formulate policies to achieve economic advance and financial stability. It took on critical, long-term ideological functions in the works of the Physiocrats and of Turgot and Condorcet in France and of Adam Smith in Britain, and it became a central feature of early French socialism. Once again, this tradition was embodied in Condorcet's educational scheme in a single lecturer who would cover the elements of (1) commerce, (2) political economy, and (3) legislation.[5]

James Harrington's political anatomy embodied yet a third tradition in political science. Like Hobbes, Harrington was vastly more concerned with institutional arrangements than Petty was; but like Petty's, his analysis dealt with social aggregates rather than individuals. His concerns had focused on the competing interests of different orders or groups (we might now say classes) in society and how those interests could be shaped to ensure that no special interest could override the common interest in a nation or society. Finally, Harrington emphasized historical evidence, the fact that the relative balance of power among different orders was dynamic rather than static and the fact that political arrangements must ultimately reflect economic arrangements. Because of its focus on structural and functional relationships among different groups in society, I will call this tradition "sociological," but it might equally be labeled "anthropological"; for it came increasingly to depend on the comparative study of social arrangements in cultures widely distributed in space and time as it was developed by Charles Secondat, Baron de la Montesquieu in France, and by Adam Ferguson, Adam Smith, and John Millar in Scotland. This tradition was critical in giving liberalism the progressive cast that can be seen in Jefferson's letter to Madison and in concentrating attention on economic issues. It played a major role in the class analysis of the French Revolution initiated by Saint-Simonian socialists, and through Saint-Simon it ultimately became a central feature of Marxist thought. Condorcet acknowledged this field by suggesting that the third of the three lecturers in the moral and political sciences be responsible for geography and the "philosophical history" of peoples.

As the eighteenth century progressed, psychological, economic, and sociological approaches to social issues often interpenetrated; for nothing like the modern professional compartmentalization of academic fields developed until well into the nineteenth century. Yet, these categories are useful for developing an understanding of changing political consciousness during the period preceding the French Revolution.

THE CONTEXT OF EARLY POLITICAL ECONOMY AND THE EMERGENCE OF *HOMO ECONOMICUS*

During the past two decades, a group of extremely able authors, including William Letwin,[6] Peter Buck,[7] and Joyce Oldham Appleby,[8] have shown how special interests and special local historical conditions shaped the basic assumptions and even the categories of analysis formulated by early English proponents of "scientific" economics. Our major concern will be directed not at the question of whether biases entered science from without (they did), but rather at that of whether science exported biases of its own. That is, was there anything about the character of science that made it easier for political economists to accept some assumptions and concepts rather than others? Granted that any treatment of political economy was bound to incorporate "ideological" components, did the attitudes and activities associated with the scientific community act to discriminate in favor of one specific set of biases and concepts over any other available one? Additionally, we will be concerned with how a set of assumptions and concepts that were clearly recognized in the seventeenth century as embodying special interests came to take on ideological standing during the eighteenth century. In particular, did their incorporation into bodies of scientific analysis have any bearing on that process?

Whether we accept the liberal or the Marxist notion of ideology, concepts and assumptions can serve an ideological function only to the extent that they seem to their users to reflect objective reality rather than to reflect the biases of any special group. If they seem to be problematic and open to doubt and analysis, they cannot carry that psycho-social authority that binds a society together in cohesive action. Even if we argue—as the Marxists seem to—that assumptions that serve the economic interests of a dominant class will somehow automatically appear as objectively true to members of that class, we must still try to discover how those assumptions also come to be accepted by others whose interests they oppose. It is this last process—that by which some special-interest arguments or assumptions come to be accepted beyond the class or group to which they might be expected to seem most congenial—that we will be specially concerned with in what follows.

Nothing is more important for the emergence of political economy as a science and as a central focus of modern ideology than the initial isolation of economic issues from broader issues relating to religion and social concerns. At least until the second decade of the seventeenth century, it was universally assumed that political authorities had both the right and the obligation to regulate the production, pricing, and distribution of goods in order to protect the interests of producers and consumers and to increase the power and prestige of the state. Economic activity was understood as serving basic material and *social* needs of the local communities in which it took place. The notion of a *just price,* controlled by guild councils, town councils, local nobles, or royal representatives, and calculated to provide goods at a cost to consumers which allowed producers just to maintain their households in a manner suitable to their proper station in the community dominated all economic ideas.

In a medieval and early modern economy characterized by effectively isolated local markets, such a conception had much to recommend it. In years of poor harvests it ensured, at least ideally, that local farmers could not raise prices at the expense of town dwellers and that the rich could not deprive the poor of food by driving up the price of commodities like bread. Similarly, it allowed for price and wage controls to eliminate competition within a trade and to ensure relative equity in employment. The amount of detailed regulation in such economies boggles the modern mind. In the baking industry, for example, not only was the price of bread and the quantity and quality of ingredients regulated, but the wages paid to apprentices, the number of holidays allowed, and the free time allowed for schooling were all strictly stipulated; and any master who overcharged his customers or underpaid his help was likely to find himself facing the local town magistrate or the village justice of the peace. Such a system functioned to maintain a stable social structure, to spread the wealth in good times and misery in poor times, and to ensure that each member of society contributed in a positive way to the well being of the whole. For just that reason, usury—the taking of interest for the use of one's money without sharing the risk—was generally forbidden because it seemed to grant income to those who were doing no productive service to the community.

Of course, the economic regulation of local communities did not always work perfectly. In hard times farmers and bakers sometimes withheld their produce and sold at higher than allowed prices on a black market, and tradesmen undersold their fellows in order to get adequate business. But it would be very hard to sustain an argument that this kind of subordination of economic activities to social goals was inappropriate. Neither did it encourage capital formation, and with it, economic growth and rising standards of living, nor did it offer much room for the expression of individuality or for upward social mobility. For the most part, however, it was capable of achieving what it sought—stable, orderly social and economic life throughout a primarily agrarian society of local and, in effect, isolated communities.

The constant and active intervention of human authority in the economic sphere seemed perfectly appropriate in the medieval intellectual scheme that assigned economic activities to the sphere of Aristotelian practical sciences as opposed to the sphere of theoretical sciences (vol. I, chap. 5). That is, economic life, like politics and law, was part of the domain of phenomena subject to human control, unlike the domains of physics and astronomy, which were totally beyond human influence and therefore subject only to natural rather than to positive as well as natural law.

During the early modern period a series of concurrent developments served to undermine both the social realities served by traditional economic thought and the intellectual framework within which that thought made sense. First, increasingly important translocal factors—both national and international—emerged to supplement and supplant the dominant concern with localized economies in the early modern period. Even in the Middle Ages a kind of international economy centered on the great fairs of the champagne region in France, and the regions around Frankfurt and Hamburg in Germany began to grow up *almost* independent of local agrarian economies. At these fairs or special markets, the merchant entrepreneurs of Florence and Flanders traded finished manufactured goods and luxury items from the Italian trade with the Orient for surplus raw wool produced in Britain and surplus grains of North Central Europe. With rare exceptions, parties to these translocal exchanges were free to strike their own bargains for the highest profits; and they developed systems of letters of credit, discounted to take account of risks and demand, which gave rise to systematic lending at interest. During the fifteenth and sixteenth centuries, the huge medieval seasonal fairs were replaced by more efficient ongoing commercial "markets" usually located at transportation nodes like London, Amsterdam, and Hamburg, which became centers of a growing international trade.

For present purposes the growing translocal trade was important both because it superimposed a dynamic element on the relatively stable economic and social structure of Europe and because it involved economic relationships that could not be accommodated within the older structure of economic thought. Since the new trade was between members of different communities rather than among members of a single community, the whole concept of just price, with its emphasis on social stability and responsibility was inapplicable—theoretically, because the merchants' social responsibility was to their own community rather than to that of their trading partners, and practically, because no *single* authority held the power to regulate international trade. The international trader was free to sell to the highest bidder and to buy where he could get the best price.

A refined commonsense means for interpreting international

trade did develop in the so-called Mercantilist theory which, in effect, treated the state as a direct party in the trade of its citizens. The goal of Mercantilist policy was thus to increase the net wealth of the state by securing a favorable balance in the exchange of specie (gold or silver coin) and to position the state and its citizens for increasingly profitable business by monopolizing both markets and cheap sources of raw materials.

Unfortunately, not only did Mercantilism treat all trade as a zero sum game, but it failed to suggest any new policy tools, and in the new economic context, the traditional expedients of regulating prices and quantities of goods sold and of imposing duties did not have the traditional effects because individual authorities could not control the entire process of production, exchange, and consumption. The increasingly complex international economy responded in complicated and often unexpected ways to any intervention. When taxes were imposed on imports, they only sometimes produced income. Just as often they drove the taxed goods away entirely, or into a growing black market in smuggled goods that evaded all control. When the regulated price of a locally produced commodity was lowered to make it accessible to local consumers, it sometimes had the undesired effect of driving the commodity into foreign markets where consumers were willing to pay more.

By the early seventeenth century, international trade, operating in ways that were often beyond the comprehension and control of governmental authorities, had come to dominate at least the Dutch, Italian, and British economies. A series of seventeenth-century crises, beginning with a major British depression during the 1620s, gave rise to a new series of attempts to understand and manage the economy. From the standpoint of the British government, the crisis of the 1620s was especially critical. Not only did it affect the wealth of the citizens, including members of Parliament who had a huge stake in the economy, but also the attendant inflation drove royal income far below expenses, demanding new sources of money for the Crown and leading to the imposition of unpopular customs duties and to the wholesale sale of unpopular monopoly licenses. In turn these measures distorted patterns of trade in largely unforeseen ways.

Both Crown and Parliament sought expert advice, and a body of acrimonious literature, filled with special pleading of the merchants, manufacturers, and bankers, developed. Some writers— usually proponents of strong monarchical regulation of trade—

attempted to integrate the new elements into the older tradition of moral economy, with focus on the ties between social and economic concerns. But as the century went on, there was a growing confluence of ideas put forth both by spokesmen for the large-scale merchants who had grown accustomed to operating with little concern for the social impact of their actions, and by men like William Petty, whose scientific orientation led them to seek to isolate economic phenomena for analytic purposes and to hunt for deterministic economic laws operating independently of intentional human interference. Joyce Oldham Appleby has characterized the late seventeenth-century trend very well:

> Two widely shared beliefs about commercial life enabled men to convert economic observations into a predictable [*sic*] science: the motive of gain and what could be called the theory of interchangeable participants. A theme with many variations, this theory was given its pithiest expression by Charles Davenant [one of Petty's closest followers], who explained that the market price of a bushel of corn would prevail over any legislative interferences: "because if B will not give it, the same may be had from C and D, or if from neither of them, it will yield such a price in foreign Countries; and from hence arises what we commonly call Intrinsic value. Nor can any law hinder B, C, and D from supplying their wants [for in the] Natural Course of Trade, Each Commodity will find its Price. . . . The supreme power can do many things, but it cannot alter the Laws of Nature, of which the most original is, "That every man shall preserve himself." Davenant had laid bare what gave consistent force to a system of apparent free choice; if one person did not follow his self interest to drive down a price or drive up a rate, someone else would. With this realization, economic writers had discovered the underlying regularity in free market activity. . . . The reality was that individuals making decisions about their own persons and property were the determiners of price in the market. The possibility was that the economic rationalism of market participants could supply the order to the economy formerly secured through authority.[9]

Some of the economic authors of the seventeenth century undoubtedly adopted the notion that pure self-interest was the sole driving force of economic life because it served to legitimize their own search for private profit regardless of the cost in terms of social justice and stability. Sir Josiah Child, governor of the East India Company and tireless economic propagandist, provides one of the most blatant examples. A need to justify merchant greed does not, however, adequately account for the enthusiasm with which men such as William Petty, Charles Davenant, and later, Adam Smith adopted the principle that rational calculation of private gain alone accounts for the economic decisions that we make.

It seems fairly clear that the *occasion* for Petty's acceptance of the private gain theory of economic interests was his familiarity with Hobbes's arguments based on the notion of purely and perfectly self-centered individuals; but the *reason* for his acceptance of this theory seems to me to lie at least partially in his passion for analytic simplification and for quantification—i.e., in his scientific biases. Focusing on a single principle of action, open by definition to calculation, made it possible to begin to establish a *science* of political economy "whose foundations are sense and the superstructures, mathematical reasoning."[10]

Whatever the combination of reasons might have been, those who initiated a science of political economy were unusual in their open acceptance of an egocentric, rationally calculative, interpretation of man—*Homo economicus.* Whereas Hobbes was loudly and frequently vilified by theologians and moralists alike and was largely unacknowledged even by those psychologizing thinkers such as John Locke and David Hartley, for whom he provided fundamental insights, he found open admiration among economic writers who not only used his emphasis on rationally calculated self-interest, but who also accepted his insistence on the equivalence of all human actors. Sir Josiah Child spoke for the whole enterprise of political economy when he insisted that "All men are by nature alike . . . [as] Mr. Hobbes has truly asserted."[11] This principle—variously called the theory of interchangeable participants by Appleby, analytic equality (to distinguish it from the Christian ideal of normative equality) by economists such as Joseph Shumpeter,[12] and "species equality" by political theorists like Michael Walzer[13]— is of immense importance in the development of a quantitative science of political economy and in the development of both liberal and socialist ideologies; for it allows the economist and political thinker alike to proceed with calculations or arguments without taking into account such complicating considerations as physical and psychological needs, social stability, and traditional class privileges.

In much the same way, although the assumption of a market "mechanism" subject to regular behavior and resistant to regulatory influences appealed to merchants and traders who wanted to be free to maximize their profits, it also had tremendous intrinsic appeal for those empirical and mechanical philosophers who sought to understand phenomena in terms of natural laws that were shorn of their old moral and normative implications. For such men the existence of "objective" economic laws was assumed because they had abandoned the old Aristotelian distinction between practical and theoretic sciences. "Trade is in its Nature free, finds its own Channel and best directeth its own Course," insisted Charles Davenant. "Wisdom is most commonly in the Wrong when it pretends to *direct* Nature."[14]

This last comment suggests an apparent paradox in the thought of the political economists of the seventeenth century. On the one hand, they insisted that economic phenomena, like other natural phenomena, must be subject to inviolable natural laws. On the other hand, all of them wrote with the intention of offering policy advice, which implied that intervention in economic affairs was both possible and desirable. On this score the Baconian dictates that "Human knowledge and human power meet in one" and that "Nature to be commanded must be obeyed" were of central guiding import.[15]

In seventeenth-century England, medicine became the paradigmatic field for Baconian scientists; for in medicine, intervention based on knowledge of anatomical and physiological phenomena seemed to offer great opportunities to bring forth practical fruits from natural knowledge.[16] In medicine the goal was to encourage and assist those natural processes conducing to health and to intervene only to block abnormal and pathological developments. Similarly, the political economists advocated that legislators and policy makers accept the natural human desire for gain and the natural operation of market forces and that they intervene in the economy primarily to encourage the natural growth of the emerging "free market" economy and to block pathological developments. Thus, when Petty wrote his *Political Arithmetic,* he emphasized the removal of "the impediments of England's Greatness,"[17] and in a letter to Lord Anglesea he spoke of his *Political Anatomy of Ireland* (written in 1672, but not published until 1691) as establishing the basis for the "political Medicine" of the country, "without passion or interest, faction or party; but, as I think, according to the Eternal Laws and Measures of Truth."[18] In this context, it is probably not coincidental that three of the most important early scientific political economists—William Petty and John Locke in England, and François Quesnay in France—were initially trained as physicians.

POLITICAL ECONOMY AND
SOCIAL STATISTICS:
THE RELATIONSHIP BETWEEN
INTEREST AND INFORMATION

If, as Petty hoped, a science of *Political Arithmetic* was to replace the "Mutable Minds, Opinions, and Passions of Particular Men" and the "passion of interest, faction, or party" as a foundation for making social decisions, then two things were absolutely necessary: 1) an adequate fund of information or experience expressed in quantitative form, and 2) a set of mathematical techniques for drawing policy inferences. These two issues were addressed in a fascinating document, *Natural and Political Observations . . . Made upon the Bills of Mortality,* by John Graunt, published in 1662. The work was dedicated jointly to the Lord Privie-Seal of England and to the President of the Royal Society of London, indicating its simultaneous emphases on science and policy. That Graunt at least partially achieved his goal of successfully linking science with political advice is suggested by the fact that he was voted into the Royal Society on the strength of his performance and that he was sponsored for election by Charles II himself.

Graunt began with a previously published source of information, the Weekly Bills of Mortality for London, which had been continuously recorded since 1603. Each parish clerk delivered a weekly account of the christenings and deaths—including cause of death and sex (but not age) of the deceased—to a central official, the Clerk of the Hall, who assembled and printed them for sale to those who wanted to purchase them—presumably, Graunt suggests, to find out when sickness was increasing so the rich could leave town, or so that "tradesmen might conjecture what doings they were likely to have in their dealings."[19] Graunt then sought to discover what "abstruse and unexpected inferences" might be drawn from "these poor despised Bills of Mortality."[20] The whole point, cleverly revealed only at the conclusion of the tract, seems to be that if a careful reasoner using simple mathematics (Graunt calls it mere "Shop-Arithmetique")[21] can tease extremely important political intelligence even from such an unpromising source as this, then wonderful things indeed might be expected from the more sophisticated mathematical treatment of a more carefully and systematically collected body of economic data.

Graunt begins his analysis by assessing the reliability of his raw data and by arguing that some adjustments need to be made, thereby showing a subtlety that his close friend, Petty, never achieved. Just a few examples will illustrate the shrewd procedures he used. In 1634 a new disease, rickets, appeared among the causes of death listed in the Bills. In that year there were fourteen deaths attributed to rickets; by 1659, the number rose to 441. Graunt asks whether it is merely a new name given to cases that had previously been reported as something else. First he asked physicians what other disease was most like rickets in its symptoms, then he showed that reported cases of that disease, "livergrown," declined as reported cases of rickets increased—indicating that diagnoses of the two diseases were at times confused. Further study showed that rickets and livergrown together did not remain a constant fraction of all deaths in ordinary (nonplague) years. This fraction grew very rapidly after 1634 although the ratio of livergrown to all deaths had previously been very stable. Thus Graunt concluded that though some cases of rickets might be the old livergrown under a new name, there really was a new disease as well.[22] In a much more important case, Graunt used the constancy of mortality from endemic diseases to show that in plague years plague deaths were initially underreported, probably in order to avoid panicking the people. In this case he showed that in initial plague years the reported non-plague mortality more than doubled, though as the plague went on, reported ordinary mortality rates rapidly returned to normal. He concluded that plague deaths—reported already as nearly double the nonplague deaths—were in fact initially underreported by about twenty-five percent.[23]

Turning from adjustments to mortality figures, Graunt argued that a perusal of causes of death and their relative numbers had important policy implications. There was a widespread feeling in the mid-seventeenth century that both starvation and violent crime, including murder, were becoming so serious that new policies for feeding the poor or decreasing the population and for repressing violence were desperately needed. Graunt purported to show that fewer than one in 4,500 adult Londoners died of starvation, so there was no need to suppress population or to develop any special new plans for feeding the poor. Similarly, fewer than one in 1,700 Londoners died by criminal act, so there was no reason to panic or to consider replacing the volunteer citizen guard of London to control crime, which was vastly worse in other countries that had professional police forces.[24]

Turning to christening figures and their relation to deaths, Graunt was able to show that London was growing, not by an excess of births over deaths, but by inmigration from the countryside. He could also show that the total population of England was well below that which could be supported by the food-producing ability of the land.[25] Comparing male and female birth and mortality figures, he could argue that the constant surplus of male over female births was so compensated by the higher job-related mortality of males that it ensured an almost perfect one-to-one ratio of breeding-age males and females, thus justifying naturally the custom of monogamous marriage.[26] By adding a small sampling to establish the average number of members of each household (8) and the average number of deaths per household (3/11), Graunt was able to estimate the total population of London at about 384,000, a number far below the six million which had been guessed by others.[27]

By using the frequency of childhood diseases to establish that about 36 out of 100 persons died before age 6, by using the number of reports of those who simply died of old age to estimate mortality beyond age 70 at six percent, and by assuming that between ages 6 and 70, mortality was independent of age, so that the number of annual deaths at any age was proportional to the number of persons alive at that age, Graunt was able to produce what has come to be called the first life table (see fig. 7) and to estimate the number of military age males in London to be about 70,000.[28]

Finally, by comparing the London figures with figures from "a certain parish in Hampshire" (figures from Romsey supplied by Petty), Graunt was able to demonstrate that London was more "unhealthful," especially for males, and that both London and the colonies were being populated from emigration from the country.[29]

"The Conclusion" of Graunt's masterpiece deserves extensive quotation and comment. "It may now be asked," he begins, "to what purpose tends all this laborious buzzling, and groping?"[30] To this question he gives four answers, the first and last of which concern us. First, he simply lists thirteen obvious results related to policy and derived from the Bills, and says that "those who cannot apprehend the reason of these enquiries, are unfit to trouble themselves to ask them."[31] The significance of Graunt's limited results was immediately obvious to almost all mathematicians and policy makers who saw them. The work was reviewed in the *Journal des Savants* of the Paris Academy of Sciences for August 1666; and within nine months the French Crown had created a bureaucracy

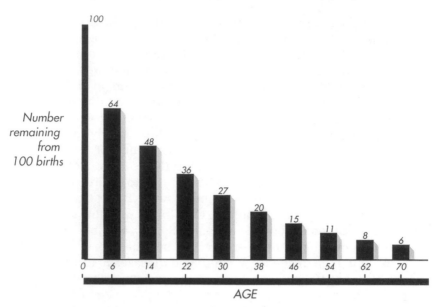

Fig. 7. Graphical representation of Graunt's "life table."

to collect information on births, deaths, and marriages in Paris because "it is of public importance for the health and subsistence of the people to know the state in all ways and to observe the causes which augment or diminish the people in each quarter of Paris."[32]

Graunt's life table was recognized as specially important because both the English and French crowns had begun to depend heavily on the sale of life annuities and the creation of Tontines (schemes in which participants paid into a fund, the income from which would be paid out to the last survivor of the group) to finance their activities. In order simultaneously to draw participants and to ensure that these schemes would not produce long-term drains on state finances, it was necessary to estimate realistically life expectancies. Prior to Graunt's tables, virtually all estimates favored annuitants, producing serious debt problems, especially for the French monarchy. At the same time, private insurance cooperatives, intended to provide retirement income for middle-class merchants and tradespeople, were having a desperate time because they systematically underestimated life expectancy for older members, leading to collapse and bankruptcy after the first generation.

Almost everyone realized that Graunt's work pointed the way to establishing accurate life tables; but they also recognized that his numbers were biased both because they were not taken from a stable population and because they made unverified assumptions about mortality rates at any given age. The first major advance on Graunt's table was made by Edmund Halley. Halley, who knew of Graunt's work through Petty, was aware that more accurate tables demanded records from a stable population and with information about the age at death.[33] In 1689, a German minister and amateur scientist from Breslau, Caspar Neuman, forwarded a Table of Births and Deaths for the city of Breslau to Leibniz. Without recognizing the significance of the materials, Leibniz let Henri Justell, who was interested in political economy, know of their existence. Justell wrote to Neuman, asking for his materials so that their suitability for constructing life tables might be judged. Neuman sent the materials in 1692 to Justell who passed them on to the most able mathematician available, Edmund Halley, who published the results as "An Estimate of the Degrees of Mortality of Mankind, drawn from Curious Tables of Births and Funerals at the City of Breslau, with an Attempt to Ascertain the Price of Annuities upon Lives," in the *Philosophical Transactions of the Royal Society of London* in the following year (17: 596–610). Improvement of Halley's work involved little but adjustments for special conditions (locale, class), which were undertaken in the eighteenth century by the French Bureau of Statistics and by private individuals in England, and the development of mathematical techniques to model and simplify the calculation of multiple life annuities initiated by Abraham de Moivre, one of Newton's protégés and Halley's colleagues in the Royal Society.

We will return to discuss the ideological importance of life tables shortly, but to set the stage we must return to the final justification for all his "laborious buzzling and groping" given by Graunt:

> I answer more seriously, [he begins,] by complaining that whereas the art of governing and the true *Politics* is how to preserve the state in *peace* and *plenty,* that men study only that part of it which teacheth how to supplant and over-reach one another, and how, not by fair out-running, but by tripping up each other's heels to win the prize.
>
> Now the foundation, or elements of harmless *policy* is to understand the land, and the hands of the territory to be governed according to all their intrinsic and accidental differences: as for example; It were good to know the geometrical content, figure, and situation of all the lands of the kingdom, especially according to its most natural, permanent, and conspicuous bounds. It were good to

know how much hay an acre of every sort of meadow will bear?
How many cattle the same weight of each sort of hay will feed and
fatten? What quantity of Grain, and other commodities the same
acre will bear in one, three, or seven years *communibus Annis*?
Unto what use each soil is most proper? All which particulars I call
the intrinsic value: for there is also another value merely accidental,
or extrinsic, consisting of the causes why a parcel of land, lying
near a good Market, may be worth double to another parcel, though
but of the same intrinsic goodness. . . . It is no less necessary to
know how many people there be of each Sex, state, age, religion,
trade, rank, or degree, etcetera by the knowledge of which Trade and
Government may be made more certain, and regular; for if men
knew the people as aforesaid, they might know the consumption
they would make, so as Trade might not be hoped for where it is
impossible. . . .

I conclude that a clear knowledge of all these particulars, and
many more, whereat I have shot but at rovers is necessary in order
to [provide] good, certain, and easier Government, and even to balance Parties and factions both in church and state. *But whether
knowledge thereof be necessary to many, or fit for others than the
sovereign and his chief ministers, I leave to consideration.*[34]

In France, which had a well established royal bureaucracy,
Graunt's calls for the collection of agricultural and social statistics
found a receptive audience, leading to the first great compilation
in the *Mémoires des intendants* of 1697–1700. But in England the
road leading to the first census in 1802 was much rockier for
reasons associated with Graunt's final sentence. The key political
question—which could not be ignored in England—was who was
to control and have access to information in order to do what to
whom.

Almost no one doubted that it would be extremely useful to
have at least some of the kinds of information suggested by Graunt
and Petty; and several private individuals collected data and were
spectacularly successful in interpreting it. Gregory King, for example, was able to produce an excellent estimate of the total population of England and of the distribution of the population into a 26-
tiered schedule of socioeconomic categories; and he was able to
construct a way of projecting the price of grain from historical data
on the size of grain harvests and prices.[35] Similarly, Richard Price
was able to produce his *Northampton Life Table* because men associated with Philip Doddridge's dissenting academy maintained
comprehensive accounts of births and deaths from 1735 to 1746.[36]

In spite of the political economists' great enthusiasm for
statistical information, however, there was strong opposition to

creating a national census in Britain, and this opposition crystallized against bills that would have created a national registry of vital statistics in 1753 and again in 1758. In both cases, bitter opposition came from the most conservative and the most liberal groups in British society. The traditional aristocracy opposed the bill as one which was "subversive of the last remains of English liberty" (1) because it would expose the property of every individual to government scrutiny, (2) because it undermined traditional distinctions of rank by opening the homes of rich and poor alike to invasion by government agents, thus leaving "all distinction . . . destroyed by universal coercion,"[37] (3) because it was certain to encourage the Crown to divert more of the public wealth to its own use, and (4) because the specific plans for administering the registry would have co-opted relatively autonomous local officials and made them subservient to the Crown. All in all, one vehement opponent wrote that such a plan threatened to turn freeborn Englishmen into the "numbered vassals of indiscriminating power."[38] On this topic such radical thinkers as Richard Price, who would become one of the most vociferous English spokesmen for the American and French revolutions, were in full accord with the old Tories. Whereas the latter feared becoming vassals, Price feared that the poor would become the "king's chattel" if centralized power was increased and if economic decisions were turned over to a class of persons whose own interests did not keep them "constantly jealous and watchful" of state power.[39]

Both conservative and radical opposition to the establishment of a national registry to collect social and economic information was based on the unstated admission that indeed knowledge *is* power and that the concentration of accurate information in the hands of the state was bound to increase the power of the state relative to that of private citizens. In addition, the conservatives rightly recognized the tendency of those who saw "political economy" as the proper foundation for policy to be insensitive to privileges of rank. And the radicals offered a penetrating insight when they suggested that locating economic decision making in a bureaucratic class—no matter what its pretensions to objectivity—was to disturb the traditional balance of interests by which the English Parliament had managed to limit the expansion of Royal power.

Price did not deny the ability of mathematicians to provide objective evaluations of economic issues. Indeed, he pushed for the establishment of private annuity societies, through which "persons in the lower stations of life might make provisions for themselves in the more advanced parts of life" and pass on property to their descendants. Furthermore, he insisted that such societies must be "guided by the authority of mathematicians," who alone could establish fair premiums to keep the societies solvent.[40] The question he was grappling with, not in a very clear and explicit way, was that of the conditions under which scientific expertise might serve the interests of the populous as a whole and those under which it must serve some special interests.

This question of the appropriate social role of expertise was dealt with much more extensively in a very different context by a group of French intellectuals connected with the *Académie des Sciences* in Paris. This group, for which Jean d'Alembert and Marie-Jean-Antoine Condorcet were the outstanding spokesmen, was obsessed with establishing a new and independent social role for scientists. Their arguments took on long-term ideological importance because Condorcet argued strenuously that those social conditions which maximize the scientific experts' opportunities to generate new scientific knowledge and to apply it—free of interference—to social problems, must also be those conditions which alone ensure progress and the maximum happiness for all mankind.

Condorcet was aware of Price's fears that knowledge might be co-opted by some authority and used to injure rather than serve the majority of mankind. His response, however, was not to try to withhold knowledge from any element of the political elite but rather to insist that the whole society must be organized to make information and scientific expertise available to the majority of men and women. To this end he urged that the organization of what d'Alembert had called "the republic of science" become the model for the organization of the state and that the state take upon itself the responsibility to ensure that all citizens receive as much scientific education as they were capable of understanding. On the face of it, the claim of d'Alembert and Condorcet that what is good for the scientist must be good for everyone has no more plausibility than the famous "What's good for General Bullmoose is good for everyone," the slogan of Al Capp's crass cartoon capitalist. But the former undoubtedly *did* become a central feature of early liberal ideology, and we must consider precisely what the claim entailed, why it was made, and how it took its authority.

The place of science in French society during the late seventeenth and early eighteenth centuries differed in important ways from its place in English society. Though the Royal Society of London had been chartered by Charles II in 1662, it neither received government funds nor exercised any governmental responsibilities. It was, in effect, a private club of largely amateur scientists. Even the *Philosophical Transactions* was the private undertaking of successive secretaries of the society until 1752. Membership in the Royal Society—which carried a substantial financial obligation—was open both to those who showed some scientific merit, like Graunt and Newton, and to wealthy amateurs who had frequently never done any original scientific work whatsoever.

When Colbert arranged for the establishment of the Académie des Sciences, by Louis XIV, however, he created a publicly funded body with clear-cut state responsibilities. Members were paid salaries and given additional income-generating posts which were intended to ensure that they made science a full-time occupation. The Académie was responsible for examining machines and inventions that had been submitted to the government for the awarding of patent rights; they evaluated scientific papers and treatises, and determined which should be published under sponsorship of the Académie or some other branch of the state bureaucracy; and increasingly as the eighteenth century progressed, they were asked to supply technically based policy advice. The famous chemist Antoine Lavoisier, for example, first came to public notice through his plan to provide street lighting in Paris in order to diminish night crime, and he served the monarchy as administrator in charge of gunpowder.[41]

At the same time that the Académie des Sciences was a creature of the state, it was one whose membership was strictly limited to those who demonstrated high levels of scientific achievement. Academicians insisted that they alone had the ability and the right to choose members of the Académie. Admission was by vote of the membership, and though factionalism within the Académie often played a role in who got elected, all agreed (at least publicly) that extrinsic concerns, including royal favoritism or opposition, should not play a role in such decisions.[42]

The problem that d'Alembert sought to explore in 1753 in his *Essai sur la société des gens de lettres et des grands* was the flip side of that articulated by Price. Price wanted to protect the citizenry against the exercise of additional state power which he saw as the logical outcome of supplying the crown with scientific intelligence.

D'Alembert was more concerned with protecting the scientists' independence of judgment in the face of a state that was willing to reward men with both prestige and money if they would only say what "the great" wanted to hear. The responsibility of the scientist to truth could only be ensured in a setting which (1) left the scientist at liberty to explore phenomena and report results free of outside interference and which (2) created a reward structure based solely on the professional judgment of scientific peers. Moreover in those professional judgments, there must be no place for inequalities based on rank or other personal characteristics. Where the scientist was left free to express his opinions to all who might want to hear and free from the pressure of others to direct his work, where he could be freed from the passion for wealth that opened him up to outside influence, where an internally "democratic" mechanism for judging the value of scientific contributions could be established—there the scientist could fully and appropriately "legislate for the rest of the nation in matters of philosophy."[43]

D'Alembert did not believe all of these conditions obtained in the Académie des Sciences. His *Essai sur la société des gens de lettres* was, in fact, a plea for reform and an ideal vision of the scientific community. When d'Alembert tried to initiate even such a simple reform as giving equal voting privileges to junior and senior academicians, he was outvoted within the Académie itself. But his vision of academic freedom in a society of equals was a powerful influence on his protégé, Condorcet, who became permanent secretary of the Académie in 1773. To d'Alembert's ideal of an autonomous and internally democratic community of science dedicated to searching for nothing less than Truth, Condorcet and Anne Robert Jacques Turgot added two critical elements: (1) an unquestioning faith in the Baconian notion that improvement in the human condition is tied to the improvement and application of scientific knowledge, and (2) a vigorous assurance—linked to their fascination with political economy—that the truths of natural science extended beyond the domains of mathematics, physics, chemistry, and anatomy into those of human society as well.

Without the dual assurances that scientific knowledge was the foundation of human improvement and that such knowledge extended even to questions of economics and politics, d'Alembert's concern for the liberty of intellectuals would literally have been of nothing but "academic" interest. But if one truly believed with Turgot and Condorcet that "to know the truth in order to make the social order conform to it, that is the *sole* source of human happiness," then it followed that "it is therefore useful, even necessary, to extend the limits of human knowledge."[44] Under this assumption, the liberty of the scientist to pursue truth became the guarantor of increased well being for everyone. Furthermore, the liberty extended to academicians could not be limited to those areas that could be presumed beyond the sphere of the conflict of human interests. They had to be free as well to study and openly discuss topics like the regulation of trade, the collection of taxes, and even restructuring the organization of the state itself.

Initially, both Turgot and Condorcet agreed with the implicit assumption of Petty and the widespread opinion among French *philosophes* that monarchy was the form of government best adapted to offer scientists free reign to develop and utilize their superior knowledge. "The great advantage[s] of monarchies," Condorcet wrote in his *Vie de Turgot*, "[are] that of being able to destroy the edifice of prejudice before it collapses under its own weight, and of carrying out useful reforms even when the crowd of rich and powerful men protect abuses; and finally, of following a regular system without being obliged to sacrifice part of it to the necessity of winning votes."[45] But a sequence of events that took place during late 1774 through early 1776 convinced Condorcet that a monarchy was impotent to protect scientific policy makers from the destructive power of vested interests.

For a short time—a twenty-month period beginning in August of 1774—it seemed as though Louis XVI was willing to offer the political economists virtually free rein to initiate a new era in scientific decision making for policy purposes. Turgot, who had been made Intendant of Limoges in 1761 largely because of his family's history of service to the state, had proved to be a superb administrator; so in 1774 he was appointed Contrôleur Général des Finances to straighten out the deplorable financial situation of the French monarchy, and he was effectively given carte blanche to do his job.[46]

One of Turgot's first acts was to establish freedom of trade in grain within France. Convinced that the long-term prosperity of the leading sector of the French economy—and therefore the long-term hope for a stable financial base for the monarchy—depended on allowing market forces to set prices and distribute grain throughout France, Turgot pressed on with the proposal in the face of opposition from his ministerial colleagues.

The harvests of 1774 were terrible throughout France, leading to major grain shortages and to deep opposition to Turgot's policies. Jacques Necker, who was to follow Turgot as Contrôleur Général, published *La Legislation et le commerce des grains* early in 1775, arguing in favor of the old doctrine of just price and citing the traditional paternalist grounds that the monarch needed both to protect his people for their own sake and to protect the stability of the realm. Within a few weeks of the appearance of Necker's pamphlet, bread riots spread throughout the Paris region, reaching Versailles on 2 May. Turgot was forced to adopt repressive measures, leading to hundreds of arrests and to the execution of two rioters. Ultimately, Louis was forced by his advisors to abandon Turgot and request his resignation.

Condorcet's reaction to this episode was deeply bitter, and he attacked Necker as a mouthpiece for the wealthy, who violated the rights of the people under the pretext of defending them.[47] Indeed, Turgot's experience convinced Condorcet that d'Alembert had understated the problem of maintaining autonomy for the scientist. D'Alembert had not recognized the fundamental truth that because new knowledge is new power, those who seek to expand and use their knowledge must always and inevitably be seen as a threat to those whose authority is grounded in the pre-existent distribution of power. "In general," he later wrote, "any power, in whatever way it is conferred, is naturally the enemy of enlightenment. It will sometimes be seen to flatter talents, if they abuse themselves to become instruments of its projects or its vanity: but any man who makes a profession of seeking and announcing the truth will always be odious to those who exercise authority."[48] Unfortunately, the response of the ignorant masses to Turgot's attempts to improve their lot certainly seemed to offer little hope that any appeal to the people could help the cause of rational policy making.

Condorcet's solution to this problem of implementing rationally established policy emerged in connection with his analysis of another reform that Turgot had proposed. Local power in France during the *Ancien Régime* was concentrated in the *parlements,* judicial councils dominated by the aristocracy, which exercised authority by registering edicts proposed by the monarchy, by judging cases at law, and by offering advice through correspondence with the king and his ministers. In order to break the stranglehold of the aristocracy on local politics, Turgot proposed the institution of elective municipal assemblies from the village to the national levels. These assemblies would establish local tax assessments and advise the central government on a variety of issues. All who held property would be eligible to vote for members of the local assemblies; their votes would be weighted to reflect the property they held. Under this plan the interests of all citizens would be appropriately reflected. Turgot recognized that the quality of advice that such assemblies could provide would depend on the intelligence and level of education of their members; so as part of his *Mémoire sur les municipalities* he also proposed the institution of a broad-based educational system for all citizens—an educational system that would make "the study of the duties of the citizen . . . the foundation of all other studies, which would be arranged in the order of their patented utility for society."[49]

In a strange and ground-breaking work, the *Essai sur l'application de l'analyse à la probabilité des décisions rendues à la pluralité des voix,* written in 1786, Condorcet attempted to discover the conditions under which the majority decisions of representative assemblies like those proposed by Turgot would warrant acceptance by the rest of society. In his words: "Our principal task here is to discover the probability that assures the validity of a law passed by the smallest possible majority; such that one can believe that is not unjust to subject others to this law and that it is useful for oneself to submit to it."[50]

Suppose that we want to know how probable it is that a simple majority decision is correct. If the probability that each individual voter will correctly decide the issue is greater than 0.5—i.e., if each member of an assembly is more likely to decide the issue correctly than incorrectly—then, Condorcet proved, the probability that a simple majority decision will be correct is proportional to the size of the assembly. If, however, each voter is more likely to be incorrect, a simple majority is more likely to be wrong as the assembly increases in size. From this consequence, Condorcet derived two critical political conclusions. Pure democracy is dangerous in an unenlightened society (i.e., one in which citizens possess neither the information upon which to make correct judgments nor the education to allow them to reason correctly regarding that information); for where few are enlightened, when more people are added to an assembly, it becomes increasingly difficult to find persons whose prior probability of making a correct decision is greater than 0.5. In the short run, then, political decisions should be entrusted to an enlightened elite, and representative assemblies should be kept small, for eighteenth-century France was a largely unenlightened society. On the other hand, the greatest possible probability of making correct decisions will exist in a pure democracy of fully enlightened persons. Thus the long-term optimization of collective decisions demands the creation of a well-informed, well-educated citizenry and the creation of a fully democratic process for making decisions. In other words, Condorcet had demonstrated mathetically that the best possible political decision-making process was one in which the whole society was modeled on d'Alembert's vision of an ideal scientific community.

Both sides of Condorcet's argument were eventually incorporated into continental liberal and socialist ideologies. For the time being, governmental decisions should be made by those best prepared to analyze the issues; but it should be the long-term goal of all states fully to educate their members and to extend participation in social decisions to all. Indeed, Condorcet even suggested that as enlightenment spreads, the very need for any special governmental authority would disappear:

> The more men are enlightened, the less those with authority can abuse it, and the less necessary it will be to give social powers energy and intent. Thus the truth is the enemy of power, as of those who exercise it. The more it spreads, the less they will be able to mislead men; the more force it acquires, the less societies need to be governed.[51]

Here we have one of the first intimations of an egalitarian social order in which the coercive power of governments becomes unnecessary and in which the state must eventually simply wither away.

Condorcet's emphases on education, democracy, and rational decision-making were, as we shall see, strongly implied by arguments associated with Enlightenment psychology; but they found one of their important sources in the policy oriented tradition of the political economists of the eighteenth century.

THE FIRST GREAT ECONOMIC SYSTEMATIZERS: CANTILLON AND THE PHYSIOCRATS

As long as claims for the efficacy of free market mechanisms occurred almost exclusively in tracts by men like Josiah Child, who used them in limited ways to argue for policies that advanced the interests of the merchant class to which they belonged, and as long as claims on behalf of the whole enterprise of political arithmetic occurred almost exclusively in tracts by men like William Petty, who used them to explain how the Crown might most effectively increase its income, such claims were bound to be viewed as grounded in class interests or as *ad hoc* assumptions made to justify some special policy. They could come to be seen as "natural" and unproblematic only as they came to be used as fundamental principles in broad interpretive works that explored a wide range of economic phenomena and that subordinated specific policy recommendations to less obviously biased claims of explanatory generality.

To clarify what I mean, let me suggest an analogy. Leibniz and many of his colleagues had complained bitterly that Newton reintroduced "occult" forces into natural philosophy in the *Principia* when he appealed to a gravitational force whose cause he could not specify. That complaint became irrelevant as a factor in accepting or rejecting Newtonian science as a result of Newton's success in accounting for the entire system of the world through the use of gravitational forces. Initially Newton's success simply allowed his followers to ignore and dismiss Leibniz' claims. But eventually it allowed them to launch a rhetorically successful assertion that the forces of gravitational attraction were "true," whereas the old Renaissance naturalists' claims on behalf of occult sympathies and antipathies were not. Through a similar process, claims that the centrality of self-interest and self-adjusting market mechanisms were merely expressions of special interests gradually became irrelevant as theorists like Richard Cantillon succeeded in creating coherent systems of political economy to account for a wide range of phenomena through their use. Cantillon's accomplishments, in turn, allowed succeeding generations of economic theorists, including the Physiocrats and Adam Smith, to argue for the "truth" of their premises and to authorize the rejection of such notions as that of the just price, with its social rationale, on grounds that they were false and misleading. That there is a logical fallacy in inferring the truth of a set of premises from the truth of a limited set of consequences drawn from them is irrelevant for historical purposes; but it should be kept in mind by anyone who thinks critically about the significance of these historical developments.

One of the most impressive early attempts to fuse the scattered insights and the quantitative techniques of the political economists into a general theory of national economies was Richard Cantillon's *Essai sur la nature du commerce en général,* written between 1730 and 1734. The work circulated in manuscript, serving to shape the thought of almost all major French economic theorists of the eighteenth century, before it was finally published in 1755, twenty-one years after its author's death. Deeply indebted to Petty, Graunt, and Halley for the central themes of political economy, Cantillon was much more sensitive to the historical and sociological tradition that we will discuss in detail later; so he was able to place his analysis of commercial society into a broader context than that of the earlier political economists. Moreover, Cantillon was more cautious than his predecessors regarding his quantitative analyses, chiding his predecessors for their "fanciful" calculations based on too few "detailed facts."[52] By promising less than his enthusiastic predecessors and delivering much more, he gained a more sympathetic hearing for their ideas and for his own refinements and extensions of them.

Cantillon's *Essai* is divided into three major sections. Each deserves serious attention for the important conclusions reached and for the kind of analysis initiated. But since Part III only refines central arguments initiated in Parts I and II, we will restrict our attention to the first two parts.

Cantillon begins Part I by defining wealth as "nothing but the Maintenance, Conveniences and Superfluities of Life." To emphasize the inseparability of two special factors in creating Wealth, he appeals to the Aristotelian notion that all substances must have both matter and form and says that "The land is the source or Matter from whence all wealth is produced," while, "The labour of man is the Form which produces it."[53] As we shall soon see, Cantillon is not an Aristotelian thinker. He believes, for example, that land and labor can be given a common measure, and he is perfectly happy to count not only services which involve no material object but also unworked raw materials as part of the "Wealth" of the nation. His use of the matter/form metaphor is simply intended to emphasize that human well-being depends both on the capacity of the environment to provide food and raw materials and on the human effort which must be applied to extract and process the products of the environment.

In an extremely brief chapter, the substance of which would provide problems for a number of subsequent social theorists, Cantillon argues that, leaving aside questions of justice, it is the observed case that in all societies that have advanced beyond the level of hunter-gatherers and nomadic animal tenders, systems of private property ownership are established which *invariably* lead to the concentration of land in the hands of a few, whether they start out with equal or unequal distributions. "Howsoever people came to the property and possession of land," he reflects, "it always falls into the hands of a few in proportion to the total inhabitants."[54]

Given the fact that land is the fundamental precondition for wealth and that a small class of landowners controls access to the land, the landed proprietary class takes on a uniquely important role for Cantillon. Ultimately, "all the classes and inhabitants of a state live at the expense of the Proprietors of Land."[55] The preferences and customs of the landowners determine the shape of the whole economy. On the one hand, where they live rustically, more interested in horses and hunting than in elegance and luxuries, more land will be given up to woods for hunting and fields for pasturage, there will be few servants and artisans, little land will be given to food production, and the region will support only a small population of unskilled farmers and laborers. On the other hand, where the landowners congregate in cities and demand fine goods and elegant living, jobs for servants and craftsmen will be plentiful, more land will be given over to food production, and a larger population with more skilled workers can be supported.[56]

Next, Cantillon turns to an initial analysis of the second major factor in wealth: labor. He begins by discussing the reasons that wages for some kinds of labor are higher than for others. He assumes a base wage for unskilled labor, then asks under what conditions the father of a son might choose to set him to some skilled trade. If the son could begin unskilled labor at age twelve, but must undertake an unpaid apprenticeship of several years to enter some trade, then the family's loss of income during the apprenticeship must be expected to be reimbursed during the tradesman's subsequent working life.[57] Given an average working life of twelve years and an average apprenticeship of seven years, one would expect wages in skilled trades to average about 19/12 those of unskilled workers if only the training time were considered. But other factors enter into consideration as well. Where serious risk is involved in a trade, wages should increase as expected working life decreases, and where special skills or trustworthiness are demanded, wages will increase either because of demand for the skill or as a way to diminish losses due to theft.

There is a market mechanism that determines the distribution of workers into the various trades. This mechanism restricts the number of workers of each kind to those who can just meet the demand for their labor. Consider the circumstances of laborers or tradesmen in rural villages. If too many children are born to a laborer or too many apprentices train in some trade, then those for whom there is not enough employment must either leave to seek work in the cities or be underemployed and either die of starvation or remain unmarried so as to bring the number of laborers into equilibrium with available work during the next generation. Thus, "if the Village continues in the same situation as regards employment, and derives its living from cultivating the same portion of land, it will not increase in a thousand years."[58] In this way Cantillon explained the remarkable stability of rural population which Petty, Graunt, King, and others had observed.

By the same token, wherever too many enter a given trade, the work will be so distributed as to drive income down. This process will either force emigration or lower the wage differential between unskilled labor and wages in the overcrowded trade enough to inhibit entry into the trade, again producing equilibrium. Where too few enter a trade, however, wages will increase, drawing tradesmen from elsewhere or making the trade attractive enough to draw more recruits from the local population.[59] From these considerations, Cantillon argued, it follows that English proposals to train more craftsmen in charity schools, or French proposals to subsidize apprentice seamen, are pointless, both as attempts to increase national resources and as attempts to help the individuals being trained, unless more jobs are created for them. In general, market forces alone will ensure that "there will never be a lack of Craftsmen in a State when there is enough work for their constant employment."[60]

Given an awareness that both land and labor vary in quality as well as quantity, Cantillon now calculates the "intrinsic value" of any object. It will be "the measure of the quantity of Land and of Labour entering into its production, having regard to the fertility or produce of the Land and to the quality of the Labour."[61] Market prices will not in general be the same as the intrinsic value of commodities, for they depend upon "the Humours and fancies of Men and on their consumption" and are fixed in relationship to the "quantity of Produce or Merchandise offered for sale in proportion to the demand or number of buyers," prices being higher as the quantity for sale is lower in proportion to demand.[62] The market price of most commodities, however, will fluctuate around the intrinsic value "in well-organized societies."[63] Cantillon understands what factors raise and lower both supply and demand; and in the special cases of labor, farm produce, and money, he recognizes how the market pushes supplies up and down in response to demand. He claims that market prices will hover near the intrinsic value of most commodities; but he never offers a general statement about how equilibrium prices are reached by the market mechanism.

Cantillon realizes, following Petty, that it would be good to discover a common measure to relate the value of land to that of labor. He rejects Petty's calculations—which assumed that the rate of return on monies invested in labor will be the same as that invested in land—as superficial.[64] In order to discover the "par" between land and labor, Cantillon performs a thought experiment. He considers the owner of a large estate worked by slaves and asks what must be provided to the least skilled slaves in order to maintain a stable estate economy. Each slave will have to be allocated the produce of enough land to provide food and other "necessities" for himself and enough land to raise one child to working age. From Halley's Breslau tables, Cantillon discovers that mortality prior to age seventeen is about one-half, so that each worker must have the allocation of land enough to support himself and two children. But because each child consumes about half as much as an adult, each slave will require double the land required to maintain a single adult. Each worker in this equilibrium state may thus be said to cost the proprietor twice the land needed to support himself. Free laborers must be paid slightly more than slave labor, and skilled persons more than unskilled; but in each case the cost of the worker will continue to be double the land needed to maintain that laborer in the customary mode of living for the particular kind of worker under consideration.

Joseph Shumpeter, among others, argues that this discussion of Cantillon's is the foundation of virtually all subsequent liberal claims—acknowledged, for example, by Adam Smith and David Ricardo—that the wages of unskilled labor in any static economy will be driven to a bare subsistence level. That is, it is the first expression of the so-called "iron law of wages."[65] In fairness to Cantillon, we should point out that he tempers this conclusion with an interesting empirical argument to show that wage levels vary from place to place depending on local modes of production and customary patterns of consumption.[66]

Since land would be unproductive if it were not cultivated, landowners have as much need of workers as workers do of the landowners; however, since landowners have the power to withhold all or part of their land from production, whereas laborers must either work or starve, Cantillon insists that only landowners are truly independent and able to shape the whole economy.[67] This notion was essential for the economic theories developed by the French physiocratic school in the eighteenth century. Cantillon's analysis of the role of "Undertakers," or "enterpreneurs," in the European economy provided suggestions that eventually led to subsequent liberal tendencies to see capital, whether in land or in money or other goods, as a more general and fundamental category than land alone.

Cantillon argued that eighteenth-century Europe is characterized by an economy in which a huge portion of the circulation of goods is in the hands of undertakers or entrepreneurs who operate at some risk in order to gain a profit. Most farmers, who do not own their own land, pay a fixed rent and buy seed at fixed prices in order to grow crops whose market price is uncertain. Merchants, both wholesale and retail, must advance capital to purchase goods before they can know what prices they can get from consumers. Tradesmen and manufacturers must invest in tools and raw materials, then try to sell the goods they have produced. Even professionals like artists, physicians, and lawyers must invest substantially in their education before they can go out and compete for clients.[68] Though such persons may live very well, they nonetheless remain fundamentally dependent on the whims of the large landowners who control access to the land—except, Cantillon says, "if some person on high wages or some large Undertaker has saved capital or wealth; that is if he have stores of corn, wool, copper, gold, silver, or some produce or merchandise in constant use or vent in a state, having an intrinsic, or real value, he may be justly considered independent in so far as this capital goes."[69] Thus Cantillon suggests that other forms of capital may function much like land. He continues to see land as more stable and less liable to accidental loss than other forms of capital. At the same time he recognizes that the proprietors of money may in some cases even rival the proprietors of land as a powerful interest group within the state. What keeps this from happening ordinarily is that the largest proprietors of money are usually also landowners. When non-landed entrepreneurs accumulate capital, they "always seek to become landowners themselves."[70]

It is very hard to fault Cantillon's analysis, for in the early eighteenth century, both in England and France, land and money were still closely linked. Most entrepreneurial capital came from landed families, and when some London or Paris merchant amassed enough wealth, one of his first moves was to invest in land. In any event, though Cantillon still found the division between independent landowners and the dependent remainder of society the most important economic division in society, he superimposed on this division one into landowners, undertakers (capitalists), and hired people (labor); the income of the first was primarily rents; the income of the second, profits; and the income of the third, wages.[71] This categorization became central to liberal ideas in the writings of Adam Smith.

Cantillon now turns to another issue which had been of central concern to earlier political economists: population. He is interested, like them, in understanding the conditions under which populations will increase and decrease, and he does believe, following Petty, that maximizing the population is, all other things being equal, a good thing because labor is a form of wealth. But Cantillon's interest lies in understanding the relationships between population and a range of factors which Petty had ignored.

The population of all plant and animal species will increase to the maximum number supportable by the land allocated for their support. Humans are no different from other species; they will "multiply like mice in a barn if they have unlimited means of subsistence."[72] For this reason, Cantillon is convinced that the population of the American colonies is bound to exceed that of England within three generations. Human population, however, is generally controlled not simply by subsistence-level food supplies, for humans develop expectations regarding their standard of living and want to "bring up children who can live like themselves."[73] For this reason, humans in some societies practice infanticide[74] to control population, whereas in Europe, they choose to delay or forego marriage. Drawing again from Halley's Breslau tables, Cantillon shows that recorded births are held to about one-sixth of what the population might be expected to generate, largely by delaying the marriage age of women until long after they become capable of bearing children. In this way the birth rate is kept down near the mortality rate, and a stable population is maintained.[75]

In England, where Cantillon insists there is a tendency to allocate increasing amounts of land to the sustenance of each inhabitant—i.e., where the standard of living is rising—one must expect a decrease rather than an increase in population.[76] Unlike Petty, Cantillon does not necessarily view as bad the implicit decision to decrease population. He explicitly says that it is "a question outside my subject whether it is better to have a great multitude of inhabitants, poor and badly provided, than a smaller number, much more at their ease: a million who consume the produce of six acres per head or 4 million who live on the produce of an acre and a half."[77] What *does* lie within his scope is to explain how specific policy decisions respecting economic activities will affect the wealth of a nation—including its ability to sustain the maximum population for any given allocation of land to food production.

Given the current state of European agriculture, Cantillon argues that only about twenty-five percent of the population is needed to work the land to provide "all the necessities of life according to the European Standard . . . [in which] Food, Clothing, Housing, etc. are course and rather elementary, but there is ease and plenty."[78] About fifty percent of the population will be too old or young to work, will be sick, or will be Proprietors of land or capital who do not do labor of the traditional manual kind; so the key economic question is how to allocate the efforts of the remaining twenty-five percent of the population. Some will become soldiers; some domestic servants; and some will use their labor to produce goods which are finer and "more nicely wrought" than is absolutely necessary.[79] These persons "add nothing to the *quantity* of things needed for the subsistence and maintenance of Man," but their labors do give "an additional relish" to the enjoyment of food, drink, and objects of use. Even the production of mere ornaments and amusements is worthwhile, for it provides work for those who need it, and "the state is not considered less rich for a thousand toys which serve to trick out the ladies or even men, or are used in games and diversions, than it is for useful and serviceable objects."[80] Indeed, the demand of the wealthy for luxury goods and for services is important in creating a wealthy state.

There are, however, some employments which increase national wealth even beyond the level which demand for luxury goods alone produces. Persons employed to produce "permanent commodities" (consumer durables) including "Tools and instruments" (productive capacity), and "Gold and Silver" (specie) all add to the permanent stock of wealth of the country. Moreover, they do this whether they mine or produce these commodities directly or whether they manufacture consumable goods, like fine linens, which are sold in foreign markets in return for consumer durables, specie, or productive capacity.[81] The key point is that the wealth of a nation is measured principally by what Cantillon calls the "reserve stock above the yearly consumption" or what we call accumulated capital;[82] so those engaged in activities that lead to capital accumulation are especially valuable to the nation.

Since one of the most desirable forms of capital, in view of its virtual indestructibility, ease of transport, ease of division, and effectively universal use as a medium of exchange, is specie, the issue of capital accumulation inevitably brings Cantillon to one of the most tendentious topics of debate among early political economists. What is the role of foreign trade and "the balance of payments"? Even though gold and silver are only one form of wealth, they are undoubtedly the preferred form, Cantillon admits. So all other things being equal, "the larger or smaller quantity of this stock necessarily determines the comparative greatness of kingdoms and states."[83] But once again, Cantillon is less interested in supporting the so-called "mercantilist" claim that the object of foreign trade is to increase the nation's supply of gold and silver than he is in exploring those factors of exchange which belie the "other things being equal" assumptions made by mercantilist theorists.

Whenever a nation exports products that involve a high ratio of land to labor in their production—items like grain, wool, or wine—it does so only by decreasing the land available to support its own population; so its wealth in specie is purchased at a cost in terms of its wealth in population. If it exports goods—like watch springs—in which almost all the value is due to the labor expended, then specie is added without any cost to the ability of the exporting state to support population.[84] It follows that where trade between two nations involves no net balance of trade—i.e., no net exchange of specie—a nation that exports finished consumable goods and imports raw materials and food stuffs will enrich itself at the expense of its trading partners.[85] It further follows that every state should seek to encourage the development of the manufacturing sector of its economy.

What makes Cantillon's comments on this issue particularly

interesting is his insistence that the best way to encourage manufacturers is to create a large consumption in the interior of the
state. Only secondarily should the state worry about discouraging
foreign manufactures or seeking foreign markets.[86] This insistence
is just one more example of a principle that divides Cantillon radically from his predecessors in economic thought. He not only sees
the market as providing the mechanism for equilibrating supply
and demand, but in every case where the supply/demand relationship is raised, Cantillon gives priority to considerations of demand.
Commercial society becomes possible only as a consequence of the
demand of landowners for luxury goods and services. Supplies of
laborers and tradesmen always follow the demand for their service;
thus, the training of skilled workers is pointless unless demand for
their skills exists. The allocation of land to the production of grain,
wool, or vegetables follows the demand. When Cantillon comes to
consider how one should encourage manufacture, he naturally falls
into his old habit of suggesting that one consider first increasing
the internal demand for manufactured goods. Never before, and not
again until the writings of Keynes in the twentieth century, was an
economic thinker so focused on the priority of demand in determining the fundamental characteristics of an economy.

In the second part of the *Essai,* Cantillon turns to a more extensive analysis of the growth and operation of a market economy dependent on the use of money and credit. In doing so he creates the
first easily identifiable macroeconomic model capable of estimat-
ing demands for money and credit. Once again, Cantillon proposes
a thought-experiment: In a certain closed economy we assume that
the annual produce of the state is worth 15,000 ounces of silver and
that all debts are settled annually. One-third of the total will be paid
in cash to the landowner, who will spend his whole share in the
cities on his life of luxury. One-third will go to the farmer; but some
of the farmer's share will be in kind and some in cash. Since the
farmer can barter for food and rural products locally, he won't need
cash for these materials. He will, however, have to pay for luxury
items and tools that he gets from the city; so we suppose that he
needs about half of his money in cash. Similarly about one-third of
the annual produce will go to support the farm—i.e., to pay hired
help, buy tools. Once more, about half of this amount will be spent
in the cities, and half may be taken in kind. So assuming an annual
settling of accounts, a country with an annual income from land

amounting to 15,000 ounces of silver will need 10,000 ounces to settle debts.

Now suppose that landlords, workers, and merchants agree to settle debts every six months. The money sent to the cities by the landlord and farmer to settle their accounts in the first half of the year must be paid back to the farmers and other land-based producers in order for the city dwellers to get their food and raw materials; so only 5,000 ounces of silver will be enough cash to keep the economy going—that amount circulating twice during the year. In practice, rents are paid quarterly, and wages even more frequently; and city dwellers return money to the country almost daily or weekly to get food. Thus the business of a closed economy can be done with about one-third of the rents due to the proprietors alone, or "the ninth part of the annual produce of the soil."[87] Cantillon admits that this calculation is very crude; but his results conform to estimates made by others based on experience; and all he wants to do at this point, he says, is to "prevent the governors of states from forming extravagant ideas of the amount of money in circulation."[88] Any government which sought to require the annual *cash* payment of taxes amounting to ten percent of the produce of the land, for example, would stop the economy almost instantly.[89]

Foreign trade demands no extra cash when trade is equal, except that by adding some additional loops in the circulation, it may slow the circulation a small amount, leading to a need for slightly more cash, or to bills of exchange or other forms of credit. In fact, Cantillon argues, commercial credit may play the role of money and further reduce the nation's need for specie.[90]

In general, large cities will collect in themselves a large fraction of the money of a country; for landowners, professionals such as lawyers and physicians, and large merchants congregate there. Outlying areas will thus be short of money. Consequently the prices of raw material, commodities, and even labor will be higher in the cities. The price differentials will pay for "the cost and risks of transport" of commodities to the city. If city prices drop below the rural cost plus transportation allowances, "Merchants will not move goods into the city; if they rise above rural cost plus transportation, merchants will flood the city markets," driving the price back down.[91] From these considerations it follows that the price of raw materials will always rise as one gets nearer to large cities. Consequently, "so far as possible Manufactures of Cloth, Linen, Lace, etc. ought to be set up in the remote provinces,"[92] and manufactures of metal tools ought to be located near sources of fuel and mines rather than in large urban centers.[93]

Up to this point Cantillon has accepted and illustrated in several contexts a maxim which he attributes to John Locke, "that the quantity of produce and mechandise in proportion to the quantity of money serves as the regulator of market price," and that therefore, where money is abundant prices will be high.[94] Now Cantillon seeks to explain precisely *how* this happens so that he can understand what groups benefit or suffer in economies whose inflationary trends are fueled in different ways. If inflation arises from mining, all associated with that industry will increase their consumption of goods; so merchants, tradesmen, manufacturers, and farmers will all be able to share in the wealth, whereas landowners who received fixed rents and wage laborers operating on fixed wages must suffer because their incomes cannot keep up with their expenses. In the long run, however, rents and wages must also rise. At some point the costs of locally produced goods will become so high that it will be cheaper to import foreign goods, causing the collapse of local "Mechanics and Manufactures" and sending money to foreign manufacturers. "The great circulation of money, which was general at the beginning, ceases; poverty and misery follow and the labor of the Mines appears to be only to the advantage of those employed upon them and the Foreigners who profit thereby."[95] According to Cantillon, this is what seems to have happened to Spain as the profits from precious metals extracted in the New World found their way into the pockets of English and French manufacturers, merchants, and farmers.

If, however, the increase of money comes from "a balance of foreign trade (i.e., from sending abroad articles and manufactures in greater value and quantity than is imported and consequently receiving the surplus in money)," as long as the favorable balance continues, the merchants, mechanics, and manufacturers will be enriched, and employment will remain high. Rents and wages will again increase, the prices of everything will rise, but if the "facility [efficiency]" of the state's shipping and manufacturing remain high, that may make up for high labor costs, continuing to allow sales of goods to foreigners. In such a case the nation in question may continue a long time with a high standard of living. This, Cantillon says, has been the case for England recently.[96] Eventually, however, it is inevitable that poorer nations will establish competing manufactures and that increasing efficiency coupled with lower labor costs must overcome the advantages of the initial exporter nation. It will become a net debtor nation; money will flow out, and "it will inevitably fall into poverty by the ordinary course of things."[97] "Such," Cantillon argues, "is approximately the circle which may be run by a considerable state which has both capital and industrial inhabitants."[98] Disaster may be delayed by governmental policies to keep money out of circulation; but in the long run no economic advantage can come to a state merely by increasing the amount of money it has.

A third method of increasing the cash position of a national economy is even more ruinous than the first. Capitalists or even the government itself may borrow money from foreigners at interest in order to establish new manufactures or to cover the expenditures of the state. In either case, short-term advantages are purchased at the cost of a huge debt service expense and "these borrowings which give a present ease comes [*sic*] to a bad end and is a fire of straw." In the short run, there is no question that a state which has an abundance of money has an advantage over its neighbors, "so long as it maintains its abundance of money."[99] For that state will find it easier to raise taxes and to support a strong military force. Ultimately, however, the wealth of a state remains proportional to its land and its labor weighted by their "quality."

The final general topic broached by Cantillon in Part II of his *Essai* is another central problem of earlier economic battles. It is the dual question of how interest rates are established and how those rates affect and are affected by the rest of the economy. For this discussion Cantillon treats money as any other commodity—one whose market price is established at some meeting ground "proportionate to the needs of the borrowers and the fear and avarice of the lenders."[100] Once more Cantillon begins with a thought-experiment. Suppose that a farmer who normally rents land for one-third of the annual crop has the opportunity to purchase his own land but has no capital with which to do so. If he could borrow money at a rate which would cost him one third of the produce of the land that he had been paying in rent, then he would lose nothing by purchasing the land. By scrimping and saving, he might eventually be able to pay off his debt.[101] If the land produces annually about one-sixth of its price, the farmer should then be willing to pay about one-eighteenth, or between five and six percent interest, on his mortgage. Consider the lender's position. Since loans on land are secured by mortgages on the land, they involve almost no risk; for if the farmer should fail, the lender simply forecloses and begins to collect rent rather than interest. Thus interest rates on mortgage loans, the lowest risk loans made, must ordinarily hover just slightly above the rents from land, as Petty had supposed without any justification.

Suppose now a journeyman hatter wants to go into business for himself. If he can borrow money at a rate that allows him to maintain himself as he was accustomed to as a journeyman and to improve his income marginally, then he will be willing to turn over the rest of the profit from the business to a lender. In this case, there is substantial risk of failure; but since the capital needed is much greater than the money which the journeyman usually consumes in a year, even if the journeyman fails, the lender might sell off some of the hats and raw materials. In such cases interest charges might settle at around twenty to thirty percent per annum.[102] Smaller entrepreneurs, who might easily squander the money they borrow, must pay much higher interest rates. At the same time, their return on capital is usually vastly higher than that for larger enterprises. In some cases, Cantillon argues, interest rates on small high-risk/high-profit ventures rises to 400 to 500 percent per annum.[103]

Turning to the causes for the raising and lowering of interest rates, most previous political economists claimed that when the amount of money in a state increases, interest rates go down; but Cantillon argued that that is only true under certain restricted conditions:

> If the abundance of money in the State comes from the hands of money lenders it will doubtless bring down the current rate of interest by increasing the number of money lenders: but if it comes from the intervention of spenders it will have just the opposite effect and will raise the rate of interest by increasing the number of undertakers who will have employment from this increased expense, and will need to borrow to equip their business in all classes of interest. Plenty or Scarcity of Money in a State always raises or lowers the prices of everything—without any necessary connection with the rate of interest, which may well be high in States where there is plenty of money and lower in those where money is scarcer.[104]

Two factors do tend to drive interest rates up. Where the nobility lives in luxury, they keep the profits of venturers high and make it possible for them to pay high interest rates on cash advances. Second, whenever there is instability and extra risk—especially in times of war, when both demand and risk increase—interest rates will be raised.[105] Given this argument that interest rates are naturally regulated, Cantillon insists that it is pointless to try to regulate them without consideration for natural factors:

> If the Prince or Administrator of the state wish to regulate the current rate of interest by law, the regulation must be fixed on the basis of the current market rate in the highest class, or thereabout. Otherwise the law will be futile, because the contracting parties, obedient to the force of competition or the current price settled by the proportion of Lenders to Borrowers, will make secret bargains, and this legal constraint will only embarrass trade and raise the rate of interest instead of settling it.[106]

Let us step back to reflect on Cantillon's *Essai* and what it might have done to hide the problematic character of its most fundamental assumptions. There is no question that Cantillon assumes throughout his essay that every economic agent acts according to some rationally calculated self-interest. In each of his many thought experiments, he begins by asking a question like "how much higher must skilled labor wages be than unskilled wages in order to make it worthwhile for a person to forego income during a period of training?" or "how much interest should a farmer be willing to pay on a mortgage in order to improve his situation?" He never raises this issue in a general form; so he never makes it possible to suggest that any other motives might be involved in economic decisions. The old Hobbesian aggressive assertion of the centrality of self-interest has apparently become so "natural" that it no longer occurs to Cantillon to try to justify it.

By the same token there is no question that he assumes that natural market mechanisms will optimize economic outcomes by adjusting supplies to demands. Again, he simply *uses* this general assumption to account for the allocation of land and labor resources to different uses, to account for aggregate population figures in different societies, and to account for interest rates. Repeatedly he *asserts* the futility of interfering with the market— whether to regulate interest, to train more skilled workers, or to encourage population growth. Once again, he never states his general assumption; so he never acknowledges the possibility of effective intervention in economic affairs except to facilitate or utilize the workings of the market mechanisms.

What is most impressive about Cantillon's *Essai,* however, is its success in accounting for such a large number of detailed and well-documented phenomena on the basis of a very small number of central assumptions. Whether one wishes to understand something as general as why there are differential wage rates in different occupations, why commodities are cheaper in the country than in the city, why market prices differ from the costs of producing goods, and why manufacturing countries like Holland have higher standards of living than agrarian countries like France, or something as specific as how an economy can operate with an amount of money that is less than ten percent of its annual produce, why mortgage interest rates are about five percent in England, but only about two percent in Holland, or why the birth rate in Breslau is only one-sixth of that which the population is biologically capable of producing, Cantillon can tell without appealing to any obviously *ad hoc* assumptions. It is hardly any wonder both that Cantillon's ideas should have had tremendous appeal to the French political economists who first noticed them and that the system seemed in some way to legitimize those assumptions that it incorporated.

At this point it is important to consider whether the central assumptions of Cantillon's work merely reflected a new emerging market-dominated society or whether their expression by Cantillon and other like-minded theorists actually played a significant role in constituting or accelerating the development of the new social reality. If the first is the case, then there would be little justification in claiming that scientific attitudes, methods, and theories were—in this instance—a key factor in the emergence of modern liberal ideology. We would have to see such scientific political economists as Petty and Cantillon as holding up a mirror to reflect society rather than as holding up a lamp for it to follow.

If a market-dominated national economy constitutes a central feature of the new order, then we have already suggested, in connection with Turgot's attempts to establish free trade in grain in France late in the eighteenth century, that not only was there nothing approaching a market-dominated grain trade in France up to this time, but also that producers, consumers, and the government itself had to be dragged, kicking and screaming, into the new order. Moreover, as we shall soon see, the prophets, architects, and agents of transformation in this case were not those who stood to reap great private profit from free trade in grain, but the "Economists," or "Physiocrats," a group of intellectuals and administrators who built a program for French economic reform on the theoretical foundations created by Cantillon. As Henry Higgs, one of the chief historians of the school has claimed, "the Physiocrats were not merely a school of economic thought, they were a school of political action"[107] through whom the science of political economy became a central foundation for the practice of politics.

Cantillon's *Essai* was not blatantly polemical or political, but it did incorporate a number of recommendations, many of which simply enjoined governments to stop interfering with the free play of the market, some of which offered general advice about encouraging rural industry, and some of which were aimed at rejecting specific French policy proposals, like Vauban's proposed monetary tithe. Like Petty, Cantillon assumed that his recommendations would be implemented—if at all—by a monarchical government; but the recommendations were independent of any assumptions about the form of government. Though he did not coin the word, his ideas certainly played a role in the French movement that its adherents labeled "Physiocracy," the rule of *nature*, to distinguish it from traditional categories like "Democracy," the rule of the people, "Aristocracy," the rule of the elite, or "Monarchy," the rule of the one. For Cantillon, as for his predecessor, Petty, and for his immediate successors, the Physiocrats, traditionally *political* decisions were increasingly seen as administrative decisions that ought to be grounded in rational and empirical knowledge of the phenomena in question—i.e., in scientific knowledge of society.

The Physiocrats, like their sometime admirers, Turgot and Condorcet (prior to 1774), did have a preference for monarchical governments because monarchy seemingly offered the fewest obstacles to rational decision making and the greatest protection against the intervention of special class interests into policy. Ultimately, government involved nothing but creating and administering policies founded in the natural laws governing society; and all of those laws relate directly or indirectly to the material conditions of life— i.e., to economics.[108]

In order to understand how Cantillon's interpretation of society was transformed into a program for liberal economic and social reform, we need to understand something of the background of François Quesnay, the leader of the Physiocratic School. Born in 1694 into a peasant family, Quesnay never lost sight of his agrarian roots even though he left the farm in 1707, never to return. Madame de Hausset, a court gossip from whom we know most of the few personal details we have of Quesnay's later life, tells us that half a century after he had moved into Paris to apprentice as an engraver, he still "loved to chat with me about the countryside . . . the wealth of the farmers, and the method of cultivation. . . . He was much more concerned at court with the best method of cultivating the land than with anything that went on there."[109] While an apprentice engraver, Quesnay took courses in medicine and surgery at St. Come. In 1717 he married a Parisian shopkeeper's daughter and moved to Nantes where he set up as a surgeon, gradually improving his social position in the town. Soon Quesnay began to write on medical and surgical topics; he became physician to a member of the nobility. Because of his competence and discretion, he was recommended to Madame de Pompadour, eventually moving to Versailles in 1749 as first ordinary physician to the king. In the meantime, Quesnay had entered the Parisian *Académie des Chirurgiens* and become its perpetual secretary. By 1751 his work in medicine and animal physiology earned his appointment to the Académie des Sciences and friendship with intellectuals like Jean d'Alembert and Etienne Condillac. The latter two were especially drawn to Quesnay because his medical works demonstrated a fascination with questions of epistemology and scientific method.

Within Court circles Quesnay had gravitated into a close association with a group of men deeply concerned with agricultural reform and the improvement of French governmental finances.[110] Given his interests and connections, it was natural that his friends Diderot and d'Alembert invited Quesnay to write a series of articles on agriculture and finance for Diderot's *Encyclopédie.* So in 1756, at age 62, François Quesnay first set pen to paper on economic topics from a background of special concerns with agricultural improvement, state finance, and scientific method.

His training as a physician provided Quesnay with an important perspective. He viewed the French economy as an organism with a serious disease of which the clearest symptom was the ruinous state of governmental finances. In order to cure the patient, however, one had to understand the causes of the disease and not merely treat its symptoms. "We must not lose heart," he wrote to his friend Mirabeau, in a letter attached to one of his most important economic manuscripts, "for the appalling crisis will come, and it will be necessary to have recourse to medical knowledge."[111]

Even before he developed any economic sophistication, Quesnay was convinced by his discussions with the *agronomes* of the Court that the fundamental problems of the French economy were tied to the relatively backward state of its agricultural sector. In his first economic article on "Farmers" for the *Encyclopédie,* he developed a notion that allowed him to modify Cantillon's theories when he came into contact with them. Cantillon had generally assumed that the produce of land was proportional to its intrinsic goodness and to the labor expended on it. But Quesnay, much more familiar with recent agricultural developments in England and with the arguments of agricultural improvers, realized that agricultural output could be greatly increased through economies of scale and by investment in new tools, livestock, rolling stock, fencing, and barns. Even more than Cantillon, Quesnay believed that the wealth of the entire state depended on the productivity of the farmers which in turn depended on their ability to generate a "net product" or surplus over and above their costs and the rents they paid, for "it is the sale of the surplus which enriches the subjects and the sovereign."[112]

At this point Quesnay is most interested in the part of the surplus which can be reinvested in agricultural improvement; for this part of the surplus creates an *expanding* economy by generating yet greater surpluses for the next year. If the farmers are to maximize their surplus and lead a French economic recovery and advance, then the government must intervene by "removing all the other causes prejudicial to agriculture."[113] In particular, the government must repeal laws controlling the production, distribution, and price of grain and bread; for such laws often force the farmers to grow grain at a loss and lock them into a grossly inefficient system of production. Similarly, the state should intervene to make certain that farmers will have incentives to invest in their farms. Thus patterns of seignorial control which drained the fruits of higher productivity away from the working farmers into the hands of the proprietors had to be abandoned, as did patterns of forced labor— the *corvée*—which disrupted the efficient allocation of agricultural labor.

Between the writing of "Fermiers" in 1755 or 1756 and that of his subsequent *Encyclopédie* articles on "Grains," "Men," and "Taxation," Quesnay came into contact with Cantillon's *Essai.* He was not sympathetic with Cantillon in enthusiasm for export manufacturing or the positive economic role of financiers; and he insisted to his friend, Mirabeau, whose *Ami du peuple* of 1756 was based on a manuscript of Cantillon's *Essai,* that "Cantillon, as a teacher of the public is nothing but a fool."[114] In spite of his differences with Cantillon, however, the latter's work provided Quesnay with an analytic perspective which he developed into the central doctrines of Physiocracy. Even when he disagreed with Cantillon, Quesnay's arguments almost inevitably developed as the amplification of one of Cantillon's insights to confute another. Given his assurance that agricultural productivity could be greatly increased by improving agricultural practices (his calculations showed that a move from small-scale, labor-intensive to large-scale, capital-intensive cultivation might roughly double costs while tripling productivity),[115] Quesnay was able to discount all of Cantillon's supposed negative consequences of exporting agricultural goods and raw materials. At least in the near future, he argued, agricultural export could have no negative impact on population; for the land is capable of feeding and clothing a vastly greater population than it does at present. He seized, however, on Cantillon's analysis of the trade cycle to argue that the allocation of resources to export manufacture in a state is pointless in the long run because "its trade can be taken away from it by other rival nations which devote themselves more successfully to the same trade."[116]

In much the same way, Quesnay also used Cantillon's analyses to attack preference for the monetary form of wealth. In a section devoted to illustrating the proposition that "The advantages of external trade do not consist in the increase in monetary wealth," Quesnay argued that money can in effect be supplied by an increase in the velocity of circulation—an argument borrowed from Cantillon; and he draws from Cantillon's discussion of the fate of Spain to show that an increase of money in the state does not lead to long-term wealth, which is dependent ultimately on nothing but quantity and quality of land and labor.[117] Finally, Quesnay adopts Cantillon's argument that internal trade—by increasing the market for agricultural and manufactured goods and the market values of all goods—is vastly more valuable to a state than external trade.[118]

On one other important topic Quesnay turns Cantillon's own analysis against him. For Cantillon, whose central concern was with consumption, it is the taste for luxury of the proprietary class which drives economic growth. It is necessary to support an idle wealthy class in order to create the demand for goods which provides jobs for the whole population. Only incidentally does he suggest that those who increase productive capacity might be especially important. Quesnay is willing to admit that wealthy consumers are at least not actively harmful to the state: "It is perfectly true that wealthy men who contribute nothing to the production of wealth, would be very harmful individuals if they did not spend their revenue. But . . . they are not harmful so long as they do that."[119] Because he is more concerned wih the mechanism of economic *growth* and issues of productive efficiency, Quesnay sees that wealth could be used more effectively in other ways: "It would be advantageous if they [the wealthy proprietors] also consumed their capital, which, by passing into the hands of industrious men, would be rendered still more useful."[120] He vastly expands Cantillon's suggestion that wealth applied to increasing productive capacity is of greatest long-term benefit to the state; for such wealth allows a drop in price without a drop in profits; and that in turn stimulates demand. Pure consumers do nothing in the long run but drive prices up, without adding to national wealth, whereas those who use their capital to improve production by investing in "machines which can contribute to reduce the cost of men's labor," or "canals . . . which avoid costs which are paid to carriers," also generate economic growth. They "bring about a price which is favorable . . . , [and] they encourage sales and production, which increases wealth and consequently population, for the increase in wealth produces an increase in expenditure, which yields an increase in the gains of all remunerative occupations and which attracts to them a greater number of men."[121]

A final analytic technique that Quesnay developed from his reading of Cantillon led in turn to one more policy recommendation that we must consider. In analyzing the need for money in an economy, Cantillon had verbally presented an economic model that followed the circulation of money and goods among three "classes" in society: the proprietors, the farmers, and the urban "mechanics, manufactures, and tradesmen." In his famous "Tableau Economique," Quesnay developed a graphic representation of the circulation of economic commodities among those same three groups, depersonalized into three classes of "expenditures"—expenditures of "Revenue" by the landed proprietors, "Productive expenditures" by the farmers, and "Sterile expenditures" by the manufacturers, merchants, and servants[122]—which allowed him and his followers to envision and analyze quantitatively critical details of economic exchange (fig. 8). Though subsequent economic theorists emphasize the critical importance of the "Tableau" as an analytic tool,[123] its interpretation is problematic and far too complex to detail here.[124] It is enough for us to note that Quesnay's analysis of the "Tableau" provided ammunition to support a specific tax policy.

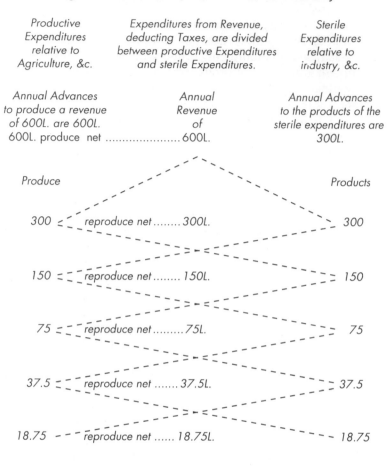

Fig. 8. François Quesnay's "Tableau Economique."

Though wealth may appear in a variety of forms, analysis of the "Tableau" demonstrates that all forms are ultimately reducible to three: "(1) the revenue of landed property, (2) the wealth which restores the costs or expenditures incurred to generate the revenue, [and] (3) the wealth produced by industrial work."[125] We are interested in how this knowledge can be used to suggest a tax policy, where taxes are understood as dues which subjects pay to a sovereign and which must somehow be "laid on the annual wealth of a nation."[126] Since the second form of wealth cannot be reduced without correspondingly reducing the produce of land in the next cycle, it "ought not to bear the burden of any taxes at all."[127] By like arguments, Quesnay purports to demonstrate that the wealth produced by industrial work cannot support taxation without severe repercussions in the industrial sector. The revenue of landed proprietors, however, is useful to the state only through consumption, and it might as easily be spent by the state as by the proprietors. It is this form of wealth, then, which must ultimately "pay the taxes levied to meet the expenditure which is necessary for the government and the defense of the state."[128] Since a tax on revenue could be much more easily collected at its source than at its point of consumption, Quesnay and the Physiocrats argued in favor of a tax paid by the farmers to the state in lieu of part of the traditional rents paid to the landowners. In this case, as Quesnay fantasized, the proprietors would never see the money; thus, they would hardly miss it at all.

Leaving aside Quesnay's intellectually blinding obsession with agriculture, what is most fascinating about Physiocracy is the way in which assumptions regarding the efficiency of market forces in allocating labor and goods and regarding the rational basis for all economic decisions have now come to play a manifestly ideological role. Quesnay had no desire to undermine the social order of the old regime, and his first and most effective disciple, Mirabeau, was not only a member of the traditional landed aristocracy but throughout much of his early career was an explicit apologist for noble privilege and for the paternalist vision of society with which defense of aristocratic privileges was linked. Yet the two men and their disciples were pushed by a scientific analysis of political economy into a series of recommendations that implied a radical reconstitution of the social order, one which presupposed the willing abandonment of feudal privileges by the landed nobility and renunciation of the traditional paternalist justification for the monarchy itself.

That the cure for France's economic ills would be more easily prescribed than administered was brought home when Mirabeau was jailed by the king in 1760 for his "Theory of Taxation" which expressed Quesnay's views. And the Physiocrats slowly seemed to become aware that it might take some kind of great social upheaval to lead to the implementation of their proposals.

The Physiocrats were *not* political radicals, and they tried to downplay the social implications of their economic doctrines.

They could not and did not retreat, however, from their efforts to rationalize the French economy by giving free rein to rationally calculated self-interest and market mechanisms. They recruited converts, gained control of two publications (the *Ephémérides du citoyen* and the *Journal de l'Agriculture, du Commerce, et des Finances*) during the 1760s, and propagandized on behalf of free grain trade, abandonment of the corvée, and a tax on the revenue of the landed nobility. In 1762, Henri-Leonard-Jean-Baptiste Bertin, Contrôleur-Général des Finances, and close follower of Physiocracy, attempted to establish freedom of competition in the cloth trade, encouraging rural manufactures by cancelling the royal privilege which gave the Lille guild of wool weavers the exclusive right to produce certain kinds of cloth. But the town council of Lille fought the plan and convinced the local intendant to refuse to implement the central government's orders, demonstrating a typical response to attempts at liberal economic reform. Finally, in 1763 the Physiocrats saw their first minor success when Bertin's Declaration of 25 May authorized the free movement of grain across provincial borders within France; but the order was rescinded after a series of local protests. When Turgot, a revisionist member of the school, became Minister of Finance, a number of their pet reform projects were briefly attempted; but as several of the group had foreseen, it took a great cataclysm—the French Revolution—to destroy the privileges that had blocked French economic growth.

The Ideological Implications of Enlightenment Social Science II: The Sociological Tradition

5

Before we turn to Adam Smith, in whose *Wealth of Nations* scientific economics and liberal ideological concerns found one of their most powerful and lasting fusions, we need to backtrack to pick up a second thread of eighteenth-century social science, the sociological/historical/anthropological thread which found its most celebrated early manifestation in Charles-Louis de Secondat, Baron de Montesquieu's *Esprit des lois (Spirit of the Laws)*, published in 1748.

MONTESQUIEU AND THE COMPARATIVE STUDY OF SOCIAL INSTITUTIONS

Montesquieu's *Esprit des lois* spawned at least two separable and powerful traditions. On the one hand, the eleventh book, which developed the classical republican notions of separation of powers in a constitutional government, took on a life of its own and played a critical role in the formulation of the American Constitution.[1] We will virtually ignore this tradition. Instead, we will be interested in the tradition of theoretical or conjectural history that found its way into Condorcet's scheme for educational reform and that found its most powerful proponents in Turgot, Adam Smith, Adam Ferguson, and Johann Herder.

When Dugald Stewart wrote his biographical memoir of Adam Smith for delivery to the Royal Society of Edinburgh in 1793, he placed Montesquieu at the beginning of a new approach to history which had been central to Smith's work on the history of government, jurisprudence, and economic institutions. Referring to governmental institutions and laws in particular, Stewart wrote:

> It is but lately . . . that these important subjects have been considered in this point of view; the greater part of politicians before this time of Montesquieu having contented themselves with an historical statement of facts, and with a vague reference of laws to the wisdom of particular legislators, or to accidental circumstances, which it is now impossible to ascertain. Montesquieu, on the contrary, considered laws as originating chiefly from the circumstances of society, and attempted to account from the changes in the condition of mankind, which take place in the different stages of their progress, for the corresponding alterations which their institutions undergo.[2]

In this statement, Stewart was guilty of foisting some of the doctrines of his successors upon Montesquieu; for there was little if anything in Montesquieu that would justify a claim that he thought of society in terms of "stages" in a "progress." But Stewart was stating a very commonly held view in seeing Montesquieu's work as seminal in the creation of a new theoretical, dynamic, and self-consciously scientific approach to government and law.

From our present perspective, Montesquieu's central accomplishment was to fuse the traditional humanistic subjects of history and law with insights and information drawn from the recent fad for travel literature into a new and systematic comparative study of social institutions. "Thus," writes Stewart, "we frequently find him borrowing his lights from the most remote and unconnected quarters of the globe, and combining the casual observations of illiterate travelers and navigators, into a philosophical commentary on the history of law and of manners."[3] Stewart, like other Enlightenment figures, undoubtedly overestimated Montesquieu's originality. In his scientific attitudes and his emphasis on the relation of various governmental forms to their related "principles," Montesquieu owed a deep debt to the tradition of Machiavelli and Harrington. In his discussion of environmental influences on patterns of social behavior and governments, he built upon a longstanding classical tradition initiated in the Hippocratic corpus.[4] But by bringing these traditions together and enriching them by reference to newly acquired knowledge about non-European cultures, Montesquieu revitalized and reformed the traditions from which he drew. Perhaps even more important, because he developed his analysis in the context of eighteenth-century conflicts between the centralized French monarchy and the traditional nobility, because he developed a defense of the parlements' privileges, because he presented his analysis in an extremely accessible form, and not least because he brought to them a reputation as a master of literary style, his arguments were ensured both a wide audience and a critical attention. As was the case with Hobbes, subsequent Enlightenment social theorists could admire or hate Montesquieu; but they could not ignore him.

Montesquieu had been born into the French landed nobility at the medieval chateau of La Brède, near Bordeaux, in 1689. At age eleven, he was sent off to study at an Oratorian college just north of Paris. Here he spent five years gaining the most progressive education available on the continent. To the study of the Latin classics was added the study of French history, mathematics, and Cartesian natural philosophy. Back home in Bordeaux, Montesquieu studied law with relatively little enthusiasm. In 1713, when he was only twenty-four years old, he became one of the judges of the Bordeaux Parlement; and when his uncle died just three years later, he became president and chief justice of the parlement of Bordeaux, inheriting his uncle's title, Baron de Montesquieu. Meanwhile he was developing as an amateur scientist, becoming a member of the Academy of Bordeaux, publishing works on acoustics, mechanics, anatomy, and geology, and corresponding with acquaintances throughout the world in preparation for writing a geological history of the earth—a project he never completed. Among his studies was one on the effects of heat and cold on animal tissues which was to have an important bearing on his later work, *Spirit of the Laws*; for it led him into speculating about the relations between climate and human life.

But Montesquieu's scientific interests were paralleled by an almost equally intense fascination with literature. His greatest admiration was for those who linked scholarly and literary attainments. The admirer of Cartesian philosophy argued that "most people know all they know of Descartes through Fontenelle's graceful exposition of his ideas," and he insisted in general that "Ideas must be gracefully dressed to survive; thus, science and belles-lettres touch."[5] To explore the literary side of his talents, the young lawyer began a fascinating work of social criticism which was to appear in 1721 as *The Persian Letters*.

Like many others in the eighteenth century, Montesquieu was an avid reader of exotic travel literature. Tavernier's *Travels in*

Turkey, Persia, and the Indes and Jean Chardin's *Account of a Journey Through Persia and the Orient* fascinated him, and a recent but very poorly written book, Charles Dufresny's *Serious and Comical Amusements of a Siamese at Paris* gave him a great idea. Montesquieu was amazed, appalled, and puzzled by the variety and absurdity of beliefs and customs in the world. He would satirize some of the disturbing features of French life through the fictional correspondence of two Persian diplomats visiting Paris. Often cynical, often slightly racy, and always entertaining, *The Persian Letters* appeared just at the end of a spectacular Paris visit of the Ambassador of the Grand Turk, and they were a smash hit, going through ten editions in the first year and spawning literary imitations— *Turkish Letters, Iroquois Letters,* and even a set of *Peruvian Letters* by Madame de Graffigny. Encouraged by the reception of the *Persian Letters,* Montesquieu sold his Presidency of the Parlement of Bordeaux for something over half a million livres to become a full-time savant.

Shortly after admission into the *Académie Française* in 1728, Montesquieu began a three-year voyage to study the political and social institutions of other European countries, staying nearly two years in England where he visited and studied the workings of Parliament and accepted membership in the Royal Society of London. In 1731 Montesquieu returned home to La Brède to work on his two major studies, *Considerations on the Causes of the Grandeur of the Romans and Their Decadence,* published in 1734, and the *Spirit of the Laws,* published in 1748. The title of the first of these two works signals Montesquieu's changed perspective on history. Like his first scientific paper on "Causes of the Echo," *Causes of the Grandeur of the Romans and Their Decadence* would focus on a search for the underlying causes of phenomena, subordinating detailed narrative to a search for general laws which might be applicable to contemporary life. "As men in all times have the same passions," he insisted, "the *occasions* which produce great changes are different, but the causes are always the same."[6]

Since all of the issues we shall be interested in are more fully developed in the *Spirit of the Laws,* we pass over the details of Montesquieu's analysis of Roman history. In the preface and early portions of Book I of the *Spirit of the Laws,* Montesquieu exposes both his Cartesian and his more broadly scientific orientation. His aim is to understand the spirit, or causes underlying the great variety of human laws which exist in the world; and his method is

the deductive method of Descartes. "I have laid down first principles," he writes, "and have found that the particular cases follow
naturally from them; that the histories of all nations are only consequences of them; and that every particular law is connected with
another law, or depends on some other of a more general extent."[7]

Laws in general, Montesquieu asserts, "are the necessary relations arising from the nature of things."[8] To suggest—with Hobbes
in his more nominalist moments—that human laws *create* justice,
for example, is to make a horrible mistake. "To say that there is
nothing just or unjust but what is commanded or forbidden by positive laws is the same as saying that before the describing of a circle
all the radii were not equal."[9] But when we consider those laws
created by intelligent beings to govern their interactions in society,
certain special problems do arise. Like the Greek Sophists, Montesquieu argues not that all human laws *are* built upon the invariable
and necessary relations of things, but that they *ought* to be. To understand the divergence between positive laws as they exist and as
they ought to exist, Montesquieu points to the finitude and tendency
to error of humans and to the uniquely human notion of free agency:

> The intelligent world is far from being so well governed as the phys-
> ical. For although the former has also its laws, which of their own
> nature are invariable, it does not conform to them so exactly as the
> physical world. This is because, on the one hand, particular intelli-
> gent beings are of a finite nature, and consequently liable to error;
> and on the other, their nature requires them to be free agents. Hence
> they do not steadily conform to their primitive laws; and even those
> of their own instituting they frequently infringe.[10]

Montesquieu at least hoped to reduce the negative consequences of
human error by providing a deeper knowledge of the invariable
"primitive" laws underlying human legal systems.

So far, except for this brief bow to human freedom, Montesquieu's aims sound very much like those of Hobbes or Harrington,
but his historical and humanistic studies led him into a strategy
which seemed to some to make his scientific approach to society
much richer than those of his predecessors, while it seemed to
others to belie his claim to be developing a science of society at all.
Far from offering a simple derivation of human laws from human
nature, as Hobbes had hoped to do, Montesquieu listed at least
twelve different classes of factors that needed to be taken into
account in understanding legal systems. First was the nature of the
"principle of each government," whose meaning we will return to

consider; second, "the climate"; third, "the quality of the soil, its situation and extent"; fourth, "the occupation of the natives, whether husbandmen, huntsmen, or shepherds"; fifth, "the degree of liberty the constitution will bear"; sixth, "the religion of the inhabitants"; seventh, "their inclinations"; eighth, their "riches"; ninth, their "numbers"; tenth, their "commerce"; eleventh, their "manners"; and twelfth, their "customs."[11] It was to Montesquieu's insistence that there must be lawful relationships among this wide range of features of society as much as to some of his subsequent specific analyses of those features that the Enlightenment tradition of theoretical history or sociology or anthropology owed its greatest debt.

As we consider the complex tradition following from Montesquieu, however, we should keep in mind that his inclusive concerns generated impetus for a countertradition as well, especially among those for whom the Newtonian success in understanding the material universe through a *single, universal* principle had dramatic appeal. Condorcet, for example, was appalled that anyone could see in Montesquieu's collection of historical examples anything but evidence of "the infinite variety of bad laws."[12] Since it was one of the great aims of science to free men from superstition and prejudice, it was unthinkable that laws should *accommodate* themselves to religion, customs, and manners, which were born of superstition and prejudices. Montesquieu's work was bitterly attacked by many intellectuals for political reasons that we will soon discuss; but it was also anathema to many because of its stubborn refusal to make the kind of analytic simplifications that had allowed mathematicians, natural philosophers, and even political economists to get on with their work. On the other hand, it had a very special appeal to German intellectuals such as Johann Herder, for its emphasis on local and nonuniversal factors seemed to legitimize a rejection of the hated French attempts to create a universal European state and culture based on French hegemony.

WAS THE *SPIRIT OF THE LAWS* SCIENTIFIC?

Because Montesquieu's work came under attack as unscientific by men like Condorcet, because it clearly did play a role in stimulating German Romantic opposition to the universalist attitudes that we have consistently identified as central to science, and because it did incorporate a distinction between the operation of laws in the "intelligent" world and in the "physical" world which certainly violated the notions of epistemic unity on which many eighteenth-century authors based their notions of social science, it is worth considering why Montesquieu's work should be identified as scientific. This is particularly important because the *Spirit of the Laws* became a model study for a powerful tradition of men who certainly understood themselves to be engaged in a *scientific* study of society.

Though Montesquieu persisted in believing in the freedom of the human to make choices, his almost constant emphasis was on considering the determinate relationships among phenomena which inevitably linked legal codes to the material conditions of societies. In this way his habits of mind are clearly different from those humanistic historians who viewed positive laws solely as the free creation of the human mind, or as combining Divine and human considerations alone. It is true that with rare exceptions Montesquieu was not sympathetic to the kind of analytic simplifications which characterized the mechanical philosophy or the extension of that philosophy to social phenomena by men like Petty and Hobbes; but such a position does not make him unscientific; it merely makes him a spokesman for a different kind of social science than that practiced by the early political economists or psychologists, and links him with Harrington and others who tended to emphasize the empirical rather than the rational aspects of scientific studies.

The two issues on which the scientific character of Montesquieu's work might seem problematic relate to whether Montesquieu developed any knowledge which can be said to be "testable" and whether he was engaged in "activities" that are recognizably scientific. I would like to approach both of these questions by considering a characterization of science offered by Turgot to an audience at the Sorbonne in 1750, just two years after the *Spirit of the Laws* appeared. In "A Philosophical Review of the Successive Advances of the Human Mind," Turgot attempts at one point to explore the natural sciences by comparing and contrasting them with pure mathematics. The passage is interesting enough to quote at length:

> To discover and verify truth [in the natural sciences], it is no longer a question of establishing a small number of simple principles and then merely allowing the mind to be borne along by the current of their consequences. One must start with nature as it is,

and from that infinite variety of effects which so many causes, coun-
terbalanced by one another, have combined to produce . . . and
when a man starts to seek for truth he finds himself in the midst of
a labyrinth which he has entered blindfolded. Should we be surprised at his errors?

Spectator of the universe, his senses show him the effects but
leave him ignorant of the causes. And to examine effects in an
endeavour to find their causes is like trying to guess an enigma. . . .

The natural philosopher erects hypotheses, follows them
through to their consequences, and brings them to bear upon the
enigma of nature. He tries them out, so to speak, on the facts, just
as one verifies a seal by applying it to its impression. Suppositions
which are arrived at on the basis of a small number of poorly understood facts yield to suppositions which are less absurd, although no
more true. Time, research, and chance result in the accumulation of
observations, and unveil the hidden connections which link a
number of phenomena together.

Ever restless, incapable of finding tranquility elsewhere than
in the truth, ever stimulated by the image of truth which it believes
to be within its grasp but which flies before it, the curiosity of man
leads to a multiplication of the number of questions and doubts,
and obliges him to analyze ideas and facts in a manner which
grows ever more exact and profound.[13]

If one accepts Turgot's characterization of the natural
philosopher or scientist as one who generates hypotheses and then
tries them out or "tests" them against observations—an accurate
reflection of the vast bulk of work in the exact sciences in the early
eighteenth century in spite of Newton's disclaimer—then there can
be no question that Montesquieu was acting like a scientist and
"testing" his social hypotheses in much the same way that an elec-
trician or chemist tested his physical ones: i.e., by seeing whether
detailed observations are consistent with their consequences. He
may not have been doing precisely the kind of quantitative social
science preferred by Petty or Condorcet, but he was unquestionably
doing science, as even Condorcet admitted implicitly when he constructed his educational schema in the 1790s.

THE CONSERVATIVE THRUST OF
THE *SPIRIT OF THE LAWS*

Drawing from a long tradition of political philosophy, Montesquieu
argues that each form of government is dependent on some dominant principle which must be supported by education and legisla-

tion if the form of government is to survive. Then he claims that the corruption of any government almost always follows from a corruption of its principles. In particular, Montesquieu argues that since a monarchy depends on the principal of honor, whenever a monarch acts in such a way as to attack the functions and privileges of the nobility or corporate bodies in the state by undermining the honor of those involved, he undermines the very foundation of the state. Montesquieu illustrates his point from a recent work on China:

> "The destruction of the dynasties of Tsin and Sou," says a Chinese author, "was owing to this: the princes, instead of confining themselves, like their ancestors, to a general inspection, the only one worthy of a sovereign, wanted to govern everything immediately by themselves."[14]

The Chinese author gives us in this instance the cause of the corruption of almost all monarchies. When the monarch takes too much of the day-to-day governing to himself, he dishonors or betrays a lack of trust and confidence in his nobility which ordinarily represents the crown in local matters. Similarly, for Montesquieu, the French Crown's attempt to centralize authority at the expense of noble privilege was certain to drive France into a corrupt and despotic regime.

It was this defense of traditional noble privilege at a time when the king and the parlements were struggling for power that set Montesquieu apart from most French intellectuals, who, we have seen, favored the monarchy because it seemed to offer the savants their best opportunity to rationalize politics and the economy. In a bitter letter to Montesquieu, Claude Adrien Helvetius spoke on behalf of many who otherwise admired Montesquieu's innovative work, but who detested his emphasis on the role of noble and corporate privileges in sustaining monarchical government:

> Our Aristocrats and our petty tyrants of all grades, if they understand you . . . cannot praise you too much, and this is the fault I have ever found with the principles of your works. . . . *L'esprit de corps* assails us on all sides; it is a power erected at the expense of the great mass of society. It is by these hereditary usurpations we are ruled. . . .[15]

This circumstance may help to explain why Montesquieu's relatively conservative approach to a science of society found its most enthusiastic adherents in Scotland and Germany, both countries

with a tradition of strong local nobles and weak central governments, whereas the French generally found the intellectually more
radical approach to society through individual psychology much
more appealing.

LAWS AND THE MODES OF SUBSISTENCE

One of Montesquieu's most important departures from the classical
tradition of political philosophy involved his arguments that (1)
there are basically four different types of society, based on four
fundamental modes of procuring subsistence—hunting, herding,
agriculture, and trade or commerce; and (2) that the complexity of
legal systems must increase from the first of these to the fourth:

> The laws have a very great relation to the manner in which the
> several nations procure their subsistence. There should be a code of
> laws of a much larger extent for a nation attached to trade and navi-
> gation than for people who are content with cultivating the earth.
> There should be a much greater for the latter than for those who
> subsist by their flocks and herds. There must be a still greater for
> these than for such as live by hunting.[16]

Like Cantillon, but so far as I know, independent of him,
Montesquieu argues that the first great difference among nations
based on different patterns of subsistence is the population that
they can support. Hunting societies will have the smallest population, dependent on the game they can hunt. Each small troop will
constitute its own nation, and their most frequent contact with
others will be in the form of conflicts over hunting and fishing
rights. Such groups will wander, without fixed habitation, and there
will be little reason even for forming long-term marriages. Under
such conditions, "The institutions of these people may be called
'Manners' rather than laws," and, "the old men, who remember
things past, have great authority; they cannot there be distinguished by wealth, but by wisdom and valor."[17] Herding nations
will need only slightly less country to furnish subsistence. They too
will frequently quarrel over rights to the use of land, but because
they can move their stock, they may combine—at least briefly, into
large bands under a single strong chief, as the Tatars do—in order
to raid other groups. In such nations, herdsmen may accumulate
stock and are likely to remain attached to their wives, who may care
for the stock when they are away. Their laws usually "regulate the

division of plunder, and give . . . a particular attention to theft."[18] Where people cultivate the land, much larger populations can be supported. Men come to possess "landed property," and "the division of lands is what principally increases the civil code."[19] Moreover, agricultural societies soon give rise to money, and where money comes into use, "injustice . . . may be exercised in a thousand ways. Hence they are forced to have good civil laws, which spring up with the new practices of iniquity."[20] First, and most generally, Montesquieu argues that the system of laws in a commercial society will be even more complex than those in an agricultural society with money:

> Commerce brings into the same country different kinds of people; it introduces also a great number of contracts and species of wealth, with various ways of acquiring it. Thus in a trading city there are . . . more laws.[21]

Much more interesting than this observation were a handful of comments on the "spirit of commerce" that served as a springboard for detailed analyses by subsequent Enlightenment speculative historians, including Adam Ferguson and Adam Smith. Commerce spawns peace, interdependence, and tolerance. Whereas hunting and herding nations usually interacted among themselves hostilely, trading nations "become reciprocally dependent," making peace "the natural effect of trade."[22] On the one hand, because commerce, by bringing members of different societies into contact with one another, makes men aware of different sets of customs and manners, it provides "a cure for the most destructive prejudices," and leads to the formulation of "agreeable manners," i.e., to *civility*.[23] On the other hand, Montesquieu suggests, the peace and unity among nations that commerce brings may be purchased only at a very substantial cost, for "if the spirit of commerce unites nations, it does not in the same way unite individuals."[24] The passage in which he explains this remark deserves special attention:

> We see that in countries where the people move only by the spirit of commerce, they make a traffic of all the humane, all the moral virtues; the most trifling things, those which humanity would demand, are those done, or there given, only for money.
> The spirit of trade produces in the mind of man a certain sense of exact justice, opposite, on the one hand, to robbery, and on the other, to those moral virtues which forbid our always adhering rigidly to the rules of private interest. . . . Hospitality, for instance,

is most rare in trading countries, while it is found in the most
admirable perfection among nations of nomadic peoples.[25]

To illustrate his point Montesquieu offers a fascinating example of the conflict between Germanic custom and Roman law.
Among the Germans, Tacitus had emphasized the importance of
hospitality. "It is a sacrilege," says Tacitus, "for a German to shut
his door against any man whomsoever, whether known or unknown.
He who has behaved with hospitality to a stranger goes to show
him another where this hospitality is also practiced." But where the
commercially minded Romans went, such a practice was unacceptable, argues Montesquieu: "This appears by two laws of the Bur-
gundians; one of which inflicted a penalty on every barbarian who
presumed to show a stranger the house of a Roman; and the other
decreed, that whoever received a stranger should be indemnified by
the inhabitants, every one being obliged to pay his proper proportion."[26] This set of comments is particularly fascinating because
precisely at the time that the political economists were elevating
Hobbesian self-interest into the universal motive for virtually all
significant human behavior and announcing that it was *the* central
feature of human nature, Montesquieu and his follower Adam Ferguson were beginning to develop an understanding of the extent
to which acceptance of self-interest as a legitimate preeminent
concern was not only socially determined but also peculiarly
characteristic of "commercial" societies.

MONTESQUIEU AND THE LAW OF UNINTENDED CONSEQUENCES

In the last few pages we have emphasized the feature of Montesquieu's *Spirit of the Laws*—his emphasis on the way that laws
and customs depend upon modes of subsistence—which places
him at the head of a materialistic-determinist school of historiography that has had a major impact on both liberal and socialist
ideologies. At least two other themes from the *Spirit of the Laws*
also shaped the sociological tradition in ways that have influenced
both conservative ideology and the *laissez-faire* side of modern
liberalism.

The first of these themes was very consciously borrowed from
Bernard Mandeville's "Fable of the Bees" but was generalized by

Montesquieu into what Ronald Meek has called the "law of unintended consequences."[27] Mandeville's strange and sardonic poem/essay, which had appeared first in 1705 and in vastly expanded versions through 1728, took as its theme the notion that private vices might be public virtues—i.e., that the general welfare might—in fact, must—be served by individuals who only intended to seek their own self-interest. Concluding his "Fable," Mandeville wrote:

> Fools only Strive
> To make a Great an Honest Hive.
> 'T enjoy the World's Conveniences,
> Be famed in War, yet Live in Ease,
> Without great Vice, is a vain
> Eutopia seated in the Brain.
> Bare Virtue can't make nations
> live in Splendor.[28]

Montesquieu's initial appeal to Mandeville drew directly from Mandeville's examples. In societies where vanity is encouraged and men and women vie in pleasing one another, "the desire of giving greater pleasure than others establishes the embellishments of dress, and the desires of pleasing others more than ourselves gives rise to fashions. This fashion is a subject of importance; by encouraging a trifling turn of mind, it continually increases the branches of . . . commerce."[29] Thus the vice of vanity increases the desire for luxury goods which in turn is the foundation of commerce. It is not the intent of the vain man or woman to increase trade and the public welfare by providing jobs, but this increase is nonetheless an important consequence of their acts. Montesquieu goes on, however, to expand the argument to include the unintended and to some extent unexpected social evils that grow out of pride; so he is not merely exploring the ironic twists that make the consequences of some "vices" beneficial but rather the general social implications of mixtures of "good and bad" qualities from which *unsuspected* goods and evils often arise. Montesquieu claims, for example, that the Spanish combination of pride and honesty interact to make them unsuited to trade, while the Chinese combination of industry and avarice make them the greatest trading nation of all.[30] At the hands of Adam Ferguson and Adam Smith, this notion—that the actions which humans undertake for one set of reasons may have unintended but extremely important social consequences—was raised to the level of a central dogma of the social sciences and liberal ideology. At the hands of Edmund Burke, it became a central feature of European conservative thought.

MONTESQUIEU ON THE INTERRELATIONSHIPS AMONG INSTITUTIONS IN A SOCIETY

A final theme of the *Spirit of the Laws* has a bearing on the central concerns of the present study. Insistent as he was on the close relationships among religion, customs, manners, laws, modes of subsistence, and forms of government, Montesquieu nonetheless argued that except under a despotic government and in a totally static society, some degree of separation and autonomy must be maintained. Laws should not attempt to dictate customs, for example,[31] nor should we attempt either to "decide by divine laws what should be decided by human laws; nor determine by human what should be determined by divine laws."[32] In a nation like that of China, where religion, laws, manners, and customs are "confounded"—i.e., so intimately connected that they constitute a single system of rituals—several consequences follow. As long as the system remains intact, there can be virtually no conflict in the society; so there will be tranquility and not even the possibility of any change.[33] Moreover, the integrity of such a system makes it virtually impervious to the influences of conquering peoples:

> Their customs, manners, laws, and religion being the same thing, they cannot change all these at once; and as it will happen that either the conqueror or the conquered must change, it has always been the conqueror.[34]

But the rigidity of such a system creates severe instability in the face of even the smallest internal change. "If you diminish the parental authority," for example, writes Montesquieu, "or even if you retrench the ceremonies which express your respect for it, you weaken the reverence due to magistrates, who are considered as fathers; nor would the magistrates have the same care of the people, whom they ought to look upon as their children; and that tender relation which subsists between the prince and his subjects would insensibly be lost. Retrench but one of these habits and you overturn the state."[35] Under this set of circumstances it can hardly be surprising that in China, each time a new dynasty emerged and initiated even the smallest changes in custom, "the state fell into anarchy, and revolutions succeeded."[36]

Montesquieu's political ideal of a parliamentary monarchy with its separation of powers and system of checks and balances that serves both to protect the liberties of subjects and to ensure the long-term adaptability of the government suggests the optimal pattern of relationships among laws, customs, and manners. Just as the legislative, executive, and judicial powers are adapted to one another without being totally subordinate in the ideal case, so too should the various aspects of society—the religion, manners, customs, and laws—be related to one another without being conflated. Only then will the society be both stable and capable of absorbing elements of change.

My own views regarding the relationship between science and other aspects of Western culture are derived from the sociological tradition initiated by Montesquieu. It should be clear that virtually every aspect of modern Western culture is influenced by scientific attitudes, activities, and concepts. At the same time it would seem to be dangerous to allow science to become so pervasive that we create a rigid culture in which the values of science are not checked and balanced in some meaningful way.

ADAM FERGUSON'S
HISTORY OF CIVIL SOCIETY

We will return shortly to consider Adam Smith's fusion of the economic and sociological traditions, but first I want to consider the development of Montesquieu's concerns by another Scot and close associate of Adam Smith, Adam Ferguson. Though Ferguson was initiated into the Montesquieu tradition of social science by Smith,[37] his most important work, *An Essay on the History of Civil Society* (1767) preceded Smith's *Wealth of Nations* by nine years. It offered a more comprehensive and subtle development of Montesquieu's ideas and a more self-conscious exploration of the scientific character of its own methods. Finally, and perhaps most important, where it diverged from Montesquieu and from Smith, it defined a set of concerns that were to become central in both the socialist and conservative ideologies that provide the primary alternatives to modern liberalism.

Although Ferguson deeply admired Montesquieu's *Spirit of*

the Laws and cited it frequently, one can hardly imagine a person whose background, training, and temperament diverged more from those of Montesquieu; and all of these factors led him to modify Montesquieu's insights at almost every turn. Whereas Montesquieu was a wealthy aristocrat, relatively irreligious, witty, and urbane, a Cartesian rationalist, a lover of peace and stability, Ferguson was none of these. The youngest son of a large family, he was born to a minister in the Highland village of Logierait, about thirty miles northeast of Perth in Scotland in 1723. After distinguishing himself as a student of classics, mathematics, and philosophy at the University of St. Andrews, he studied Divinity at Edinburgh and then served from 1745 to 1754 as chaplain to the First Highland Regiment of Foot, with which he traveled to the continent and to Quebec for a time, gaining a reputation for both piety and bravery.

Returning to Edinburgh in 1757, Ferguson briefly succeeded David Hume as librarian of the Advocates Library. In 1759, he became Professor of Natural Philosophy at Edinburgh, focusing especially on Newtonian experimental philosophy. Deeply committed to "Newtonian" empiricism, as it had been adapted by Hume to the study of man, Ferguson had a much greater sensitivity to the "instinctual" or nonrational elements of human nature than Montesquieu had shown. Finally, in 1764, Ferguson succeeded to the coveted chair of moral philosophy at Edinburgh. He held this position—with brief periods of absence to travel in Russia and to serve on the British team sent to negotiate the settlement following the American Revolution—until 1785. Retiring to a remote farm in Perthshire, Ferguson continued as a working farmer and writer on moral philosophy until his death in 1816 at age 92.

The *Essay on the History of Civil Society*, Ferguson's first major work, published in 1767, was rapidly translated into French and German. Though it enjoyed only modest success in its English version, the German academic community embraced it, and it remained popular in Germany throughout the nineteenth century, becoming central in both socialist theory (because of Marx's enthusiasm for its arguments)[38] and in the most conservative tradition of German sociology (because of Ferguson's positive attitude toward war and because of his focus on the positive role of unintended consequences).[39]

Ferguson begins his *Essay* with a section devoted to "The General Characteristics of Human Nature" as if he were going to follow the Hobbesian pattern of constructing a social theory on the basis of individual psychology. But Ferguson immediately shows that his commitment to the Newtonian/Humian experimental method demands a very different approach. He starts with a methodological critique of Hobbes and all others who posit the existence of men in a state of nature or independent of some social context.[40]

It is a simple fact, insists Ferguson, that all of our experiences of humans are experiences of beings "assembled in troops and companies," and it is absolutely pointless to speak about a condition or a time in which humans had not banded together and adopted the use of language; for even if there had been such a time it is beyond the pale of scientific investigation: "It is a time of which we have no record, and in relation to which our opinions can serve no purpose, and are supported by no evidence."[41] If, then, we are to ground our understanding of society in the "general characteristics of human nature," we will have to appeal to a "nature" which is unavoidably social. That is, we must begin with what we might today call "social psychology" rather than with some imaginary individual psychology which abstracts human beings from their social setting. "Any experiment relative to this subject [human nature]," Ferguson writes, "should be made with entire societies, not with single men."[42]

One consequence of the social character of humans, coupled with their use of language, is that they are capable of a long-term progress unavailable to any other species:

> In other classes of animals, the individual advances from infancy to
> age or maturity and he attains in the compass of a single life, to all
> the perfection his nature can reach: but in the human kind, the
> species has a progress as well as the individual; they build in every
> subsequent age on foundations formerly laid; and in a succession
> of years, tend to a perfection in the application of their faculties, to
> which the aid of long experience is required, and to which many
> generations must have combined their endeavours.[43]

Thus does Ferguson provide a foundation for the dynamic view of society which he shared with Smith and Turgot, among others. If his view of human society is dynamic rather than static, as was Montesquieu's, it is also distinctively evolutionary rather than revolutionary. The scene of human affairs was not a stagnant pool, but it was not a raging torrent either.[44] More important yet, for Ferguson the process of change in human society was in a sense inevitable, unavoidable, and not completely controllable. He might have said providential—for he was a thoroughgoing believer in Christian natural theology—but he does not. Instead he insists that the deterministic dynamic of social development arose from the nature of humanity reacting to its material environment.

Returning to the theme of social change from a slightly different perspective, Ferguson argues that political institutions are not the product of individual, rational constitution building, like that attributed to Lycurgus and advocated by most political theorists, but are rather the complex result of complex natural factors:

> The crowd of mankind are directed in their establishments and measures, by the circumstances in which they are placed . . . and nations stumble upon establishments, which are indeed the result of human action, but not the execution of any human designs.[45]

FERGUSON ON THE DIFFERENCE BETWEEN NATURAL AND MORAL LAWS

One important argument follows immediately from Ferguson's insistence upon the social character of man and the dynamic character of society. The distinction that moralists such as Rousseau, among others, had drawn between a virtuous "natural" man and a corrupted man made "artificial" by society could make no sense. "Art itself is natural to man," Ferguson insisted, so it makes no sense to talk about men quitting the state of their nature in advanced civil societies. The state of nature is equally to be found in contemporary Britain and among the savages living at the Straits of Magellan. Thus, the words "*natural* and *unnatural* . . . employed to specify a conduct which proceeds from the nature of man, can serve to distinguish nothing: for all the actions of man are equally the result of their nature."[46]

It is certainly appropriate for moral philosophers to make value judgments regarding what is just or unjust, good or bad, in various societies. But Ferguson is insistent that the norms they use cannot be simply discovered in nature itself. Like Montesquieu, Ferguson distinguished between "physical" laws and "moral" laws which apply respectively to the realms of determinist science and of human choice; and like Montesquieu, Ferguson finds in human social behaviors and institutions a confluence of physical and moral considerations, but he does not view moral laws as simply natural laws imperfectly followed. Moral laws are those "which we *desire* to have uniformly observed," and a moral law is said to be law "in consequence of its rectitude, or the authority from which it proceeds . . . not in consequence of its being the fact."[47] What is "natural" has reference to physical laws and carries no normative implications. Natural laws simply describe what *is*. In order to discuss what *ought to be,* one must appeal beyond the domain of nature to that of morality. Ferguson does not, of course, believe that morality and nature are independent; for moral choices have to be made from among complex natural possibilities. A social science which can discover the complex consequences of different social organizations is thus a necessary but not a sufficient precondition for an effective moral philosophy.

Having laid out his basic methodological principles and some of their most immediate implications at the outset of his *Essay*, Ferguson turns in the remainder to explore some key features of human nature as it is observed. He admits that self-preservation and "interest" in the narrow Hobbesian sense of "selfishness" do form a part of human nature, but even before he goes on to explore what he considers some of the more admirable but equally "natural" aspects of men in society, he points out that rationally calculated interest certainly cannot be supposed to encompass all the motives of human conduct. Even if we were to deny disinterested benevolence to men, he points out, "hatred, indignation, and rage, frequently urge them to act in opposition to their known interest, and even to hazard their lives, without any hopes of compensation in any future returns of preferment of profit."[48]

Just as natural to man and woman as their pursuit of private interest is "a propensity to mix with the herd and, without reflection, to follow the crowd of his specie [*sic*],"[49] which can be totally divorced from any expectation of personal advantage by association with the group:

> Men are so far from valuing society on account of its mere external conveniences, that they are commonly most attached where those conveniences are least frequent; and are the most faithful, where the tribute of allegiance is paid in blood. Affection operates with the greatest force, where it meets with the greatest difficulties: In the breast of the parent, it is most solicitous amidst the dangers and distresses of the child: In the breast of a man, its flame redoubles where the wrongs or sufferings of his friend, or his country, require his aid.[50]

In extensive sections on "Moral Sentiment," "Happiness," and "The Felicity of Nations," Ferguson constantly downplays the importance of mere private economic profit and loss in favor of more "vehement" passions centered in the human heart and focused on admiration or contempt for the actions of others. Humans act out of malice, jealousy, and envy, and out of kindness, love, generosity, and friendship; and though we cannot understand why this should be so, he followed Hume in arguing that "We must, in the result of every inquiry, encounter facts which we cannot explain."[51] If anyone doubts that we act for these noneconomic and nonrational reasons, he says, just "ask those who have been in love."[52] It is the *social* passions and not mere private greed which offer the greatest scope for human happiness:

> It should seem . . . to be the happiness of man to make his social disposition the ruling spring of his occupations; to state himself as a member of a community, for whose general good his heart may glow with an ardent zeal, to the suppression of those personal cares which are the foundation of painful anxieties, fear, jealousy, and envy. . . .[53]

Like Montesquieu before and Marx after, Ferguson was inclined to see the focus on economic competition within a commercial society that offers great economic benefits to individual members as peculiarly destructive of the social and "affectionate" side of human nature:

> It is here indeed, if ever, that man is sometimes found a detached and solitary being: he has found an object which sets him in competition with his fellow creatures, and he deals with them as he does with his cattle and his soil, for the sake of the profits they bring. The mighty engine which we suppose to have formed [commercial] society, only tends to set its members at variance, or to continue their intercourse after the bonds of affection are broken.[54]

The major difference between Ferguson and Montesquieu or Marx on this point lies in the greater distinction that Ferguson draws between physical or "natural" and "moral" laws. For Ferguson, the recognition that commercial societies may unintentionally lead to greater isolation of individuals and to a decline of affection does not lead to acquiescence in the fact. Instead, it points to an area in which moral choices can be made to intervene in the natural process to salvage strongly desired characteristics. It is only when we realize that commercial society encourages selfishness that we can fight that tendency and strive to keep the social affections alive.

One way of expressing the relationship between physical and

moral laws—i.e., between science and morality—in the formation of society as Ferguson understood the situation is to say that social systems are "softly" deterministic. Left alone, they will inevitably develop along certain lines; but the possibility of changing those lines by conscious and intentional intervention does exist. The whole point of a "social science," then, is to explore the opportunities for and likely consequences of intentional moral action. Without the science, morality is blind; but without the morality, science is useless, pointless, and paralytic.

FERGUSON ON THE POSITIVE SIDE OF CONFLICT

One of the most intriguing examples of how our scientific observations help to guide and inform our moral life is found in Ferguson's discussion of "War and Dissention." Unlike both Hobbes and Montesquieu, who view peace as an unalloyed blessing and the most central goal of the creation of governments, Ferguson accepts a huge accumulation of evidence that men may be naturally competitive and even combative:

> Every animal is made to delight in the exercise of his natural talents and forces. . . . Man too is disposed to opposition, and to employ the forces of his nature against an equal antagonist; he loves to bring his reason, his eloquence, his courage, even his bodily strength to the proof. His sports are frequently an image of war; sweat and blood are frequently expended in play; and fractures and death are often made to terminate the pastimes of idleness and festivity.[55]

Moreover, the instinct for opposition or hostility may well be useful to nations by uniting them in the face of an enemy,[56] and it may lead to the development of man's highest virtues:

> To overawe, or intimidate, or when we cannot persuade with reason to resist with fortitude, are the occupations which give its most animating exercise, and its greatest triumphs, to a vigorous mind; and he who has never struggled with his fellow creatures is a stranger to half the sentiments of mankind.[57]

Ferguson's participation in the games and battles of his Highland regiment no doubt disposed him to consider the positive side of conflict and to take seriously an emotional side of mankind which those whose knowledge of humanity came largely from "hours of retirement and cold reflection"[58] could not fathom. Moreover, this disposition had a major impact on how he viewed the limits placed on morality by natural laws. Thus, he argued that we may well seek to pacify and unite all men by ties of affection, but in doing so we are bound to lose something of great value:

> We may hope, in some instances to disarm the angry passions of jealousy and envy; we may hope to instill into the breasts of private men sentiments of candor toward their fellow-creatures, and a disposition to humanity and justice. But it is vain to expect that we can give to the multitude of people a sense of union among themselves, without admitting hostility to those who oppose them. Could we at once, in the case of any nation, extinguish the emulation which is excited from abroad, we should probably break or weaken the bonds of society at home, and close the busiest scenes of national occupations and virtues.[59]

Indeed, Ferguson insists that conflict among nations is as important to humans as cooperation and affection within a given nation:

> Their wars, and their treatises, their mutual jealousies, and the establishments which they devise with a view to each other, constitute more than half the occupations of mankind, and furnish materials for their greatest and most improving exertions.[60]

FERGUSON ON THE LIMITATIONS OF POLITICAL ECONOMY

Just as Ferguson seems to have broken from Montesquieu on the potential value of conflict between social groups, initiating the modern sociological emphasis on conflict, he diverges radically from the political economists in his estimate of the importance of wealth and population for the well-being of a nation. Humans are happiest not where they are merely secure in their economic existence, but where they have the greatest opportunity to exercise their social passions. By the same token, though wealth, commerce, and population all give a nation the power to preserve itself, "their tendency is to maintain numbers of men, but not to constitute happiness. They will . . . maintain the wretched as well as the happy."[61]

On most specific economic topics, Ferguson reflects the liberal line initiated by men like Cantillon and raised to its high point by his friend Adam Smith. But when he comes to consider the overall importance of "commerce and wealth," Ferguson insists that recent able writers left one important thing to be expressed: "the general caution, not to consider these articles as making the sum of national felicity, or the principle object of any state."[62] Indeed, Ferguson goes farther than this. He attacks those "who would have nations, like a company of merchants, think of nothing but their stock"; and he insists that "whatever may be the actual conduct of nations in this matter, . . . [economic theorists] would hurry us, for the sake of wealth and of population into a science where mankind, being exposed to corruption . . . are in the end, subject to oppression and ruin."[63]

In general, though special circumstances, including the size of surrounding states, may demand consideration, people are happiest "where nations remain independent and are of a small extent," for each citizen finds his own role in the community diminished as the community is enlarged. In any event, the search for a complete unification of peoples "is a ruinous error; and in no instance, perhaps, is the real interest of mankind more entirely mistaken."[64] Marx, too, especially in his early works, views self-fulfillment through the exercise of one's passions in a community as a far higher goal than mere wealth. But unlike Ferguson, Marx does not identify the community with the state, and he does look forward to a time of total unity among humankind.

In the process of understanding what interests and advantages will motivate men, we must remember their social nature and recognize that the interests of "classes" or "orders" of men are likely to be even more powerful than private interest. Indeed, Ferguson argues that advanced society exists primarily as "the very scene in which parties [a synonym for "classes" or "orders," terms used earlier in the paragraph by Ferguson] contend for power, for privilege, or equality."[65] Both Machiavelli and Harrington had emphasized the importance of balancing group interests in order to establish stable political institutions. Montesquieu had explored and defended hierarchical distinctions in connection with monarchical government. And Ferguson's friend and colleague, John Millar, engaged in the first extended sociological analysis of what we would call class distinctions or social stratification which would appear as *The Origin of the Distinction of Ranks; or an Inquiry into the Circumstances Which Give Rise to Influence and Authority in the Different Members of Society* (London, 1771). But Ferguson was among the first theorists to view class conflict and the mediation of class interests as a central mechanism for legal and social progress rather than as something to be avoided in order to achieve stability and avert social corruption:

> In free states, therefore, the wisest laws are never, perhaps, dictated by the interest and spirit of any order of men: they are moved, they are opposed, or amended, by different hands; and come at last to express that medium and composition which contending parties have forced one another to adopt.[66]

It is certainly reasonable to ask whether the traditional notions of "orders" and "ranks," with which Ferguson and Millar worked in exploring the dynamic implications of social conflict, ought to be identified with subsequent notions of "class," even if Ferguson did use the latter term. I follow a number of recent historians of sociology who concur in viewing Ferguson's and Millar's discussions of social stratification and its economic foundations as among the first attempts to replace traditional notions of sociopolitical hierarchies with modern notions of socioeconomic classes.[67]

Turning to an analysis of "civil liberty," Ferguson argues that because different people assign values differently, the term *liberty* will have different meanings in different contexts. Some will define freedom or liberty in terms of protecting their economic goods; others, in terms of guaranteeing their political participation; still others, in terms of ensuring their hierarchical standing or their economic equality. Because of these differences, people "are led to differ in the interpretation of the term [liberty]; and every people is apt to imagine, that its signification is to be found only among themselves."[68] Ferguson focuses on analyzing two particular notions of "liberty" which might serve as models for subsequent scientific socialist and liberal ideologies. "Some," he writes, "having thought that the unequal distribution of wealth is unjust, required a new division of property, as the foundation of freedom."[69] Taking Sparta as an example of an egalitarian state—which later socialists could not do because of its dependence on the helot class—Ferguson describes the consequences of the Spartan idea of freedom:

> The citizen was made to consider himself as the property of his country, not as the owner of a private estate. . . . The individual was relieved from every solicitude that could arise on the head of his fortune; he was educated and he was employed for life in the service of the public; he was fed at a place of common resort to which he

could carry no distinction but that of his talents and virtues; his children were the wards and the pupils of the state; he himself was taught to be a parent, and a director to the youth of his country, not the anxious father of a separate family. . . . On this plan, they had senators, magistrates, leaders of armies, and ministers of state; but no men of fortune. . . . A citizen, who in his political capacity, was the arbiter of Greece, thought himself honored by receiving a double portion of plain entertainment at supper. He was active, penetrating, brave, disinterested, and generous; but his estate, his table, his furniture, might, in our esteem, have marred the lustre of all his virtues.[70]

Wherever, as in Sparta, Ferguson continues, wealth is not allowed to become a sign of distinction and a source of corruption,

the citizen is dutiful, the magistrate upright; any form of government may be wisely administered; places of trust are likely to be well supplied; and by whatever rule office and power are bestowed, it is likely that all the capacity and force that subsists in the state will come to be employed in its service: for on this supposition, experience and abilities are the only guides and the only titles to public confidence.[71]

Contrasting this idyllic vision of an economically egalitarian state with modern commercial society, Ferguson argues that a very different conception of liberty or freedom must prevail. In what was to become the central perspective from which both conservative and socialist critics approach liberalism, Ferguson offers an essentially negative understanding of the liberal concept of freedom.

We must be contented to derive our freedom from a different source; to expect justice from the *limits* which are set to the powers of the magistrate, and to rely for protection on the laws which are made to secure the estate and the person of the subject. We live in societies where men must be rich in order to be great; . . . [and] where public justice, like fetters applied to the body, may, without inspiring the sentiment of candor and equity, prevent the actual commission of crimes.[72]

Even in such a society, human instinctive social passions will avert a complete collapse into totally selfish anarchy; but there can be no question of Ferguson's distaste for the central dogma of economic liberalism—its exclusive emphasis on self-interest.

In spite of his obvious preference for egalitarian societies and for the small-scale, face-to-face cultures which characterized earlier stages in human history, Ferguson was driven by his determinist view of historical dynamics to accept the fact that the "advance-ment" of both civil and commercial life was virtually inevitable:

> The establishments of men, like those of every animal are . . . the result of instinct, directed by the variety of situations in which man-kind are [*sic*] placed. Those establishments arose from successive improvements that were made, without any sense of their general effect; and they bring human affairs to a state of complication which the greatest reach of capacity with which human nature was ever adorned, could not have projected.[73]

Thus, whether we like it or not, the historical process, fueled pri-marily by the continuing division of labor and increasing special-ization of human activities, has placed us in complex commercial society.[74] Ferguson admits that this process has created a range and abundance of material goods which should be welcomed. Moreover, it has made possible the efficient prosecution of civil and military policies by persons of modest abilities through the bureaucratiza-tion of military and governmental institutions:

> The servants of the public, in every office, without being skillful in the affairs of state, may succeed, by observing forms which are al-ready established on the experience of others. They are made, like the parts of an engine, to concur to a purpose, without any concert of their own. . . .[75]

On the other hand, the division of labor—especially in the so-called "mechanical arts" or manufacturing sector of society—has a major pernicious impact on those who engage in it; for it tends "to contract and limit the views of the mind."[76]

Later analysts would say that industrial work under the prin-ciple of division of labor is "de-humanizing" and "alienating," by which they wanted to emphasize that it led to human beings' treat-ing one another as interchangeable objects rather than as fellow human beings with their own feelings and needs, and that it sepa-rated humans both from one another and from the products of their labor by robbing them of any understanding of the social contexts within which those products were produced. Without having this later vocabulary available, Ferguson seems to be expressing nearly the same sentiments. He bemoans the fact that as the division of labor proceeds, people tend to be isolated, "in separate cells . . . in-vented to abridge or facilitate [their] separate tasks[s]." Many arts, he reflects, "succeed best under a total suppression of sentiment and reason; and . . . Manufactures, accordingly, prosper most, where the mind is least consulted, and where the workshop may, without any great effort of imagination, be considered as an engine, the parts of which are men."[77]

Most of the key features of subsequent socialist critiques of labor under capitalism are represented here—the isolation of workers from their fellows and their products alike, the suppression of sentiment and intelligence, and the treatment of workers as mere cogs in a productive machine.

Under such circumstances the workers will be ignorant, servile and corrupt, incapable of participating in their own governance. As a consequence, the working class must be subordinated to a class which presumably has broader views than their own, and, "not withstanding any *pretention* to equal rights, the exhaltation of the few must depress the many."[78] But it is not at all clear to Ferguson that the "superior orders" in a commercial society are any less involved with "sordid cares and attentions" than the working class; for though their economic concerns may be slightly broader, they are likely to be equally as focused on generating profits and manipulating the human cogs in the productive machine as the workers are in fulfilling their own narrowly defined functions. It is true that the laws in a commercial society will give an illusion of equality by offering to protect the property and liberty of every individual and by insisting upon equality before the law. But, argues Ferguson, "If the pretention to equal justice and freedom should terminate in rendering every class equally servile and mercenary, we make a nation of helots, and have no free citizens."[79]

Within this vision of a society becoming increasingly enslaved to wealth and greed, Ferguson does offer a double suggestion regarding how the most negative impacts of commercialization might be avoided. As part of the process of specialization, a class of intellectuals and "men of science" emerges whose central function is to enliven and enlarge the domain of human discourse and to protect and promulgate those sentiments which most classes in the society would otherwise ignore. Where the society as a whole encourages the education of all or many of its members at the hands of this group, a tradition of liberal education opposes the tendencies toward narrowness implicit in commercial society.[80] In making this argument, Ferguson speaks on behalf of the powerful Scottish tradition of liberal education for which his student, Dugald Stewart, is perhaps the most insistent and effective spokesman.[81] It remains even today the most powerful and persuasive justification for liberal learning.

As long as international crises or internal class conflicts continue to draw men's attention away from themselves and into a concern with public rather than private issues, the worst dangers of commercial society can be avoided. But should that time come when "a growing indifference to objects of a public nature should prevail and . . . put an end to those disputes of party, and silence that voice of dissent, which generally accompanies the exercise of freedom. . . . The period has come, when . . . private interest and animal pleasure become the sovereign objects of care."[82] Thus, for Ferguson, a world without tension and conflict is to be shunned. For him the later Marxist vision of an end to class and state would have been anathema—not so much impossible as undesirable—for it would mean the end to all of those challenges that bring out the best in human beings.[83]

ADAM SMITH'S *WEALTH OF NATIONS* AND THE SOCIOLOGICAL TRADITION

When we turn from Ferguson's sociological approach to human institutions to the scientific approach of Adam Smith, we find an even deeper commitment to a kind of historical determinism tied to modes of subsistence. Smith tended to transform Montesquieu's four-fold division of societies into a rigid historical sequence;[84] and in his most popular and influential work, *An Inquiry into the Nature and Causes of the Wealth of Nations* (1776), he did what Ferguson could never allow himself to do: he accepted the leading assumption of political economy; i.e., that self-interest is *the* dominant motive for human activity. Moreover, he produced an analysis of human history and society in which economic determinism became the sole focus and in which free-market capitalism appeared as the inevitable and glorious high point of a dynamic process of social and economic change.

From our present perspective, The *Wealth of Nations* is of greatest interest because of the way that it virtually completed the process of turning the triune notions of self-interest, unintended consequences, and self-adjusting market mechanisms into the cornerstones of liberal ideology. It did so in part by incorporating them as basic assumptions in a scientific analysis of society that was both vastly more comprehensive and more popular and accessible than those of Cantillon or the Physiocrats. Joseph Schumpeter claims that from a technical standpoint it was "the most successful, not only of all books on economics, but . . . of all scientific books that have appeared to this day,"[85] and Ronald Meek quite rightly identifies it as "the *Bible* of the liberal bourgeois,"[86] whose authority remains almost as great in the United States in the 1980s as it was in Scotland in the 1780s. It made self-interest and the self-regulating market the bases of ideologies by explicitly arguing that they are—like Newton's law of gravity—both true and *sufficient* to account for virtually all economic phenomena. Perhaps most important, it did so in spite of the fact that Smith found some of their consequences almost as distasteful as did his colleague, Ferguson, or his successor, Karl Marx. This is perhaps the acid test of a concept that has taken on an ideological character—it appears to be true or necessary even though those who believe in and use it might often wish it were not.

That Smith viewed his own commitment to the primacy of self-interest with a kind of resignation rather than with enthusiasm, and that he saw it as like Newton's principle of gravity, with the *Wealth of Nations* playing the role of the *Principia,* is suggested by his curious return, immediately after he completed the *Wealth of Nations,* to a youthful essay entitled "On the Principles Which Lead and Direct Philosophical Inquiries: Illustrated by the History of Astronomy." In this essay, which he had completed no later than 1758, and which surveyed developments up to the time of Descartes, Smith had developed an analysis of scientific theorizing that was totally relativistic. Scientific theories and philosophical systems are created in response to powerful and painful sentiments of wonder and fear that humans have in response to uncommon or unexpected events. In order to obviate these painful and often destructive emotional responses, men seek to construct systems which link the previously unknown to the known through some kind of "invisible chains." The progress of the human mind consists in the creation of ever more inclusive systems capable of dissipating novelties or in the simplification of systems without loss of generality. Thus, Smith compares the refinement of scientific theories to the refinement of mechanical devices:

> Systems in many respects resemble machines. A machine is a little system, created to perform, as well as to connect together, in reality,

those different movements and effects which the artist has occasion for. A system is an imaginary machine invented to connect together in the fancy those different movements and effects which are already in reality performed. The machines that are first invented to perform any particular movement are always the most complex, and succeeding artists generally discover that, with fewer wheels, with fewer principles of motion than had originally been employed, the same effects may be more easily produced. The first systems, in the same manner, are always the most complex, and a particular connecting chain or principle, is generally thought necessary to unite every two seemingly disjoint appearances: but it often happens, that one great connecting principle is afterwards found to be sufficient to bind together all the discordant phenomena that occur as a whole species of things.[87]

As an undergraduate, Smith had been particularly drawn to mathematics and had carefully studied Newton's *Principia*, yet when he first approached the history of astronomy, he persisted in viewing the theory of gravity as just one more hypothetical system invented to account for phenomena. When he finished writing the *Wealth of Nations*, Smith returned to the history of astronomy, adding a concluding section on Newton from a personally changed perspective. Now he was inclined to feel that Newton had somehow managed to escape the old limitations. "Even the most skeptical cannot avoid feeling this," he wrote. "And we, even we, while we have been endeavouring to represent all philosophical systems as mere inventions of the imagination, to connect together the otherwise disjointed and discordant phenomena of nature, have inevitably been drawn in, to make use of language expressing the connecting principles of *this* one as if they were the *real* chains which Nature makes use of to bind together several operations."[88]

Given the timing and thrust of this comment, it seems likely that it was keyed by Smith's response to his own great work. In spite of his intention to maintain a healthy skepticism toward his own economic system and its fundamental hypotheses about human motives and economic mechanisms, Smith had succumbed. He could no longer doubt his own assumptions, and he was forced to admit that Newton's system could have the same seductive character. Just as Hume, Smith's closest confidant over a quarter of a century, admitted that he could not sustain his skepticism when he left his study and entered into ordinary daily activity, Smith found that he could not sustain his belief in the "imaginary" character of the connecting principles of his own and Newton's sciences in the face of their explanatory power and precision. For him they had literally become "discover[ies] of the most important and sublime truths," rather than "attempt[s] to connect, in the imagination, the phenomena," and that is precisely what we mean when we talk of something taking on an ideological function.

Though a full analysis of the *Wealth of Nations* is beyond our scope, we must spend enough time with Smith to illustrate his commitments to self-interest, unintended consequences, and market-directed economic theory, and to show that he accepted them in the face of very real qualms about their implications as a result of his commitments to identifiable scientific habits of mind. Then we must consider at least a few of the consequent doctrines that "the Bible of the liberal bourgeois" in turn authorized and legitimized.

SMITH'S MORAL PHILOSOPHY LECTURES: EMPIRICAL ETHICS

Born in 1723 at Kirkaldy, Scotland, the only child of a minor customs officer, Smith attended a local burgh school, then went to the University at Glasgow for four years between 1737 and 1741. There he studied mathematics under Robert Simson and moral philosophy under Francis Hutcheson, the founder of the Scottish Moral Sense School. In 1740 he earned a fellowship to Oxford, spending the next six years there, primarily studying Greek and the classics. In the late 1740s Smith returned to Scotland and taught a series of private courses on literature and economic topics at Edinburgh. Then in 1751 he was invited to take the chair in Logic at Glasgow, moving the next year to the chair in Moral Philosophy. In that position, according to John Millar, Smith divided his attention among four general topics: natural theology, ethics, jurisprudence, and political economy. After twelve years at Glasgow, Smith resigned his professorship to accept a lucrative post as tutor and traveling companion to the third Duke of Buccleuch with whom he spent the better part of three years in France. There he developed particularly close friendships with the Physiocrats and Turgot. In 1767 he returned to Kirkaldy to finish work on the *Wealth of Nations*, and thereafter until his death in 1790 he split his time among Kirkaldy, Edinburgh, and London, while serving as a commissioner of customs for Kirkaldy and socializing with his two closest friends, the chemist, Joseph Black, and the geologist, James Hutton.

As far as we can now tell, Smith included natural theology in his moral philosophy course primarily out of a sense of obligation rather than out of any great enthusiasm. Though he maintained an outward stance of orthodox Presbyterianism in order to avoid trouble with church and university authorities, there is little doubt that privately he had a great deal of sympathy for ancient materialist theories which attributed the origins of religion to human fears of the strange and unknown.[89] In fact, the "invisible hand" image, which plays such an important role in the *Wealth of Nations,* and which is often called upon to justify claims regarding Smith's supposed belief in a providential God underlying events, first appeared as the "invisible hand of Jupiter" in a passage that draws directly from atomist theories of the origin of polytheism.[90]

His Glasgow lectures on ethics became the basis for Smith's first major published work, *Theory of Moral Sentiments,* which appeared in 1759. This work is fascinating for the peculiar twist it gave to Hutcheson's empiricist approach to moral philosophy and for its blurring of distinctions between social and private passions through an early version of the principle of unintended consequences. We begin by considering Smith's claim about the way in which general moral rules are formed. He writes:

> They are ultimately founded upon what, in particular instances, our moral faculties, our natural sense of merit and propriety approve, or disapprove of. We do not originally approve or condemn particular actions; because, upon examination they appear to be agreeable or inconsistent with a certain general rule. The general rules, on the contrary, are formed, by finding from experience, that all actions of a certain kind, or circumstanced in a certain manner, are approved or disapproved of.[91]

Two features of this notion of moral laws are important. In the first place, what is moral or virtuous is an *empirical* question, open to scientific investigation. It is what, in fact, humans do approve of, rather than, as for Ferguson, something given by a higher authority. Equally important for Smith, our judgments of virtue or morality are judgments about *actions* and their *effects,* rather than about the *motives* of actors, which are not directly accessible to outside observation. Given his phenomenal and empirical definition of morality, Smith explicitly denies the arguments of Hutcheson and Mandeville that "self-love was a principle which could never be virtuous in any degree or in any direction." In fact, he insists that

> regard to our own happiness and interest . . . appear on many occasions very laudable principles of action. The habits of economy, industry, discretion, attention, and application of thought, are generally supposed to be cultivated from self-interested motives and at the same time are apprehended to be very praiseworthy qualities which deserve the esteem and approbation of everybody.[92]

Smith is certain that private drives for status and wealth serve as the primary stimulus to economic activity. But unlike Mandeville, Smith cannot view this fact as an ironic instance of "vices" leading to public advantage. Instead, it is a case in which truly virtuous acts—i.e., acts which receive and deserve approval—simply emerge out of the motive of private gain.

When he does turn to analyze motives, Smith's emphasis in *the Theory of Moral Sentiments* is on a curiously self-oriented social passion. Above all, humans want to be approved and admired by their fellows:

> Nature, when she formed man for society, endowed him with an original desire to please, and an original aversion to offend his brethren. She taught him to feel pleasure in their favorable, and pain in their unfavorable regard. She rendered their approbation most flattering . . . for its own sake.[93]

Even the search for wealth, it turns out, hinges on this primary human desire for approval; for the great stimulus to seek wealth is not a demand for mere subsistence, but rather the expectation that riches will bring the admiration and approval of others.

This realization brings us to the third major segment of Smith's philosophy course: jurisprudence. Justice and law have to do with establishing the boundaries within which the human drive for wealth and status must be contained if it is to be admired rather than resented. To address this issue Smith turned to the historical/sociological approach initiated by Montesquieu, transforming the latter's static division of societies into a dynamic theory of development through four stages, and seeking to understand why certain kinds of laws and rules of justice emerge in the different stages of society. Though Smith never published the material from his lectures on jurisprudence, they unquestionably had a major impact on the sociological writings of his students and colleagues, John Millar and Adam Ferguson, as well as on his own economic writings. Their content is known from a brief description by Millar who sat in on the lectures in the early 1750s, through a very spotty set of notes by John Anderson dating from about 1755, and through two extensive sets of students' notes, both from the early 1760s.[94]

Smith's lectures on jurisprudence are much more exclusively concerned with the means of subsistence as determining factors in all other features of society than were Montesquieu's or Ferguson's writings. Independent of the French Physiocrat, but at about the same time that Quesnay was insisting that "subsistence is the primary object of all societies . . . everything else is only modification," Smith was telling his Scottish students that "all the arts, the sciences, law and government, wisdom, and *even virtue itself* tend all to this one thing, the providing of meat, drink, raiment and lodging for men, which are commonly reckoned the meanest of employments."[95]

Like Marx later on, Smith had ostentatiously abandoned any but an empirical and "objective" understanding of how government, law, and morality emerged out of the changing conditions of production. Also like Marx, he could not avoid a kind of bitterness, anger, and revulsion at some of the developments he saw. First, consider Smith's introductory overview of the development of property and law—an introduction that only slightly expands Montesquieu's views while sustaining a dispassionate stance:

> Where the age of hunters subsists, theft is not too much regarded. As there is almost no property among them [i.e., hunters], the only injury that can be done is the depriving them of their game. Few laws, or regulations will be requisite in such an age of society, and these will not extend to any great length, or be very rigorous in the punishments annexed to any infringements of property. . . . But when flocks and herds come to be reared . . . many more laws and regulations must take place; theft and robbery being easily committed, will of consequence be punished with the utmost rigor. In the age of agriculture . . . there are many ways added in which property may be interrupted as the subjects of it are considerably extended. The laws therefore, tho perhaps too rigorous, will be of far greater number than amongst a nation of shepherds. In the age of commerce, as the subjects of property are greatly increased, the laws must be proportionately multiplied. The more improved any society is and the greater length the several means of supporting its inhabitants are carried, the greater will be the number of their laws and regulations necessary to maintain justice, and prevent infringements of the right of property.[96]

When Smith returned to amplify his reflections about the increasing complexity of law in the transition from hunting to herding culture and to assess some of the implications it has for the growth and power of governments, however, he was unable to retain his emotional distance; and the nature of his language signals his distress. When animals are domesticated and become the private property of individuals, the distinction between rich and poor begins to emerge. Then,

> when . . . some have great wealth and others nothing, it is necessary that the arm of authority should be continually stretched forth, and permanent laws or regulations made which may ascertain the property of the rich from the inroads of the poor. . . . *Laws and government may be considered in this and indeed in every case, as a combination of the rich to oppress the poor,* and preserve themselves in the inequality of goods which would otherwise be soon destroyed by the attacks of the poor, who, if not hindered by the government would soon reduce the others to an equality with themselves by open violence. . . .[97]

Smith had used his economic determinist view of society to demonstrate why, as Cantillon had asserted long before, it was inevitable that property should become unequally distributed in advanced societies and how law and governments come into existence to maintain socioeconomic hierarchies. But Smith could not free himself of the feeling that there was something wrong with the patterns of inequality which emerged with economic progress; thus, he uses the term *oppression* to speak of the relation of the rich to the poor.

Throughout all of Smith's writings there is an implicit acceptance of a principle which his moral philosophy teacher, Hutcheson, had insisted upon but which his own empirical approach to morality had pushed into the background. Hutcheson had argued that the highest good—the public good—was produced by providing "the greatest happiness for the greatest numbers" in society.[98] Since there were clearly so many poor, it seemed hard to understand how any system which subordinated their happiness to that of the few rich could ultimately be justified. This perspective pushed Smith into seeing the rich as oppressors and into bemoaning the fact that "the labor and time of the poor, are in civilized countries, sacrificed to the maintaining of the rich in ease and luxury."[99] Even in the *Wealth of Nations* he continued to argue that "no society can surely be flourishing and happy, of which the far greater part of the members are poor and miserable" and "It is but equity besides, that they who feed, clothe, and lodge the whole body of the people, should have such a share of the produce of their own labor as to be themselves, tolerably well fed, clothed, and lodged."[100]

In his earliest lectures on jurisprudence, Smith found a way to excuse what he himself described as the "oppression and tyranny" exercised by the rich over the poor in commercial societies, arguing that even the poorest members of modern society enjoy greater material wealth than the members of savage societies. He never grew completely comfortable, however, with this argument. In the *Wealth of Nations,* his most vigorous defense of the system of natural liberty which would lead to maximizing wealth, Smith admitted that its positive economic benefits could be purchased only through a system which left "all of the nobler parts of the human character . . . obliterated and extinguished in the great body of the people,"[101] and which was likely to leave the intellectual facilities of the inferior ranks of people "mutilated and deformed."[102]

What distinguished Smith from Ferguson, who clearly shared his dislike for many of the developments which accompanied the growth of modern economies, was that Ferguson retained a foundation for morality that lay outside his descriptive science of society, whereas Smith did not. Although Ferguson could continue to criticize the developments he saw and to argue in favor of intentional attempts to manipulate the process to bring about desired ends, Smith was forced increasingly to pin his hopes on the "invisible hand," i.e., on the law of unintended consequences.

SMITH'S "INVISIBLE HAND" RE-EXAMINED

Smith turned to compose his *Inquiry into the Nature and Causes of the Wealth of Nations* during the mid-1760s when he was resident in France; but though his analysis of the pivotal role of capital in commercial societies may have been sharpened through discussions with Turgot,[103] his major themes had been well developed in his earlier lectures on moral philosophy.

Abandoning all reference to motives other than "the uniform, constant, and uninterrupted effort of every man to better his condition,"[104] Smith writes:

> It is not from the benevolence of the butcher, the brewer or the baker that we expect our dinner, but from their regard of their own inter-

est. We address ourselves not to their humanity, but to their self-love, and never talk to them of our necessities, but of their advantage.[105]

But as in the *Theory of Moral Sentiments,* he insists that self-interest may unintentionally produce public benefits. After detailing the historical tendency of the feudal nobility to replace their support of large numbers of relatives with the purchase of luxury goods, he summarizes the transition from feudal to "commercial" society in the following way:

> A revolution of the greatest importance to the public happiness, was in this manner brought about by two different orders of people, who had not the least intention to serve the public. To gratify the most childish vanity was the whole motive of the greatest proprietors. The merchants and artificers, much less ridiculous, acted merely from a view to their own interest, and in pursuit of their own pedlar principle of turning a penny wherever a penny was to be got. Neither of them had either knowledge or foresight of that great revolution which the folly of the one, and the industry of the other, was gradually bringing about.[106]

Smith even returns to the "invisible hand" image which he had used in both his essay "History of Astronomy" and the *Theory of Moral Sentiments* to symbolize the way in which the search for private gain is turned to public benefit. But the context in which the invisible hand appears in the *Wealth of Nations* is both vastly different from that in which it had been used earlier and extremely illuminating. In the astronomy essay "the invisible hand of Jupiter" was understood as a human construct, created in response to ignorance of natural causes—a construct that was abandoned as more advanced systems of astronomy obviated its need. In the *Theory of Moral Sentiments,* the invisible hand again appeared to signify some unknown and possibly divine principle which turned private motives into public gain. But in the *Wealth of Nations* the invisible hand appears only at the conclusion of an extensive analysis of how the law of supply and demand coupled with a self-interested concern with minimizing risks, ensures that profit on capital must be maximized when it is employed "in the support of that [domestic] industry of which the produce is likely to be of greatest value"[107]—i.e., when it also serves to maximize the aggregate revenue of the society. Summarizing the natural law that directs capital allocation, Smith describes the capitalist's actions:

> He generally, indeed neither intends to promote the public interest, nor knows how much he is promoting it. By preferring the support of domestic to that of foreign industry, he intends only his own security; and by directing that of industry in such a manner as its produce may be of the greatest value, he intends only his own gain and is in this, as in many other cases, led by an *invisible hand* to promote an end which was no part of his intention.[108]

Clearly the invisible hand that links the capitalist's aversion for risk and search for profit to the public good is nothing but the self-equilibrating market mechanism. It is neither unknown nor divine, and it is invisible only because the capitalist need not have it in mind when he acts. Ironically, then, Smith appeals to the invisible hand in the *Wealth of Nations* to emphasize the fact that we no longer need to appeal to some unknown or divine intervention to account for the private source of public benefit. He believes he has discovered the mechanism which ensures that when private profits are maximized, so too are public goods. Considering any individual, we can now be assured that "the study of his own advantage *naturally,* or rather *necessarily* leads him to prefer that employment which is most advantageous to the society."[109]

As applied by Montesquieu and Ferguson (and by Smith in his early writings), the law of unintended consequences served as a kind of admission of ignorance and as a conservative warning against undue reliance on rational planning; for the consequences of our acts must always be more extensive than those that we can envision. But in the *Wealth of Nations* Smith manages to transform this expression of humility into a strange guarantee of the economic liberal dogma that the more we ignore social interests the more we will serve them.

Throughout the *Wealth of Nations,* Smith "demonstrates" that every approach to political economy which seeks to *direct* the economy to some publicly laudable end—whether it is an improved balance of trade (the mercantilist approach), the improvement of agriculture (the Physiocratic system), or even the adequate provisions for the citizenry (the scholastic or "moral economy" system)— "is in reality subversive of the great purpose which it means to promote."[110] Economic conditions are optimized when all public systems of preference or restraint are removed, and when "Every man, as long as he does not violate the laws of justice, is left perfectly free to pursue his own interest in his own way, and to bring both his industry and capital into competition with those of any other man, or order of men."[111] Under these circumstances, governments are completely absolved of all responsibility for guiding economic affairs and are limited in their aims to providing national defense, administering justice, and providing certain public works and institutions—e.g., roads and schools—whose existence demands the support of whole societies.

SOME LIBERAL POLICY IMPLICATIONS OF SMITH'S ECONOMIC ASSUMPTIONS

Smith began the *Wealth of Nations* with three chapters on the division of labor. In the first chapter he argued that it is the division of labor "which occasions, in a well governed society, that universal opulence which extends itself to the lowest ranks of the people."[112] In the second chapter he insisted that the division of labor emerges out of a natural "propensity to truck, barter, and exchange."[113] And in the third chapter he developed a new, or at least greatly expanded argument that "the division of labor is limited by the extent of the market,"[114] illustrating his point from evidence that the first and most extensive division of labor occurs where water carriage "opens the whole world for a market to [sic] the produce of every form of labor."[115] Smith did not himself explicitly link this set of comments on the division of labor to any specific policy recommendations, but his readers and economic followers during the nineteenth century derived from them a special justification for attempts to increase the international division of labor by encourging free trade.

From 1804 until 1846, when all restrictions on the import of grains were lifted, the free-trade side of the virtually continuous "Corn Law" debates in Parliament was led by the nearly sixty economists—including David Ricardo and John Stuart Mill—who sat during the period;[116] and their arguments constantly emphasized the inviolability of the basic laws of Smithian political economy. Responding to one opponent's speech, George Scrope, author of *The Principles of Political Economy* (1833), presented a typical argument:

> The hon. Member does not seem to be aware that the principle he declaims against as a cold dogma of stern political economy is the one sole vivifying principle of all commerce—the stimulus to all improvement—the mainspring of civilization—the principle, namely, of obtaining the largest and best result at the least

> cost . . . . I call on you then no longer by unwise and unjust laws to
> prevent the industrious classes of this country from availing them-
> selves in the ample means which God and nature have placed at
> their disposal for obtaining, by the exercise of their unrivaled skill
> and energy, an abundant supply of the first necessaries of life.[117]

Indeed, with rare exceptions, the economist–Parliamentarians
were so much guided by their free-trade principles that those principles overrode all other considerations. In 1846, for example, when
on the grounds of religion, justice, and humanity, Henry Brougham
spoke against allowing the import of slave-grown sugar, eleven of
the twelve economists then sitting (most of them anti-slavery) voted
against Brougham.[118]

Another feature of Smith's analysis contributed to the long-
term liberal insistence on economic expansion and involved his
treatment of the wages of labor. Smith started from Cantillon's discussion of wage rates, but he was much less hopeful that unskilled
wages could be long maintained above the *minimum* subsistence
level in a stable economy. He recognized that employers must always have a negotiating advantage by virtue of their ability to survive a period during which their capital is unemployed, whereas
workers must earn wages "for the sake of present subsistence."[119]
Given this condition, only where the demand for labor is high and
continually increasing will employers be forced to compete with one
another for workers and drive up wages. It is thus "not the actual
greatness of national wealth, but its *continual increase* which occasions a rise in the wages of labor. It is, accordingly . . . in the most
thriving states or in those which are growing rich the fastest, that
the wages of labor are highest."[120] For this reason, Smith insisted,
wages were tremendously high in the relatively poor but rapidly expanding American colonial economy, but barely above subsistence
in the rich but stable economy of China.[121] Smith's whole argument
that the public good is most fully served by maximizing aggregate
wealth depends on his assertion that the condition of "even the
poorest members" of society must improve.[122] It thus demands the
indefinite expansion of the national economy and suggests that
the rate of economic growth must be the single best indicator of a
nation's health.

The extent to which this argument became a dominant feature
of both liberal and socialist ideologies is mind-boggling. Even at
the height of Western imperialism there were at least a few voices—
socialist or humanitarian—raised against the Europocentric exploi-

tation of colonial empires; but I can think of no nation even today which does not measure its own well-being chiefly in terms of an increasing Gross National Product.

SUMMARY OF CHAPTERS 4 AND 5

Just as the extensive development of natural philosophy in the classical Greek world was followed by a period during which the new techniques of analysis were extended to moral and political considerations, the spectacular developments in the natural sciences associated with the scientific revolution of the sixteenth and seventeenth centuries were followed by a period in which the attitudes, ideas, and methods of analysis associated with mathematical, mechanical, and experimental natural philosophy were extended to the social domain. The new social sciences of the late seventeenth and eighteenth centuries initiated major changes in the way Europeans and Americans understood the domains of morality and politics, undermining the theological and Aristotelian foundations used to justify the traditional hierarchical society with its focus on reciprocal duties and obligations. Like the ancient Sophists, the new social scientists tended to replace the sacred foundations for social relations of deference and authority with secular and "natural" foundations for those modern systems of social categories, values, and arrangements which emerged in the nineteenth century under the labels of liberalism and socialism.

The first of the social sciences to emerge as an autonomous domain of concern was political economy or economics. The political economists drew heavily from the quantitative emphases and the obsession with analytic simplification associated with the seventeenth-century *esprit géométrique,* as well as from a medical tradition that viewed intervention in natural processes as justified almost exclusively to block pathological developments. They sought to divorce economic phenomena from the broader social and moral context that characterized Medieval and Renaissance attempts to understand "moral economy," and they sought to discover quantitative relationships among economic variables, developed statistical techniques, and encouraged schemes to collect demographic and economic information from which to construct and test their candidates for the status of economic laws.

Beginning with William Petty and John Graunt, a tradition of scientific analyses of economic phenomena was self-consciously developed by men such as John Locke, Edmund Halley, Richard Cantillon, the Physiocrats, Turgot, and Adam Smith. Though these men may have differed on a number of details, they collectively created an increasingly sophisticated and inclusive system of ideas that built upon and gradually seemed to legitimize the following set of assumptions which form the foundations of modern economic liberalism:

1. Economic activities are governed by natural laws which human, positive laws are powerless to modify.
2. In analyzing economic activity, we must assume that every individual acts solely in terms of his or her own rationally calculable self-interests.
3. All economic actors are analytically equivalent; i.e., for purposes of economic analysis such noneconomic factors as social status, sex, religion, race, national origin, educational background, and emotional needs may not be considered.
4. Though economic actors intend only their own private gain, there is a self-regulating market mechanism operating through relations of supply and demand which ensures that the private pursuit of economic gain will also maximize the public wealth.
5. A measurable quantity, the aggregate wealth of a society, stands as an adequate measure of the general welfare of that society so that it makes sense to equate public welfare with public wealth.
6. It follows directly from 1, 4, and 5, that any interference with the market's operation to maximize wealth leads to a diminution of the public welfare; so governmental involvements in economic life should be limited to eliminating conditions which interfere with the natural and free operation of markets.
7. If analysis focuses on maximizing public wealth through unregulated economic competition, the problem of the distribution of wealth within a society will take care of itself. (By the time of Adam Smith this rule is understood to hold only in growing economies.)

The attempts of British political economists to gain governmental support for their data-gathering activities or to base public policy on the implications of their theoretical systems were almost uniformly unsuccessful prior to the beginning of the nineteenth century in part because the British scientific community had no special political standing; thus, no mechanism for incorporating scientific advice into public policy had been developed. The incorporation of a scientific element into the operations of the British government was complicated especially by the fact that both aristocratic and democratic political actors saw that scientific intelligence was likely to augment the power of the central government.

So there was self-conscious opposition to any extension of scientific policymaking. In France, however, the scientific community acquired significant governmental functions during the seventeenth and early eighteenth centuries through the Académie des Sciences and through the involvement of scientists in the growing bureaucracy. At the same time, such scientific leaders as Fontenelle and d'Alembert had been able to formulate and popularize the notion that the scientists represented a politically disinterested body devoted to the public welfare. Thus the French political economists were in a position to gain governmental support for their policies; and, at least prior to 1791, they faced no special political opposition just because they were formally linked with the Crown. As a consequence, the French political-economist-bureaucrats quickly began to collect statistical information. During the second half of the eighteenth century, they were even able to implement schemes to establish a free grain trade and to abolish the traditional *corvée* which disturbed the free allocation of labor. This specific policy failed miserably in the face of opposition to attacks on privilege and of disastrous grain harvests which drove the price of bread beyond that which many of the poor could afford. So it was only after the French Revolution that liberal political economy became the foundation of most French economic policy.

The second approach to social science to gain a substantial number of adherents was the tradition of philosophical history that gave rise to anthropology and sociology. Though it was closely linked to the seventeenth-century political anatomy of James Harrington, subsequent members of the sociological tradition tended to view Montesquieu's *Esprit des lois* as the founding work of their science. Much less concerned with quantification and analytic simplicity than the political economists, Montesquieu and his inheritors sought to account for the immense complexity of human social arrangements and legal systems across huge spans of time and place. Culling data almost indiscriminately from classical history, traveler's descriptions of primitive cultures and the venerable civilizations of India and China, as well as from their own experiences of eighteenth-century European commercial cultures, men such as Montesquieu and Turgot in France and Adam Smith, Adam Ferguson, and John Millar in Scotland developed an increasingly compelling theory to account for the bewildering variety of laws and customs as responses to environmental conditions and patterns by which humans manage to meet their needs for food and shelter.

A static taxonomy of four fundamental modes of procuring sustenance suggested by Montesquieu—hunting, herding, agriculture, and commerce—became in the hands of Turgot, Smith, and their followers the foundation for a theory that viewed societies as dynamic, developing naturally through successive stages in which changing modes of production led to changing patterns of law and government. Though they tended to see economic activities—e.g., the way in which men satisfy their fundamental physical needs—as the most fundamental determinant of the system of customs and laws of any society, the members of the sociological tradition insisted that local conditions and inherited customs produce many variations. Furthermore, the sociologists were generally insistent that the political economists were wrong in assuming that self-interest was the sole and universal driving motive of men. Indeed Ferguson, and to a lesser extent, Montesquieu, saw the dominance of self-interest as a peculiar and lamentable consequence of the character of commercial society. Cooperation and sympathy were, they insisted, the prevailing motives for activity in primitive societies.

Members of the sociological tradition insisted that there was good cause to reject some of the dominant values of commercial society, as well as good reason to believe that now that men understood the sources of their laws, customs, and values, they might use that knowledge consciously to direct what had up to now been a process dominated by the often unintended consequences of human efforts. A series of developments that involved a mixture of desirable and undesirable characteristics might now be turned into an unambiguously progressive process. In particular, the narrowing of interests and numbing of intellect that seemed to accompany the economically valuable division of labor might be controlled by providing a genuinely liberal education to all men and women, and the increasing isolation of individuals in commercial society might be combatted by making them sensitive to the value of cooperation and sympathy. Thus the sociological tradition introduced a series of elements which became fundamental in the subsequent development of both liberal and socialist ideologies. By creating a theory of dynamic social development driven by economic conditions, it provided the foundation for optimistic liberal expectations of economic and social progress at the same time that it provided the ground work for subsequent socialist theories of dialectical materialism, with their simultaneous emphases on economic determinism and the possibility of effective human intervention to direct historical developments. In its treatment of the negative consequences of the division of labor, it provided the foundation for the Marxist analysis of alienation.

If the long-term impact of the eighteenth-century sociological/anthropological tradition was progressive and even radical, however, it was viewed by many eighteenth- and early nineteenth-century intellectuals as conservative in its political implications; for by acknowledging that particular institutions were often natural responses to local conditions, it provided a way of explaining and justifying laws, customs, and social organizations that some bitterly opposed.

The Ideological Implications of Enlightenment Social Sciences III: Sensationalist Psychology and a New Focus on Equality, Distributive Justice, and Education

6

If the early scientific traditions of political economy and philosophical history contributed in major ways to articulating and promoting most of the attitudes connected with economic liberalism—including its emphasis on private economic interest, aggregate wealth, governmental noninterference, and economic growth—both traditions were silent on issues of distributive justice. Moreover, when they addressed them at all, both were positively conservative with respect to broader social and political concerns. They accepted as natural the division of society into various orders, hierarchically related through patterns of obligation and deference, and they viewed political—though not economic—activity in terms of balancing the interests of the various orders in society, drawing heavily from Greco-Roman understandings of political life. Whenever they commented on the tendencies of "commercial" activity to break down traditional understandings of public as opposed to private responsibility, they viewed that breakdown as unfortunate and either urged that it be resisted (Montesquieu and Ferguson) or accepted it with an uneasy resignation (Smith).

In order to understand the more radical tendencies of Enlightenment thought—those which dominated both the political side of liberalism, with its greater emphasis on equality, political participation, and educational opportunity, and the emergence of early socialist emphases on distributive justice—we must explore the third major scientific attempt to understand society, an attempt grounded in the analysis of individual psychology.

In two ways this tradition was unquestionably more radical

than either of the others that I have discussed: first, it implied the need for much more extensive and fundamental modifications of the sociopolitical order; and second, it sought to go beneath the apparent complexity and variety of social arrangements to some more fundamental or primitive uniformities of human nature. Thus, for example, Helvetius and Condorcet, who provided some of the most extensive discussions of the social implications of sensationalist psychology, violently opposed Montesquieu not just because he gave aid and comfort to those "Aristocrats and petty despots" whose power is "erected at the expense of the great mass of society through hereditary usurpations"[1] but, more fundamentally, because his methods were totally inadequate to the task he set himself. The true foundations of law could not possibly be discovered by "collecting a mass of notes on the laws of all peoples, and . . . collecting them under different titles."[2] Laws are indeed the embodiment of ideas of justice, but those ideas are not to be understood as different for different societies; rather, they can and should be derived from *universal* characteristics of human nature:

> The idea of justice, of right, necessarily forms in the same manner in all sensitive beings capable of making the necessary combinations to acquire these ideas, [insisted Condorcet] . . . . *They will therefore be uniform.*[3]

If different countries have different laws, it can only be because laws are most frequently based upon superstitions or habits that need to be overturned in order to allow *true* justice to prevail. Whether one followed this line of reasoning into a publicly oriented utilitarian ethics like that of David Hartley and Jeremy Bentham or into a purely self-oriented materialist utilitarian ethics with Helvetius or into an argument on behalf of the universal rights of man with Condorcet, one was bound to reject Montesquieu's arguments as an invitation to "inherit all of the errors that have been accumulated since the origin of the human race."[4]

In its more extreme form, this doubly radical position suggested that because one of the most important properties of human nature is its capacity for indefinite change and perfectibility and because the chief aim of any government must be to protect the status quo, all governments are by their very nature evil rather than good in their long-term effects. That is, they operate to oppose progress. Thus *no* study of past governments can lead to an adequate science of morality and justice. As William Godwin, who fused the English and Continental traditions, insisted, the study of existing governments only

> gives substance and permanence to our errors. It reverses the genuine propensities of mind, and instead of suffering us to look forward, teaches us to look backward for perfection. It prompts us to seek the public welfare, not in innovation and improvement, but in a timid reverence for the decisions of our ancestors, as if it were the nature of mind always to degenerate, never to advance.[5]

Once the principles of human nature were understood, then for Condorcet, Helvetius, and the English radical philosophers like Hartley, Bentham, and Godwin, all principles of justice and morality could be directly deduced from them:

> They are in effect the *necessary* result of the properties of sensitive beings capable of reasoning; they derive from their nature; from which it follows that it is sufficient to suppose the existence of these beings for the propositions founded on these notions to be true; just as it is sufficient to suppose the existence of a circle to establish the truth of the propositions which develop its different characteristics.[6]

The first major post-Hobbesian attempts to derive morality from the fundamental characteristics of human nature emerged at the hands of David Hartley, who coined the modern term *psychology*[7] to signify that study of the human mind which was to provide the underpinnings for any subsequent understanding of ethics and morals. We will consider Hartley's development of associationist psychology—which Joseph Priestley credited with throwing "more useful light upon the theory of the mind than Newton did upon the theory of the natural world"[8]—as a general introduction to the enterprise of deriving morality from human psychological principles. This procedure will allow us to focus on distinctive characteristics of subsequent discussions of the relationship between morality and sensationalist psychology.

DAVID HARTLEY AND NEWTON'S METHODS APPLIED TO PSYCHOLOGY

David Hartley was a self-professed but highly idiosyncratic follower of Isaac Newton and William Whiston who was as admiring of the masters' work in biblical chronology and prophecy interpretation as he was of the *Principia* and the *Opticks*. Moreover, Hartley became known in Germany almost exclusively as a religious thinker, because his German translator dropped or digested almost all of Hartley's psychological backgrounding to focus on the theological content of Hartley's *Observations on Man, His Frame, His Duty, and His Expectations* (1749).[9] Hartley's major long-term impact in England and France, however, depended on his associationist psychology and its application to aesthetic, ethical, and moral issues separated from his theology. Hartley himself had begun with a strong moral emphasis, arguing in *Various Conjectures on the Perception, Motion, and Generation of Ideas* (1746) that "the principal use of the doctrine of association must be considered to be the amendment of ethics and morals."[10] Expanding on this theme, Hartley outlines the major steps of the strategy that guided not only his own work but also that of a large number of eighteenth- and early nineteenth-century thinkers whom Elie Halevy has designated "philosophical radicals":

> Having been led by this thread [i.e., associationist psychology] we are able to investigate the primary origins of mental pleasures and pains, and thereby of desires and aversions, and lastly of the voluntary and semi-voluntary power over these; the same task begins to illuminate the ways by which good motivations . . . may be fostered [and] the bad restrained . . . and, what is particularly noteworthy, by what precepts the life of tender minds of children can best be formed for virtue and piety.[11]

Note both the almost universally unacknowledged Hobbesian linkage of pleasure and desire with "good," whereas pain and aversion are linked with "bad"—a linkage that was to dominate all psychological theories of morality—as well as the central focus on education as a way of "forming" human beings for social life.

Even more important than his own emphasis in determining the long-term significance of Hartley's works was Joseph Priestley's *Hartley's Theory of the Human Mind, on the Principle of the Association of Ideas* (London, 1775), a popularization of Hartley's psychological writings which subordinated the religious arguments. As a consequence of Priestley's partisan advocacy, Hartley appeared to most subsequent English thinkers in a largely secular guise.[12]

Throughout his *Observations on Man*, Hartley develops two theories in parallel. One, a physiological doctrine, he calls the doctrine of *vibrations*; the other, a mental doctrine, he calls the doctrine of *association*. The first seeks to provide a more detailed and sophisticated version of the Hobbesian theory that ideas are caused by motions in the brain; the second seeks to explore the detailed implications of the empirical approach to mental phenomena suggested by Locke and more extensively developed by David Hume. Both Locke and Hume had realized that two or more seemingly unconnected ideas may become linked (associated) with one another merely by virtue of the fact that they habitually appear either simultaneously or in an invariable sequence. It is critical to any adequate understanding of Hartley to recognize how he understands the two doctrines to be related to one another. Hartley accepts the ordinary post-Cartesian distinction between body and mind. He insists that there is some close relationship between the two, such that bodily changes are accompanied by mental ones and vice-versa; but he is unwilling to commit himself to any particular theory about the precise relation of physical to mental events. Indeed, he insists that everything he says about both the vibrational doctrine and the associational doctrine may be equally true, whether "Motions in the medullary substance be the physical cause of sensations, according to the system of the schools; or the occasional cause, according to Malebranche, or only an adjunct, according to Leibniz."[13] Nor, Hartley insists, does he "presume to determine whether matter can be endued with sensation or no."[14]

For most purposes, then, we may treat Hartley as an advocate of psycho-physical parallelism, and his doctrine of vibrations as a physiological correlate of his doctrine of association. Strictly speaking, his associationist doctrine does not logically depend on his vibrational theory, and he is very insistent that the doctrine of associations is "unquestionable" and "a certain foundation, and a clue to direct our future inquiries, whatever becomes of that of vibrations."[15] Why, then, we might ask, did Hartley bother to encumber his work on associationist psychology with a physiological theory that he himself acknowledged to be less certain and that his friend Priestley saw as an utter embarrassment? In order to answer this question, we must consider Hartley's understanding of scientific method and the special roles played by hypotheses and analogies.

Though Hartley insisted that the most proper method of conducting science was the method of analysis and synthesis described by Sir Isaac Newton in the general scholium to Book III of the *Principia*, he acknowledged in the first chapter of *Observations of Man* that this approach was excluded from an investigation of human psychology "on account of the great intricacy, extensiveness, and novelty of the subject."[16] Instead, choosing a method that was second-best but nonetheless capable of producing genuine knowledge, he proceeded by a hypothetical method patterned on the mathematician's technique of successive approximations or "the Rule of False." In this method, wrote Hartley, "the Arithmetician supposes a certain number to be that which is sought for; treats it as if it was that; and finding the deficiency or overplus in the conclusion, rectifies the error of his first position by a proportional addition or subtraction, and thus solves the problem."[17] In precisely the same way, he argued that

> it is useful in Inquiries of all kinds, to try all such suppositions as occur with any appearance of probability, to endeavour to deduce the real phenomena from them; and, if they do not answer in some tolerable measure, to reject them at once; or if they do, to add, expunge, correct, and improve, till we have brought the Hypothesis as near as we can to an agreement with nature. After this it must be left to be farther corrected and improved or entirely disproved by the light and evidence reflected upon it from the contiguous, and even, in some measure, the remote Branches of other sciences.[18]

The doctrine of vibrations was just this kind of hypothesis for Hartley; and it was especially fruitful because it suggested a whole range of psychological phenomena which had not been previously observed. But the physiological phenomena predicted by the doctrine of vibrations could not be directly observed; so Hartley was faced with the question of how to establish the plausibility of this doctrine. At this point he followed a method recently adopted by Joseph Butler in his *Analogy of Religion* (1737). He appealed simultaneously to a set of ideas regarding the use of analogies and to the doctrine of associations. As John Locke had carefully explained, when direct evidence cannot be had on some issue, we are forced to have recourse to analogical thinking. Hartley amplified the Lockian idea of how analogy functions:

> The Analogous natures of all the things about us are a great assistance in deciphering their properties, powers, laws, etc., inasmuch as what is minute or obscure in one may be explained and illustrated by the analogous particular in another, where it is large and clear. And thus all things become comments on each other in endless reciprocation.[19]

In Hartley's case the unobservable physiological phenomena predicted by the doctrine of vibrations could all be expected to

cause or to be accompanied by analogous observable mental phenomena—i.e., those dealt with by the doctrine of associations. So,
to the extent that every prediction of the vibrational hypothesis
was paralleled by an analogous observed mental phenomenon, the
vibrational theory received support by analogy.

At this point the nature of the doctrine of association becomes
critically important. Unlike the doctrine of vibrations, the doctrine of associations, as it was understood by Hartley, was not a
theory or a hypothesis at all, but a set of descriptions of mental
phenomena established inductively. Direct evidence for it did not
depend upon physiological assumptions, and it carried a kind of
certainty that the doctrine of vibrations was incapable of. So evidence from the mental domain could be accepted as independent
evidence—though merely by analogy—for the physiological events
implied by the doctrine of vibrations.

The relationship between the doctrine of vibrations and that
of associations, however, was by no means limited to the indirect
support that the phenomena of mental associations offered for the
hypothesis of vibrations. Indeed, that support was the least important feature of their relationship to one another. We have already
seen that Hartley viewed all analogous relations as being *reciprocally* reinforcing. Even more important for Hartley, "Analogy may
also in all cases be made use of as a guide to invention."[20] For
his *Observations on Man* this possibility was important because
the doctrine of vibrations predicted physiological events that suggested—by analogy—the existence of a range of mental phenomena
which had not previously been incorporated into associationist
psychology. Thus in the long run, vibration theory—supported by
previously observed mental phenomena—was most important for
Hartley not for itself, but as a suggestive tool for extending the
explanatory domain of associationist psychology. Once new mental
relationships had been suggested by the vibrational doctrine, their
existence could be directly confirmed observationally. This is why
Hartley could insist upon the validity of his expanded doctrine of
associations, "whatever becomes of that of vibrations"; and this is
why Priestley felt justified in jettisoning the vibrational doctrine
when he sought to popularize Hartley's work. Vibration theory had
done its job, and there was no point in linking associationist psychology with the liabilities of a problematic—or at best, merely
probable—physiological doctrine that it no longer needed.

Having established this methodology, we can now turn to

Hartley's analysis of sensations and ideas. Hartley began by offering evidence that "the white medullary substance of the brain" is the physical instrument by which "ideas are presented to the mind," or more precisely, that "whatever changes are made in this substance, corresponding changes are made in our ideas and vice versa."[21] Furthermore, he explained that the nerves and spinal marrow link the brain to our sense organs, and he appealed to evidence from visual after-images to claim that "sensations remain in the mind for a short time after the sensible objects are removed."[22]

Given these facts, Hartley proposed his vibrationist doctrine as a plausible explanatory hypothesis. Objects transmit a sequence of longitudinal vibrations to our organs of sense. There they set in motion minute vibrations within the nerves. These vibrations are in turn directed through the nerves to the brain where they set in motion the small "infinitesimal" medullary particles, the vibration of which either constitutes, causes, occasions, or accompanies our sensations. The foundation of this doctrine, Hartley acknowledged, is drawn from "hints concerning the performance of sensation and motion, which Sir Isaac Newton has given at the end of his *Principia* and in the Questions Annexed to his *Opticks*";[23] but its details and implications remain to be worked out.

In order to account for the variety of our sensations, we must presume that the related medullary vibrations can be distinguished from one another in a variety of ways. Hartley suggested that there are four basic differences: degree in "vigor" or amplitude, kind or frequency, place or "region of the medullary substance," and line of direction (fig. 9).[24]

Whenever some object impresses its vibrations upon the nerves and generates a medullary vibration of a particular amplitude, frequency, location, and orientation in the brain, we have a particular and corresponding "feeling in our mind" which we call a *sensation.* In general we call any feeling in the mind an *idea;* so sensations are a special class of ideas. But clearly we have many ideas which are not sensations, and we have to figure out how these are acquired. The first and most important ideas which are not merely sensations—or rather which are not *ordinary* sensations—are pleasure and pain. Initially, Hartley argued, pleasure and pain *accompany* sensations of unusual vigor; pleasures accompany vigorous sensations, whereas pains accompany sensations that have become too vigorous; thus pleasure and pain differ only in degree and pass into one another as sensations are intensified or di-

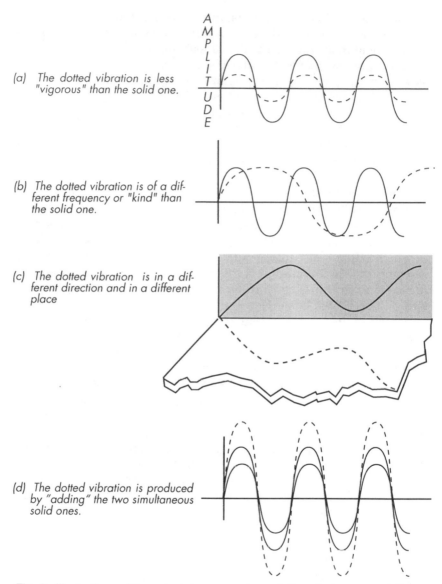

(a) The dotted vibration is less "vigorous" than the solid one.

(b) The dotted vibration is of a different frequency or "kind" than the solid one.

(c) The dotted vibration is in a different direction and in a different place

(d) The dotted vibration is produced by "adding" the two simultaneous solid ones.

Fig. 9. Illustrations of the principles governing relations between medullary vibrations according to Hartley.

minished. Pleasurable warmth, for example, passes into burning heat; pleasant sounds become painful when they are increased in loudness. Pain, Hartley suggested, occurs when vibrations become so great that they produce a "solution of continuity,"[25] or a long-term deformation. Thus, wounds are so painful because they completely disjoin parts of the nerves, and the violent release of tension is transmitted to the brain. One curious consequence of this theory is that while pains will usually be more intense than pleasures, they are unlikely to be so frequent, because they depend upon "such violent states as cannot arise from common impressions."[26]

Frequently repeated sensations give rise to "vestiges, types, or images of themselves" which we call ideas in the mind, and this occurrence is easily accounted for as a consequence or correlate of vibration theory, for any medullary vibration decays only gradually in the brain. If it is repeated, the new vibrations reinforce the old and "beget a disposition to diminutive vibrations" which Hartley calls "vibratuncles" or "miniatures," corresponding to their originals in the brain.[27] By the same principle we can account for the mental phenomenon of *association,* by which sensations or ideas become linked to one another in space or time. Understanding how associations come about is absolutely critical, for as Hartley insists, everything we know about "the power of custom, habit, example, education, authority, party-prejudice, the manner of learning the manual and liberal arts, etc." depends upon the doctrine of associations.[28]

The power of some idea, sensation, or word to excite related ideas—as the smell of a flower may give rise to an idea of its visible appearance in the mind—follows from vibratory theory. Since the repeated experiences of some object generate a "vibratuncle" which includes miniature vibrations corresponding to its smell, visual impression, and sound, when any of the corresponding impressions is made on a single sense, it stimulates the whole pattern of miniature vibrations or vibratuncles in the brain, giving rise not only to the idea of the single sense impressed, but to related ideas of sense. Similarly, whole sequences of ideas can be called up by one which usually initiates the series because the related sequence of vibrations is initiated in the brain. Or the idea of a phenomenon can be called up by a word or name because experience has linked the vibration connected with the word to those connected with the experience to which it refers.[29] In this manner, simple ideas may even be compounded into complex ones in which all awareness of the simple elements has been lost, just as simple vibrations superimposed upon one another create complex wave forms in which the initial, constituent, simple waves are no longer recognizable. Such complex ideas as those of beauty, honor, or goodness are of this kind. Though formed initially from simple sensory impressions, they may become extremely difficult to decompose into their constituent simple ideas, just as their corresponding vibratuncles cannot easily be analyzed into simpler elements.[30]

At this point Hartley first began to explore the implications of his vibrational and associationist doctrines for morality. In the first place, those complex ideas which are one or more steps from the simple ideas of sensation, and which we therefore call "intellectual," often become even more powerful and vivid than the original ideas of sensation. This phenomenon is suggested by vibration theory because intellectual ideas are created by the positive interference of many simple ideas whose additive force must be relatively great. Thus, intellectual ideas will often become the source of pleasure because their corresponding vibrations have great amplitude.[31] At the same time, argued Hartley, because such complex vibrations will not be highly localized or directional, they are not so likely to produce a "solution of continuity" at any particular place in the brain; so intellectual ideas are not so likely to produce pain as those of mere sensation.[32]

Several important features follow from the source and complex character of the intellectual ideas:

1. They must be the source of the greatest number of our unalloyed pleasures and therefore the focus of most of our desires; for the sensible, or bodily, pleasures and pains have a "greater tendency to destroy the body than the intellectual ones." It follows then, "that our ultimate happiness appears to be of a spiritual [i.e., mental], not corporeal nature."[33]
2. They are all *acquired* rather than innate, for they are constituted out of a combination of sensory ideas initially derived only from experience.
3. It follows that humans are what they are almost exclusively because of the experiences they have, rather than because of any innate and differential tendencies toward good or bad.

4. It also follows that in an important sense, all humans are created at least potentially equal; for, in Hartley's terms, "if beings of the same nature, but whose associations and passions are, at present, in different proportions to each other, be exposed for an indefinite time to the same impressions and associations, all the particular differences will, at last, be overruled and they will become perfectly similar, or even equal." They may also be made perfectly similar, in a finite time, by a proper adjustment of the impressions and associations.[34]

The whole rationale for studying associationist psychology in the first place follows from the above four characteristics of our intellectual ideas. If we analyze our affections, passions, and associated ideas into

> their simple compounding parts, . . . we may learn how to cherish and improve good ones, check and root out such as are mischievous and immoral [i.e., productive of more pain than pleasure], and how to suit our manner of life, in some tolerable measure, to our intellectual and religious wants. And as this holds in respect of all persons of all ages, so it is particularly true and worthy of consideration in respect of children and youth.[35]

That is, associationist psychology becomes the foundation of all moral understanding, improvement, and education. It provides the technique by which the mixed state of human kind can eventually be transformed "into one in which pleasure alone would be perceived" or, in religious terms, it is the means by which fallen man can be returned to the "paradisiacal" state.[36]

Hartley self-consciously followed his doctrines of vibration and association into a "necessitarian" position that denies the *traditional* philosophical and religious notion of free will, replacing it with a notion of will borrowed directly from Hobbes and termed the "popular and practical" notion of free will. Every human action, insisted Hartley, "results from the previous circumstances of body and mind, in the same manner, and with the same certainty, as other effects do from their mechanical causes; so that a person cannot do indifferently either of the actions A, and its contrary a, while the previous circumstances are the same; but is under an absolute necessity of doing one of them, and that only."[37] If we suppose otherwise, then we "destroy the foundation of all general abstract reasoning," at least as it is applicable to human behavior. Ultimately this argument seems to reduce to the assertion that unless the old notion of free will is abandoned, a completely deterministic scientific knowledge of human behavior is impossible; therefore, we must abandon the old notion of free will. Put this way, it is clear that Hartley's claim begs the central question whether a completely deterministic science of human behavior is possible. Hartley simply assumed the applicability of early eighteenth-century scientific techniques to the study of man; and if his assumption is accepted, his conclusion certainly does follow.

If we define *will* as "the desire or aversion which is strongest for the present time,"[38] following Hartley, then we can retain the notion that the will produces actions that are said to be voluntary—as opposed to automatic actions such as the withdrawal of our hand from a hot object. Under these circumstances, the claim that we have free will can indicate that our physical circumstances are such that we are not constrained from carrying out that action which we will. Hartley believed that his new-made notion of free will was completely consistent with Christianity; but he was aware that it would seem threatening to many. In England it did appeal to some Dissenters such as Priestley; but in France, as we shall see, the doctrines of associationist psychology—though initially developed in an orthodox religious context—rapidly became associated with anticlerical and purely secular trends.

As Hartley extended his analysis of sensations and ideas to ever greater levels of complexity, he assumed that the primary goal of each human was to increase his or her own happiness. At the same time, however, because each human being was to be equally valued, Hartley accepted the notion that the ultimate end of moral acts was to produce the maximum happiness not just for a single individual, but for *all* people. The major problem that Hartley and all subsequent utilitarian philosophers faced—i.e., those who sought to maximize aggregate happiness or the happiness "compounded of the happiness of the several individuals composing the body politic"[39]—was somehow to resolve the tension between a single individual's private happiness and the aggregate, or public, happiness. Just as economic liberals such as Adam Smith had somehow to explain how the private search for gain related to the issue of public wealth, Hartley and subsequent utilitarians had to explain how the private search for happiness related to the greatest public happiness, the ultimate goal of morality.

HARTLEY AND THE BRITISH DOCTRINE
OF NATURAL ACCOMMODATION

Hartley's answer to this question, which dominated early English liberal and utilitarian doctrines, was the doctrine of *natural* accommodation. That is, Hartley and such early followers as William Godwin insisted that there were self-acting psychological mechanisms which ensured the confluence of private with public happiness, much as the free market ensured the confluence of private economic gain with public wealth. Continental associationists, such as Claude Adrien Helvetius, could not accept the English doctrine of natural accommodation and were forced into arguments for an *artificial* accommodation which had much more radical political implications.

Although the various sensory and intellectual sources of pleasure and pain occur in a determinate sequence from sensations through imagination, ambition, self-interest, sympathy, theopathy, and finally, the moral sense, there is also a mechanism by which higher-order associations react upon lower order ones in a process that modifies and ultimately "perfects"[40] the ideas which initially emerged at lower levels. Thus, for example, self-interest is first generated in connection with what Hartley calls the "gross" or crude feelings associated with sex, food, the enjoyment of luxury goods, or money (which may procure these as well as the admiration of others).[41] But once one has experienced the pleasures associated with higher-order ideas, such as sympathy, those pleasures modify our notions of self-interest to provide what Hartley calls ideas of "refined" self-interest:

> A person who has had a sufficient experience of the pleasures of friendship, generosity, devotion and self-approbation cannot but desire to have a return of them, when he is not under the particular influence of any one of them, but merely on account of the pleasure which they have afforded; and will seek to excite these pleasures by the usual means, treasure up to himself such means, keep himself always in a disposition to use them, etc., not at all from any vivid love of his neighbor, or of God, or from a sense of duty, but entirely from the view of private happiness.[42]

In the long run Hartley does *not* want to claim that we follow our moral sense exclusively or primarily out of self-interest. The pleasures arising out of benevolence do not depend in the first instance on any calculation of benefit to one's self. Indeed, they arise out of conscious acts of self-denial which are directed at producing "the greatest happiness and least misery" for others, regardless of consequences to self.[43] What Hartley purports to show through his associationist doctrines is that in morality the precise converse of Smith's "invisible hand" in economics is at work; for when we act solely out of benevolence, piety, and moral sense, completely abandoning any immediate concern for our gross self-interests, then and only then will we be providing the "only true, lasting foundation of *private* happiness."[44]

Of all of Hartley's arguments about why self-interest in the ordinary sense must be modified in order to produce the maximum aggregate happiness, one took an increasingly important role among subsequent associationist psychologists. Addressing the issue of gross self-interest in particular, Hartley for the first time introduced the concept of equity or distributive justice into scientific discussions of morality, though it had long been part of religious and philosophical discussions of morality. Hartley wrote:

> Gross self-interest, has a manifest tendency to deprive us of the pleasures of sympathy and to expose us to its pains. Rapaciousness extinguishes all sparks of good-will and generosity, and begets endless resentments, jealousies, and envies. And indeed a great part of the contentions, and mutual injuries, which we see in the world arise, because either one or both of the contending parties desire more than an equitable share of the means of happiness.[45]

The political economists of the eighteenth century either ignored or artfully sidestepped issues of the "equitable" distribution of goods; and the sociologists, including Montesquieu and Ferguson, tended to accept the unequal distribution of property and other forms of wealth as a necessary, if often unpleasant, consequence of all but the hunting way of life. If one starts from the Hartlian assumptions of associationist psychology, however, there is no effective way to argue that one person deserves a greater share of the means of happiness than any other. Under these circumstances the pursuit of wealth generated what began to be recognized as both a practical and a theoretical problem, for as Hartley wrote—with only a slight and unimportant misunderstanding—"whatever riches one man obtains, another must lose."[46] Even allowing for the fact that the distribution of wealth is not, strictly speaking, a zero-sum game in which there must be a loss to balance every gain, Hartley correctly understood that the major issue involved was one of perceived equity. Actions based solely on gross self-interest must inevitably lead to the resentment rather than the approval of others, lessening our overall pleasure. When we act out of sympathy, friendship, and benevolence, however, no such negative consequences follow; and as Hartley insisted, pleasure is increased both for others and for one's self. In the end, then, Hartley claimed to be able to show that what he called "self-annihilation" and refined self-interest are naturally identical.

Only a full-scale analysis of the origin and development of our sensations and ideas can provide this knowledge; so Hartley's associationist psychology becomes the prerequisite for moral knowledge. Without such knowledge one might be subject to the pathological development of some particular associations that could block the normal development of private and public happiness alike. In Hartley's case, then, proper education plays the role of a kind of moral medicine. All experience is, of course, part of our education; for every sensation and association creates or modifies our ideas and our consequent patterns of behavior. In a sense, then, any moral error or pathology is the product of our "education," broadly conceived. But *proper* education, guided by a knowledge of associationist psychology, can be used to counteract any pathological patterns of association that might block the normal development of a benevolently oriented morality.

Subsequent associationist moralists, especially those in France, were inclined to place an even greater emphasis than Hartley on the precise details of a proper education; but I want briefly to indicate some of Hartley's ideas regarding *formal* education so they might be compared with the independently developed and more important notions of Etienne Bonnet de Condillac, Claude Adrien Helvetius, and Marie-Jean Condorcet.

What distinguishes the narrowest from the broadest sense of "education" is the central role played in the former by words, language, and formal processes of reasoning that depend on logic or syntax. At one level our linguistically oriented education is important because it offers an efficient surrogate for direct personal experience. That is, it allows us to build associations rapidly which could grow only very slowly out of extended sensory experience. It thus allows for the progressive character of culture by making it possible to incorporate the experiences of past generations into our own. Even more important, because words come to act as the most important links in most of our complex trains of association, language not only reflects, but also shapes our knowledge. Thus, as modern structuralist philosophers have rediscovered, when we understand the language of a particular society, we can better understand how the people in that time and place structured their experience.[47]

Though natural languages reflect the specific needs and interests of the special groups they serve, all languages have certain features in common. By comparing languages to mathematical systems, Hartley explained the general rules of thought or "logic" which all languages express. Indeed, he insisted that "language itself may be termed one specie [*sic*] of algebra."[48] Finally, Hartley reasserted the old Baconian expectation that it should be possible to construct a universal philosophical language capable of expressing an "accurate knowledge of things" without all of those distorting characteristics which natural languages impose on our knowledge because of the local needs that they are created to serve.[49] Only in such a language could the perfection of science and morality ultimately be expressed.

ETIENNE CONDILLAC AND FRENCH ASSOCIATIONIST DOCTRINES

In France, doctrines of associationist psychology very similar to those developed by Hartley were initiated by Etienne Bonnet de Condillac in a series of works published during the 1740s and 1750s. Beginning with his *Essay on the Origin of Human Knowledge* (1746), and in his successive *Treatise on Systems* (1749), *Treatise on Sensations* (1754), and *Treatise on Animals* (1755), Condillac presented an associationist psychology that paralleled Hartley's in many ways. Both argued for the centrality of heuristic hypotheses in scientific investigations. Both held a provisional commitment to psycho-physical parallelism. They were equally concerned with seeing associationist psychology as not only consistent with, but also positively supportive of natural theology, and they were equally insistent that proper education was the key to human happiness.

On just two major issues did Condillac diverge widely from the path taken by Hartley. Condillac was much more skeptical about accepting any particular physiological theory, such as that of vibrations, as a foundation for associationist psychology. He criticized Hartley's hypothesis of a "soft substance" in the brain, capable of preserving traces of sensory vibrations, and he acknowledged that he had "only an extremely imperfect and vague idea" about the physiological correlates of mental activity.[50] More important, Condillac was not willing to commit himself formally to a full rejection of the traditional doctrine of free will. While he insisted that we usually act in a fashion determined by our conformation and our needs, he nonetheless allowed some room for "arbitrary" behavior, thus making his doctrines acceptable to the faculty of theology at the Sorbonne.[51]

Like Hartley, Condillac viewed languages as playing a central role in shaping experience. He identified well-made languages with algebraic systems, and he insisted that the construction of a philosophical language alone would make possible the perfection of the arts and sciences.[52] Condillac, however, took the next critical step by transforming his analysis of ideas and language into a textbook designed to teach to a wider audience the "art of thinking," or "analysis," grounded in associationist principles. Condillac's *La Logique,* written in response to an invitation from the Polish government in 1778, became an immediate success. According to Dominique Joseph Garat, who was chosen to head the *Ecole normale* after the French Revolution, it created "a revolution in the teaching of all peoples."[53] Whether Condillac's teaching penetrated society as far as his enthusiastic advocate Garat hoped, it did provide the key psychological foundations upon which such *philosophes* as Claude Adrien Helvetius, Condorcet, and the whole school of "ideologues" built their educational doctrines and their moral and political theories. As Kingsley Martin has insisted, it was this new psychology "that separated the eighteenth century philosophers from their predecessors under Louis XIV. . . . [T]he philosophers had a positive doctrine to substitute for the orthodox creed. They believed they could demonstrate scientifically that knowledge was the key to happiness, and that it sufficed to enlighten men to make them perfect[able]."[54]

CLAUDE ADRIEN HELVETIUS AND THE SOCIALIST IMPLICATIONS OF ASSOCIATIONIST PSYCHOLOGY

The first person to press the moral and political implications of associationist psychology significantly beyond the very general and nonthreatening conclusions of Hartley and Condillac was Claude

Adrien Helvetius, whose major role in the emergence of a revolutionary consciousness during the late eighteenth century is generally much more fully acknowledged—though not necessarily better understood—by Marxist historians than by liberals.[55]

Helvetius's background makes him a very curious candidate for the socialist equivalence of sainthood. Born in 1715, the only son of Jean Claude Adrien Helvetius, principal physician to Louis XV's Queen, Marie Leszczynska, Helvetius was apprenticed to an uncle who was director of tax-farming in Caen. Through the queen's influence he became a farmer general of taxes in 1738 and then took up a court post as *maître d'hôtel* to the queen in 1748. Known best for his womanizing and for the financial abilities which made him one of the richest financiers of France, Helvetius began to frequent salon society in Paris, becoming closely linked with Voltaire, with the Physiocrats, and with the great naturalist and French Newtonian, Georges-Louis Leclerc, Comte de Buffon. In 1751 he married the noblewoman Catherine d'Autricourt, whom he had met at the salon of Mme. de Graffigny. From then until his death in 1771, with the exception of brief periods of travel, Helvetius and his wife split their time between Paris, where they established their own salon, and their country estate at Vore, where Helvetius turned his talents to pursue the career of a savant.

In 1758 Helvetius published *De l'Esprit (On the Mind)* in what seems to have been the innocent expectation that it would be approved by the Church, the state, and his philosophical colleagues as an eloquent and useful development of the sensationalist/associationist tradition initiated by Condillac. But Helvetius's work was more overtly critical of both the Church and the government than that of his respectable predecessors. Moreover, it appeared at a time when both the Gallican Church and Louis XV were especially sensitive to any criticism. During the 1750s a series of longstanding conflicts between Jansenist sympathizers and Jesuit factions within the Church and between the parlements, which sought greater secular control over the Gallican Church and the Gallican Bishops, who sought to preserve their own authority, came to a head on 5 January 1757 with an attempt to assassinate Louis XV by a fanatic supporter of the parlementary and Jansenist factions. As a consequence of the attempt on his life, a panic-stricken Louis XV sought the rigid repression of any writings that might be interpreted as impious or subversive:

All those who are convicted of having written or having caused to be written or printed any writing tending to attack religion, or to incite minds to impugn our authority and to trouble the order and tranquility of our state shall be punished with death.[56]

In spite of the tense mood, Helvetius's work was approved by two different censors; but when it appeared in print, the book was condemned by religious and secular authorities alike. The Archbishop of Paris presented the widespread opinion among religious and secular authorities:

We condemn the book as containing an abominable doctrine tending to overthrow natural law and destroy the foundations of Christian religion; as adopting as its principles the detestable doctrine of materialism; denying the primitive notions of virtue and justice; establishing maxims totally opposed to evangelical morality; substituting for holy doctrines a morality of interests, passions, and pleasures; tending to trouble the peace of the state and to turn the subjects against the authority and even the person of their sovereign . . . .[57]

The antagonism generated by *De l'Esprit* spilled over into a general condemnation of suspicious works, including the *Encyclopédie* and previously unchallenged works by Voltaire and Diderot. Curiously, because Helvetius and his wife had extremely close personal friendships within the court—extending to the queen, Madame de Pompadour, the Duc de Choisiel, and even the king himself—he was personally never in any serious danger. To the surprise and consternation of the other philosophes, much of the hostility generated by *De l'Esprit* fell on them rather than on its author.[58] As a consequence of its notoriety, *De l'Esprit* achieved a kind of *succès de scandale* which raised its importance and its circulation in clandestine editions far beyond that which its intellectual merits might otherwise have achieved. Moreover, its reception consequently drove its author into an increasingly bitter and radical position which was fully expressed only in his posthumously published *De l'Homme* or *Treatise on Man* of 1774. At the same time, the hostile fallout from the reception of Helvetius's *De l'Esprit* initiated a major split among philosophes between the moderates like d'Alembert, Turgot, and Voltaire, who still viewed an enlightened monarchy as the greatest hope for incorporating their ideas into social practice and who therefore sought to avoid alienating secular authorities, and the more radical and democratic like Helvetius, d'Holbach, and Rousseau (in some moods), who produced increasingly strident criticism of the religious, social, and political structures of the old regime in France.

The immediate occasion for Helvetius's first work was his reading of Montesquieu's *Spirit of the Laws,* which he saw as misguided both in terms of its method and its aims. If one seeks to understand the true principles of legislation, one must consider two different but related problems. First, one must seek "the discovery of laws proper to render men as happy as possible and consequently to procure them all the amusements and pleasures compatible with the public welfare."[59] Only after that step has been taken can one effectively approach the second problem, which is to discover the "means by which a people may be made to pass insensibly from the state of misery they suffer to the state of happiness they might enjoy."[60] Montesquieu had unfortunately put the cart before the horse. It was true that in order to figure out how to reach the desired state from the present, one must have "a knowledge of the principal laws and manners of the people whose legislation is to be insensibly changed."[61] But one needs first to know where one wants to go. In answering that question, one finds the historical approach not merely useless, but indeed harmful. When students of mechanics are given a complicated problem, they begin by making simplifying assumptions. They calculate the motions of bodies without regard to the resistance of surrounding fluids or the friction in gears and pulleys. By the same token, Helvetius insists, when we first approach the complex problem of legislation, "we should . . . pay no regard to the resistance of prejudices, or the friction of contrary and personal interests, or to manners, laws, and customs already established."[62]

For our purposes, it matters little that Helvetius's concept of scientific method was open to serious criticism. What is important is that he honestly viewed his work as an attempt to apply a more properly scientific approach to law and morality than that which had been attempted by Montesquieu. Given his preferences for simplification, universality, and starting from unimpeachable first principles, Helvetius argued that in order to understand the law, we must focus initially on the chief end of legislation, which is to provide for the greatest possible happiness of the members of a society. We must, moreover, begin by considering what constitutes human nature and human happiness. So we return once again to Condillac and Hartley's questions of how humans come to have ideas and to associate them with pleasures and pains, and of how the happiness of a single individual is related to the happiness of the other members of society.

Though Helvetius self-consciously emphasized the implications of sensationalist/associationist psychology for morality and social life, he brought to his task two perspectives which gave his work a very different thrust than associationism had at the hands of Hartley and Condillac. In the first place, he shared with his early friends, Fontenelle and Voltaire, a much deeper anticlerical attitude and an associated antagonism to all organized religion that went far beyond even Hume's thoroughgoing but benign skepticism. Indeed, Helvetius seemed to have accepted the theory set forth by Fontenelle in his *Histoire des oracles* that religions are most often an imposition on the common people by a priestly class concerned primarily with its own power and aggrandizement.[63] In *De l'Esprit* his unmistakable anticlerical stance was mitigated by a formal "admission" of the truth of Christian revelation which contains "the perfection of human morals."[64] But unlike Hartley or Condillac, Helvetius made no positive use of religious beliefs at all. He did not even accept the widely held notion, voiced by Voltaire, that were Christianity not true, its doctrine of an afterlife would have had to have been invented to keep the common folk fearful and in line. Helvetius had an abiding hatred of fraud and deceit which would not allow him to accept the two-tiered morality suggested by Voltaire's religious skepticism. He acknowledged no life after death and thus totally abandoned the notion that otherworldly rewards and punishments have a bearing on morality. Whatever rewards or punishments were to be associated with a moral or immoral act—and it was the act rather than the motive which, for Helvetius as for Smith and Hartley, was to be judged[65]—they had to be recognizably of this world.

In the second place, probably from his association with the Physiocrats, Helvetius derived a more sophisticated understanding of political economy and a greater interest in the role of economic phenomena in the pursuit of human happiness than that of his psychological predecessors. Helvetius did not accept the political economists' confidence that wealth was an adequate measure of human well-being. Nor did he admit the ability of free market mechanisms to maximize wealth. Moreover, he rejected the economists' uncritical acceptance of the notion that unequal division of property was a necessary condition for the growth of wealth.

Instead, he saw inequalities of wealth as among the most important causes of unhappiness in the world. But Helvetius *did* fully accept the economists' emphasis on egoistic self-interest as the sole driving force of human activity, placing this emphasis on a new foundation in associationist psychology. Furthermore, he did reassess an important series of economic issues in light of new insights gained from his psychology, with startling and important consequences.

Both John Locke and David Hume had contributed to political economy as well as to associationist psychology; but it is curious that for both, the domains of economics and psychologically grounded morality remained unintegrated. Helvetius managed to bring the two traditions together in a fruitful way. Like Hartley, but unlike Condillac, Helvetius accepted the fully determinist implications of associationist psychology. It was literally impossible for him to believe that any behaviors could be without antecedent motives or causes; and those causes were all passions of one kind or another. But since passion is to the moral world what motion is to the physical world[66]—i.e., the cause of all events—and since passions are, by definition, all *individual* desires to attain pleasures and avoid pains, Helvetius could not follow Hartley into an admission that any human motives could transcend self-interest. Helvetius admitted the existence of social passions; but he was convinced that all such passions, like friendship and love of country, must ultimately be grounded in self-interest or "self-love" rather than in self-denial:

> I say that all men tend only towards their happiness; that it is a tendency from which they cannot be diverted; that the attempt would be fruitless, and even the success dangerous.[67]

Given that private pleasure and pain are "the only movers of the moral universe," Helvetius was convinced that "self-love is the only basis for a useful morality."[68] But self-love is by itself neither good nor bad, neither moral nor immoral; it is simply the source of all virtues *and* vices; so how can one make it the foundation of morality and virtue? In order to understand how virtue can be based on self-love, we must have some notion of virtue, and this notion too, Helvetius shared with Hartley. "Public utility," "the general happiness," or the "public welfare" is the aim of all just and virtuous acts.[69] And this public utility consists in the aggregate private happiness of all members of the relevant society.

Unlike Hartley, Helvetius could find no reason to believe that there is a natural identification of private and public happiness. Indeed, he was intensely aware that individual and class interests within most societies constantly conflict with one another. The key to making men virtuous must then lie in establishing or legislating a set of social arrangements in which men *perceive* an identification between their self-interests and those of the society as a whole, so that "the idea of virtue is so united to that of happiness, and the idea of vice to that of contempt, that they are hurried away by a lively sensation" to perform actions that seem to be in their interest even though they may not truly be so. Thus Helvetius returned to a vision of publicly constrained private behavior much like that advocated by James Harrington. Under such a set of circumstances,

> the virtuous man is not . . . he who sacrifices his pleasure, habits, and strongest passions to the public welfare, since it is impossible that such a man should exist; but he whose strongest passion is so conformable to the general interest, that he is almost constantly necessitated to be virtuous.[70]

Throughout *De l'Esprit,* Helvetius reiterates this fundamental theme—i.e., that the goal of the moralist is not to decry the selfish nature of men but to teach the legislators how to unify private and public interests by attaching appropriate rewards to virtuous acts and punishments to vicious ones.[71]

Helvetius is even more concerned than Hartley to establish the fundamental equality of all persons—in terms of both their abilities and their needs. He purports to be able to demonstrate that the *natural* differences among normal human beings are totally insignificant in comparison with their *educations,* broadly conceived, in explaining the observed differences among them. Apparent differences in the delicacy of sensations, in the capacity of the memory, in the intensity of passions, and in the attention that different persons seem able to bring to bear on different topics, can all be accounted for in terms of the training they receive and the circumstances in which they are placed, rather than in terms of any innate differences in capacities. It follows both that there are no fundamental, innate sources of merit which might distinguish some men or women from others and that within the limits of human nature in general almost any person is capable of being educated to become almost as passionate, as docile, as scholarly, or as virtuous as any other. "The great inequality—observable in mankind," he concludes, "only depends on the different education they receive, and the unknown and varied chain of circumstances in which they are placed."[72]

One consequence of the native equality of all humans, according to Helvetius, is that "Nature, who has engraved on all hearts the *sensation* of primitive equality, has placed an eternal seed of *hatred* between the great and the little."[73] In the long run, then, no society which is divided into a dominant class and an exploited class—whether such distinctions are grounded in economic conditions, in differences between a politically active and politically passive group, or between the priesthood and the laity—can be one in which aggregate happiness is maximized. The best society must be one in which the distribution of wealth is nearly equitable,[74] in which all inhabitants have a "share in the management of public affairs,"[75] and in which there exists no dominant priesthood requiring "an arbitrary morality" in place of one based on public utility.[76]

Since Helvetius's analysis of the relationship between economic equity and human happiness was both more extensive and, in the long run, more important than his concerns with political participation and religious institutions, they deserve special consideration here. In *De l'Esprit,* Helvetius offered a brief discussion of the economic aspects of happiness in connection with a discussion of the notion of luxury, but in *De l'Homme* he launched a bitter critique of all money-based economies. Starting from the arguments of Cantillon and the Physiocrats, Helvetius first offered what he considered the best possible defense of luxury, or the employments of riches on "superfluities." The demand for luxury stimulates the economy, provides work for artisans and laborers, "softens manners," "creates new diversions," and in general, "causes life to circulate through all the members of the state."[77] At this point Helvetius argued that luxury becomes destructive only to the extent that the distribution of wealth becomes too unequal and that a nation "divides itself into two classes—one abounding in superfluities, the other wanting necessities."[78] Unfortunately, Helvetius argued, economic inequality is not only self-perpetuating but is also self-intensifying. It is always in the interest of rich proprietors to lower the wages of workers, and it is always possible for them to withhold work until laborers are willing to accept their bare subsistence wages. Thus over time it comes to be the case that "seven or eight millions of people languish in misery, and five or six thousand riot in an opulence which renders them odious, without augmenting their happiness."[79] Unlike Adam Smith, who considered the same problem but convinced himself that even the poor in commercial societies enjoy a higher standard of living than members of pre-commercial societies, Helvetius was certain that the masses in commercial society are, in fact, "more unhappy than the savage nations, which are held in such contempt by the civilized."[80] This was so, Helvetius insisted, because happiness was *not* simply measured in terms of wealth; and the savage, unlike the modern poor, feared no oppressive lord, paid no excessive taxes, led a healthy life, and enjoyed the satisfactions of equality and liberty.

If it were the case that the happiness of the few was really increased at the expense of the many, there might possibly be some justification for great divergences in wealth; but Helvetius was able to show (to his own satisfaction at least) that the rich suffer almost as much as the poor as a consequence of huge inequalities in wealth. All pleasures known to mankind are ultimately associated with food, warmth, and, above all, sex.[81] Up to the point at which a person's wealth allows him to satisfy his or her natural passions for food, protection from the elements, and sex, its increase also increases happiness. But there is a point beyond which wealth must not only offer fewer pleasures, but also generate increasing anxiety and ill-health.[82] Thus great inequalities in wealth must produce lowered happiness among the rich as well as the poor. Helvetius attacked even the long standing argument that increases in personal and public wealth can at least provide security against indigence and external threats. For an individual, he argued, excessive wealth produces enemies faster than it produces security. With respect to national wealth, Helvetius reiterated Cantillon's analysis of trade cycles to demonstrate that the accumulation of wealth in a country must inevitably lead to impoverishment and weakness in the long run. Only the poor person or nation which possesses nothing desired by others can be truly secure.[83]

When Helvetius returned to discuss the relationship between distributive justice and happiness in *De l'Homme,* his basic perspective remained the same; but his analysis was enriched by a greater emphasis on class interests, a new concern for the role of money in corrupting societies, and a new understanding of labor not merely as a source of wealth, but also as an avenue for the achievement of a newly recognized pleasure associated with what Helvetius called "the exercise of talents" and which others call "self-expression."[84] Each of these features was to be extensively expanded in the Utopian Socialist tradition which emerged out of the period of the French Revolution.

"There is no country," Helvetius insisted, "where the order of common citizens, always oppressed, and rarely oppressors, do[es] not love and esteem virtue. Their interest leads them to it."[85] Moreover, he suggested elsewhere, there is only one interest that *never* opposes that of the public interest, and that is the interest of the oppressed common people.[86] That this is so follows from the great numerical superiority of the common people coupled with the very definition of public interest. On the one hand, every improvement of the public happiness must imply an increase in the happiness of the common people. On the other hand, it seems to be in the interests of the great "to be unjust with impunity [and] . . . to stifle in the hearts of men every sentiment of equity,"[87] for only by maintaining the differentials of wealth and power which they hold can the great retain their supposedly advantageous position. Class conflict between the order of the rich and powerful and that of the common man who stands for the public is thus the inevitable consequence of an economy in which great inequalities of wealth exist.

At this point Helvetius attempted a piece of typical philosophical history to explain how societies almost inevitably come to develop inequalities of wealth and consequent divisions of class interest. We are asked to imagine a few families arriving at an uninhabited island, more than large enough to support them. At first, each family will appropriate just as much land as it can effectively cultivate; and though the stronger and more industrious may accumulate slightly more land and goods, the differences in wealth will be small.[88] Soon the families will reproduce until there are more workers than can find agricultural work. Initially, these may find work as artisans, providing implements and other goods needed in their agrarian society. At some point the population will grow beyond the labor needs of an agricultural society, and some may leave to find work elsewhere, whereas some may turn to theft to meet their needs; but a few will "invent some new article of commerce, in exchange for which his fellow citizens may supply his wants."[89]

If the desire for the product provided by one of these entrepreneurs becomes sufficiently great, he will collect together more of the surplus population at a site favored because of its easy access to raw materials and transportation and begin a manufacturing enterprise. Soon, other entrepreneurs will locate nearby because of the favorable location, and a town will form. As the commerce of the town prospers, it will draw riches to it, which in turn will draw the purveyors of pleasures and the wealthier proprietors of land who want to spend at least part of the year enjoying the greater range of luxuries available. As more and more of the surplus population migrates to the town in a search for work, the labor supply inevitably exceeds the demand. Competition for work drives the wages of labor down and the profits of the entrepreneurs up, creating an ever greater disparity in wealth. Crime becomes common; harsh and repressive laws are passed. Thus, Helvetius argued, "the unequal partition of the national wealth, and the too great increase of men without property, producing at the same time an empire of vices and sanguinary laws, at last develop those seeds of despotism, which ought to be regarded as a new effect of the same cause."[90]

The details of how political despotism arises out of economic inequality we need not follow here. It is enough to point out that despotism becomes inevitable only when nations give up the direct participation of all citizens in political decisions and turn over that function to some group of governors who can exploit the division of interests among the governed to expand their own authority. In this way Helvetius backed into his position of support for direct— not representative—democratic government.[91]

It might seem that all of the disastrous developments that emerge in society could be avoided by a simple equitable redistribution of property; but Helvetius did not see that as a practical option, both because it must go against the ingrained habits of the members of a commercial society and because, in the long run, the unequal distribution of wealth is a necessary consequence of the introduction of *money* into a state.[92] As long as there exists a money-based economy, if wealth is redistributed it must eventually become unequally distributed once again. If commerce could be carried out through the direct exchange of commodities, luxuries might be fewer; but people could still be "wholesomely fed and well clothed." Moreover, a country without money must be a country without despotism, for "the prince who raises his taxes in kind . . . can seldom raise and keep in pay a number of men sufficient to put his people in fetters."[93] In a country without money, public esteem and glory are major rewards for service, and they are at the disposal of the public at large; so there is a strong public hold over private interests. In countries where money is used to reward, however, rewards will more often be in the hands of the enemies to public happiness than in the hands of the public.[94] Finally, money economies are subject to the trade cycles which Helvetius had pointed out in *De l'Esprit,* and they make possible destructive public indebtedness. Under present forms of government, Helvetius thus concluded, money is *inevitably* the source of corruption and destruction.[95]

Since all of his analyses seem to show that widespread happiness is impossible in money-based commercial societies governed according to present forms, and since the use of money, the growth of commerce, and the development of present forms of government seem to have emerged as part of an almost inevitable historical development, Helvetius realized that he had an obligation to suggest the conditions under which men might live in happiness and to explain both what went wrong and how it could be changed. To the question of under what conditions men might be happy, Helvetius initially claimed that "When every citizen has some property, is in a certain state of ease, and can, by seven or eight hours of labor, abundantly supply his own wants and those of his family; they are then all as happy as they can be."[96] He demonstrated the truth of this claim by exploring how men can and do employ their time. Roughly half of every day is spent in eating, drinking, sexual activity, and sleep. These basic needs can all be met with very modest resources, and there is no evidence that great riches in any way increase the enjoyments associated with meeting them. Supposing that all have the ability to meet these basic needs, whether men are happy or miserable will then largely depend on how they use the remaining half of their days.[97]

Helvetius recognized two basic ways of filling the time during which we are not responding to our most basic needs: either with labor or with idleness. Labor is generally regarded as evil, Helvetius insisted, only because under most present governments, those who labor for their necessities are forced to labor excessively to support not only themselves, but also the idle classes. The very idea of labor has become associated with pain, but labor is not intrinsically painful. In fact, "when it is pursued without remarkable fatigue, it is in itself an advantage."[98] Any number of men—especially artisans—demonstrate this fact by continuing to work when there is no need because their labor is itself a source of pleasure. Helvetius's explanation of how labor becomes associated with pleasure is not particularly compelling. He suggested that it is largely a "pleasure of expectation annexed to the payment of his work" which becomes associated with a man's labor and transforms it into an intrinsically pleasurable activity.[99] But even this derivative pleasure is not available to the idle man who "is forced to wait 'til nature excites in him some fresh desire" before he can experience pleasure. It is thus a "disgust of idleness" that "fills up the interval between a satisfied want and a rising want" in the opulent and idle man. As a consequence, "the wealthy idler experiences a thousand instances of discontent, while the laboring man enjoys the continual pleasure of fresh expectations," and men find their greatest happiness when they are above indigence yet not exposed to the discontent of the idle rich, i.e., when they can provide for their wants by a moderate and pleasurable labor.[100] If the laws of the nation assigned some property to every individual, they would "snatch the poor from the horror of indigence, and the rich from the misery of discontent," rendering all more happy. But without additional dramatic changes in education, broadly conceived, such laws could not sustain an acceptable division of resources; because from their infancy almost all persons in present societies have been taught to associate the idea of riches with that of happiness.[101] Moreover, many have been taught to multiply artificially their desires beyond the ability of any attainable level of resources to satisfy them.

Why did we come to slip into circumstances in which laws and education effectively conspire against achieving happiness? If, somehow, human beings had been capable of developing the science of associationist psychology and the derivative science of utilitarian morality before they had begun to join into societies for mutual protection, then they might have guided the development of laws and customs and avoided the error-filled historical process which has created the present. The adoption of these laws and customs, however, could not wait for the science; as a consequence, legal systems and education developed in connection with the expression of class interests which almost uniformly subverted the broader public interest. According to Helvetius, the historical priority of class over public interests in guiding the behavior of individuals follows from the application of sensationalist/associationist psychology. So he devoted extensive and critically placed segments of both *De l'Esprit* and *De l'Homme* to explaining the emergence, power, and tenacity of those special group interests which displace the natural human interests in moderate amounts of food, shelter, and erotic activity, and preempt broader public interests in the minds of most humans.

In *De l'Esprit,* Helvetius considered the general issue of the relationship between the interests of the public and those of what he called "private societies," where private societies were understood to include such units as the family, occupational groupings, and the Court of the French King.[102] From the standpoint of the broad concerns of this work, Helvetius discussed the relationship between the interests of subcultures or specialist groups within a culture and the interests of the broader culture of which they are a part, rather than class interests in any narrowly economic sense. His major point is that as we grow and receive our education and training, we associate almost exclusively with members of those "private societies" or subcultures to which we belong; so the ideas and values with which we associate our own self-interest tend to be those of the relevant private societies. Moreover, our initial acceptance of the ideas and attitudes of the private group creates a situation that leads us to continue to seek the company of like-minded persons rather than to expand our social associations. Since flattery and esteem provide us with pleasures, whereas derision and contempt cause us pain, we soon learn to limit our associations to those with whom we agree. As Helvetius argued:

> There is scarcely a man so stupid, but, if he pays a certain attention to the choice of his company, may spend his life amidst a concert of praises, uttered by sincere admirers; while there is not a man of sense who, if he promiscuously joins in different companies, will not be successively treated as a fool, a wise man, as agreeable and tiresome, as stupid and a man of genius.[103]

At one point, Helvetius's emphasis on the extent to which our ideas—and our very language—is determined by the interests of the special groups we associate with led him to express the very real fear that it might be impossible to create the kind of philosophical language that Hartley and Condillac had shown must be the precondition for the objective or "disinterested" moral science that could in turn lead to the perfectibility of law and the human condition.[104] Setting aside this long-term theoretical problem, he argued that it is presently the case that almost no modern "public" society incorporates either formal or informal mechanisms to educate its members for "citizenship" to counteract the impact of special interest groups. Only a few private societies "of the academical kind" self-consciously seek to inculcate the "absolute detachment from personal interest" and the "profound study of the science of legislation" which truly allow their members to recognize and work for the public interest.[105]

In the long run, Helvetius hoped that "education for citizenship" might become such a large part of the education of all men and women that ever larger portions of the public would identify with public rather than with special class interests.[106] But he expected neither that political and religious authorities would initiate such civic education nor that if it were initiated it would soon bring about any dramatic improvements. In the first place, those in power have a vested interest in preventing the masses from recognizing their oppressed state. Even if by some accident a well-grounded moral education were offered in some isolated academic setting, it would be counteracted by the more powerful educational forces operating through the laws and manners of the nation. "When the precepts of these two parts of education are contradictory," Helvetius acknowledged, "those of the former become void."[107]

To understand why scientific moralists have not already initiated the legal and education reforms which should improve and ultimately perfect society, Helvetius returned to consider the private interests of three groups whose importance lies in the fact that they are sufficiently powerful to impose their own wills upon other groups and to keep the populace in a state of ignorance of those true principles of morality which "would have opened their eyes with respect to their own misfortunes and their rights, and have armed them against injustice."[108] The three classes or groups are the economically privileged, the politically powerful, and the religiously dominant. Members of each group "forced by their private interest to establish laws contrary to the general good, have been very sensible that their power had no other foundation than the ignorance and weakness of mankind."[109] If reform is ever to occur, Helvetius concluded, it will only be when these groups have been *unmasked* and shown to be "the most cruel enemies of human beings."[110]

In *De l'Esprit,* Helvetius reserved his most bitter criticisms for religious elites, whose obsession with preserving their own power has most frequently led not to peace and love but to intolerance and death:

> If we cast our eyes to the north, the south, the east, and the west, we everywhere see the sacred knife of religion held up to the breasts of

women, children, and old men; the earth smoking with the blood of
victims sacrificed to the false Gods or to the Supreme Being; every
place offers nothing to the sight but the vast, the horrible, carnage
caused by a want of toleration.[111]

In *De l'Homme,* Helvetius continued his anticlerical attitudes. He
argued that the religious elite has the most to lose from any general
enlightenment regarding morals because its position is grounded
solely in the maintenance of an arbitrary, false, and supposedly revealed morality which stands opposed to utilitarian morality.[112] The
religious authorities must thus be expected to repress all attempts
to explore the nature of human happiness from the standpoint of
corporeal sensibility, for "they are sensible that the people, when
enlightened by that study, will measure the esteem or contempt
due to different actions by the scale of public utility; and what
respect will they then have for bonzes, bramins and their pretended
sanctity."[113]

   While he continued his antagonism toward dogmatic religion
in *De l'Homme,* Helvetius turned to an equally hostile critique of
oppressive governments, including the French monarchy. He had
attacked despotic governments in *De l'Esprit,* but had openly denied that the French government was despotic. When he wrote *De
l'Homme,* however, he complained that "my country has at length
submitted to the yoke of despotism" and that "the disorder which I
had hoped in some measure to remedy, is become incurable."[114] Not
only are despotic governments cruel but they also deny citizens
access to that knowledge which might help them to improve their
lot. Under these conditions, Helvetius argued, the people have not
so much a right as an obligation to revolt:

> When men fall under despotism they must make efforts to shake it
> off, and those efforts are, at that period, the only property the unfor-
> tunate people have left. The height of misery is not to be able to de-
> liver ourselves from it, and to suffer without daring to complain.
> Where is the man barbarous enough to give the name of peace to
> the silence, the forced tranquility of slavery! It is indeed peace, but
> it is the peace of the tomb.[115]

Thus Helvetius suggested that some kind of insurrection against
the tyranny of the political and religious authorities was necessary
to break the pattern of powerful special interests which blocked the
emergence of an egalitarian and improved society.

   Starting from the apparently innocuous principles of associationist psychology, Helvetius managed to follow his analysis into a justification of open rebellion by the poor and weak against their rich and powerful oppressors. Moreover, this justification of rebellion was not based on any claim—like that which Locke and the social contract theorists had made—that rebellion could only be justified by the prior violation of a contract or law by the sovereign power. Helvetius realized that the rich and powerful had been able so to structure the laws in their favor that they did not need to violate them in order to oppress the poor. The revolution needed not merely to reject those who violated the laws, but the very system of legal arrangements which allowed oppression and exploitation in the first place. Unlike England in the age of the Glorious Revolution, France in the eighteenth century needed a change not of monarchs under an old and still venerated constitution, but of constitution.

In terms of their long-range impact, Helvetius's writings must be considered in three different contexts. His extreme emphases on equality, on the historical development of class conflict as the central feature of money-based commercial societies, and on the need for the consequent unmasking of class interests and their replacement by an objective understanding of social relations by a special class of "disinterested" scientific moralists played a central and acknowledged role in the emergence both of Utopian socialism after the French Revolution and of "scientific" socialism in the nineteenth century. His arguments for the development of a utilitarian morality—i.e., a morality which sought to maximize human happiness, understood in terms of the pleasures and pains associated with "corporeal sensitivity"—provided a central theoretical stimulus to the utilitarian foundations of nineteenth-century liberalism as they were developed by Jeremy Bentham and John Stuart Mill in England and by the "Ideologues" in France. Finally, and most immediately, Helvetius's writings, with their emphases on a more equitable distribution of wealth, on the oppressive character of the Catholic Church, on the "despotic" character of the French monarchy, and on the consequent need for legal and constitutional change, played a major role in creating the intellectual perspective of the Jacobins and other radical leaders during the period immediately preceding and during the French Revolution.

The role of the philosophers—especially of the scientific tendencies of their thought—in preparing for and shaping the French Revolution has been debated by historians and social theorists since the early stages of the Revolution itself.[116] There were at least some revolutionaries who viewed the new scientific approaches to man and society as fundamental to the emergence of those attitudes that led to the Revolution. Such, for example, were Condorcet and Tom Paine. Moreover, the most eloquent opponents of the Revolution and founders of the modern European conservative political tradition—men like Edmund Burke in England and Joseph de Maistre in France—tended to place the central *blame* for the Revolution on the secular scientific social theorists of the eighteenth century. We shall return to this general debate more extensively when we consider conservative reactions to the growth of scientific liberalism. At this point we are more concerned with considering the relation of Helvetian analyses to pre-Revolutionary and Revolutionary socialist tendencies and the impact of socialist thought on the Revolution. We must begin by admitting that *socialism* and *socialisme* were terms coined only during the 1820s and 1830s by groups associated with Robert Owen in England and Saint-Simon in France.[117] So by definition, socialism did not exist as a self-conscious movement until well after the French Revolution. But if we understand socialism (1) as dedicated to the abolition, or at least the more equitable distribution of private property, (2) as grounded in an insistence on human equality, (3) as opposed to emphases on individual competitive and crudely self-interested activity in favor of concern for the welfare of society as a whole, and (4) as closely linked with secular theories of human happiness, then it is clear not only that socialist ideals were articulated by French theorists beginning in the 1750s, but also that those theories played an important role in Revolutionary rhetoric and in some of the concrete acts of the successive Revolutionary governments.

For the most part the ideological thrust of the Revolution was more liberal than socialist. Membership in the various assemblies and conventions as well as in the *de facto* executive councils, including the Parisian Committee of Public Safety, was dominated by middle-class figures—lawyers, physicians, and scientists—most of whom sought to protect private property and ensure the exercise of supposedly natural rights to individual liberty in the pursuit of happiness. During a period of Jacobin dominance, however, there was a special concern for the rights of the poor and with imposing limits on private property. In 1793 the radicals pushed through a law which sought to limit the differences between the wealthy and the poor by forcing an equal division of property among all heirs to any inheritance. Moreover, the pressure to retain this law was sufficient to get it incorporated into the Napoleonic code; so it remained part of French law into the twentieth century.[118]

When most of the socialist participants in revolutionary debates—men like François Babeuf, Mallet du Pan, Brissot, François Boissel, Camille Desmoulins, St. Just, Linguet, and Barnave—appealed explicitly to earlier theorists rather than to presumably self-evident principles, they appealed more often to Rousseau's *Discourse on Inequality* (1749), André Morelly's *Code de la nature* (1755), and the abbé G. B. Mably's *De la legislation ou principes des lois* (1768), than to Helvetius's *De l'Esprit* and *De l'Homme*. The rhetoric of Rousseau, Morelly, and Mably was more polished; their opposition to private property, more insistent; and their anti-clerical, but not anti-Christian attitudes, less offensive to men who were more inclined than Helvetius to distinguish between the Gallican Catholic Church in particular and Christian religion in general. But when the Convention acted in 1793, it was to incorporate the limitations on property proposed by Helvetius rather than the abolition of property which Morelly and Mably favored. Moreover, recent work has suggested that Morelly's and Helvetius's analyses are so similar that they must have been developed together,[119] and there is direct evidence for their close personal friendship and mutual support at least after 1760.

Perhaps more important, it seems likely from the details of their analyses of class conflict and the special role of population increase in spurring the disparity between rich and poor that Mably and Barnave owed substantial direct debts to Helvetius, and it is also clear that many elements of Boissel's *Cathéchisme du genre human* (1794) are derived directly from the "Moral Catechism" presented in Helvetius's *De l'Homme* (1774).[120] So to the extent that socialist ideals did have some impact during the Revolution, Helvetius's works seem to have had a substantial role to play.

LIBERAL IDEOLOGY, EDUCATIONAL REFORM, AND ASSOCIATIONIST PSYCHOLOGY

In many ways Condorcet followed Helvetius's analysis of both the foundations of morality and the class interests which made it so difficult to ground legislation in a "true" knowledge of human interests. Thus, for example, he saw in associationist psychology the foundation for morality:

> [T]he analysis of our sentiments made it possible for us to discover, in our faculty of experiencing pleasure and pain the origin of our moral ideas; the foundation of the general truths, resulting from these ideas, which determine the immutable and necessary laws of justice and injustice; and finally the reasons for directing our conduct in conformity with these laws, reasons founded on the very nature of our sensibility, on what could in a sense be called our moral constitution.[121]

Moreover, both he and his friend Turgot fully accepted Helvetius's analysis of the interests that religious and political authorities have in being intolerant and maintaining governmental abuses. But Condorcet could not stomach the thought that humans were totally selfish. So he, with his mentor Turgot, turned away from what they termed Helvetius's "crude utilitarianism."[122]

Like Hartley, but independent of him, Condorcet argued that our interests are not limited to *self*-interests and that we have an interest in being just and virtuous, "founded on the pain necessarily inflicted on one sensitive being by the idea of evil suffered by another. . . ."[123] That is, we have a *natural* sentiment of sympathy or benevolence. Given our sense of benevolence, Helvetius's argument that the function of legislation and education is to manipulate rewards and punishments to bring self-interests into line with the common good was rejected in favor of a more expansive notion of education and a more limited and negative role for legislation. Condorcet conceded that Helvetius was right in seeing that the present legal system allowed self-interest to conflict with justice. But, according to Condorcet, laws placed too much rather than too little emphasis on self-interest. What was needed was a system that allowed more scope to our natural benevolence. The best system of legislation and education would free men to "fulfill by a natural inclination the same duties which today cost him effort and sacrifice."[124] Condorcet's perspective also involved the reversion to an optimistic faith in reason which the earlier associationists' emphasis on pleasure, pain, passion, and mere habit had abandoned. This optimistic faith in reason played its own role in the "scientific" foundations of eighteenth-century liberalism which should be acknowledged and understood, even if we can no longer sustain it.

The starting point of Helvetius's utilitarian ethics and morality had been a definition of "human" that focused on the uniformity of "corporeal sensibility" and the associated pleasures and pains.

The starting point for Condorcet's social and moral theories, in contrast, was a definition of humanity that focused on the uniformity of human *reason*:

> The idea of justice, of right, necessarily forms in the same manner in all sensitive beings capable of making the necessary combinations to acquire these ideas. They will therefore be uniform. . . . [Moral propositions] are the necessary result of the properties of sensitive beings *endowed with reason*; from which it follows that it is sufficient to suppose the existence of these beings for the propositions founded on these notions to be true.[125]

It may be reassuring to some to recognize that there are uniform moral laws which derive from the nature of man, but such a recognition is vacuous until the content of those moral laws is discovered or recognized. Condorcet began to explore their content in 1786 in a brief essay entitled *De l'influence de la révolution d'Amérique (On the Influence of the American Revolution)*. It follows, he insisted, from the "natural and primitive equality of man" that all must equally enjoy four basic natural rights—the right to personal liberty and security, the right to the free and secure enjoyment of one's property, the right to the equal protection of the laws, and the right to participate in the formulation of policy and the enactment of laws.[126] As the forerunner of Condorcet's first draft of what eventually emerged as the National Assembly's *Declaration of the Rights of Man and of the Citizen,* this initial attempt to establish moral laws on a firm basis deserves careful consideration. The first question one might reasonably ask is how Condorcet could possibly have viewed his list of rights as anything but a set of arbitrary claims. In what sense could they have seemed to him "scientific"? Rightly or wrongly, it seems that he viewed the existence of these basic rights as true in the same sense that the basic postulates of mathematics are true: once they are articulated, they are simply recognized immediately as agreeing with the ideas of humanity as sensitive and rational. But what about the claim that men have a primitive and fundamental equality? This claim is especially odd in view of the fact that Condorcet explicitly criticized Helvetius's argument that men are physically equal and made unequal only by education. Condorcet admits that men and women are born with greater or lesser physical strengths and intellectual abilities, yet they are identical as humans in much the same way that all circles

are the same. Any mathematical proposition that is true for one circle is true for all. By the same token, any moral proposition true for
one human must be true for all.[127]

Different humans may have different physical and mental
abilities, but all must be equally free to exercise those talents which
they possess without interference—so long as they do not interfere
with the liberty of others. Similarly, some people may have more
property and some less, but every person's right to enjoy and dispose of that which he owns must be the same. Thus it follows that
one of the most basic laws of civil society is that old favorite of the
liberal economists: humans must be free to enter into exchanges of
property without interference of any kind.

Equality before the law seems a relatively unproblematic consequence of the primitive equality of men, but the equal right to participate in public decisions creates a major problem for Condorcet.
He cannot deny political participation to any because of the basic
equality of men, but since laws can exercise their function of protecting the rights of citizens and leaving them free to exercise their
benevolence only if they are grounded in rational and scientific
analyses of society, political rights should belong only to the enlightened and not to the ignorant and prejudiced. In order to get out
of this bind, Condorcet offers a two-stage consideration. First, in the
*Essai sur l'application de l'analyse à la probabilité des décisions
rendues à la pluralité des voix,* he offers *indirect* political participation through the choice of representatives, who are presumed to be
sufficiently enlightened to reach correct decisions. Second, he argues that through appropriate education all men can *eventually*
come to attain that equality in fact which they have in principal at
birth.

This last sentence demands both explanation and qualification. Condorcet does not really believe that most men and women
can *ever* attain a level of knowledge and rationality sufficient to
warrant their active participation in the formulation of law and
social policy. Such direct policy formulating activities must always
remain the domain of an elite of extraordinarily talented persons
organized to be free of interference by any special interests and
trained to allow the fullest expression of their abilities. What is possible for the rest of the citizenry is that they be educated to a level
which will allow them to give truly *informed* consent to the laws
and policies established by the elite. Condorcet is convinced that
in politics, as in mathematics, though there may be only a few men

capable of generating new truths and proofs, many are capable of recognizing the validity of a demonstration when it is presented to them. Informed consent is a very real and critical form of political participation for Condorcet, for it is the existence of an enlightened and informed citizenry, constantly monitoring the decisions of the policy-making elite, that guarantees that the elite will never be tempted to substitute their own special class-interest for the interest of the public.

Condorcet's presentation of the relative roles of a policy-making elite and a consenting citizenry reflects a central characteristic of French republican governance well into the twentieth century. The sociologist Dorothy Nelkin reports in comparing mid-twentieth-century debates about nuclear power in a number of national contexts, that France is unique in the degree to which its population is still willing to consent to decisions made largely by scientific experts within the government.[128] Perhaps even more important, the contending emphases on policy formulation by a scientific (or technocratic) elite and on democratic consent has been an ongoing central feature of socialist theory and practice from the emergence of Utopian socialism in the works of Condorcet's immediate inheritor, Saint-Simon, to the recent renunciation of the Cultural Revolution in post-Maoist China. We shall return frequently to this theme in future chapters; but it is important to recognize its initial formulation and resolution at the hands of Condorcet; for the problem of how or whether meaningfully democratic governance is possible in a world in which only a small elite has the knowledge and training—i.e., the expertise—to bring their intelligence to bare on formulating policy issues is one of the most critical problems facing both modern capitalism and modern socialism.

We have seen that for Condorcet, as for all other associationist theorists, an educated citizenry becomes one of the key preconditions for long-term social and political progress. In a series of five *Mémoires* on public instruction published during 1791 and 1792, he developed a detailed plan for public education. Then, he chaired the National Assembly's Committee on Public Instruction, whose *Report on the General Organization of Public Instruction,* sought to make free public education for all citizens one of the chief responsibilities of the nation. Condorcet's proposals were far too elitist to survive the extreme radicalism of the Revolution, and his insistence on treating law and politics as subjects for critical analysis rather than blind acceptance made them anathema during the Napoleonic

Restoration. Condorcet's visions, however, did guide the development of at least higher education during the period after the Reign
of Terror. Moreover, they provided the theoretical underpinnings for
the broad scheme of liberal education that was finally instituted
in France under the Third Republic at the end of the nineteenth
century[129] and that has shaped American education into the twentieth century.[130]

The opening sentence of the National Assembly *Report* lays
out Condorcet's basic educational ideals:

> Public education is a duty that society owes to all citizens to assure
> to each one the opportunity of making himself more efficient in his
> work, of making himself more capable of performing his civic func-
> tions; and of developing, to the highest degree, the talents one has
> received from nature, thereby establishing among the citizens ac-
> tual equality in order to make real the political equality decreed by
> law.[131]

Education must be public at the lower levels in order to ensure
that access is not limited to the wealthy. Even more important, it
has to be universal to counteract the mind-numbing consequences
of industrialization and the division of labor. Condorcet was deeply
moved by Smith's fears that the division of labor would tend to
create a working class of virtually unthinking automata unless an
extensive educational system helped to mitigate the effects of the
monotony of the workplace.[132] Moreover, like Hartley, Condorcet
seems to have become convinced that the pleasures of the intellect
were greater than those associated with the senses alone. So it followed that education could help eliminate the vices which arise
from the tendency of the ignorant to escape their boredom "through
the senses rather than through ideas."[133] Above all, universal education would eliminate the blind dependency of the masses on the
educated elite which made despotism possible. Where the ordinary
citizen "knows ordinary arithmetic, that he needs in daily life, [he]
is not dependent on the learned mathematician." Under those circumstances and those alone, "the latter's talents can be of the greatest use to him, without danger to the enjoyment of his rights."[134]

Applied to political issues, this notion that education should
protect the enjoyment of rights led to a focus on critical evaluation
rather than mere indoctrination. The aim of education is not to in-
still admiration for the existing political system but to create a critical attitude toward it. Each generation should not be compelled to

submit to the opinions of its predecessors, but it should be enlightened so that it can govern itself "by its own reason."[135] It is certainly true that all citizens should learn about the constitution and laws of their country, but the constitution and laws should be taught as something to be "explained and analyzed" rather than as dogmas to be accepted without question. Anyone who seeks to "capture the imagination of the child in order to stamp on it images that time cannot erase, so the adult would be attached to the existing political system through blind sentiment" seeks to create "patriotic charlatans" rather than responsible citizens.[136]

In the second part of the *Report on Education,* Condorcet's committee proposed a set of institutions to carry out the educational mission they envisaged. There would be four levels of educational institutions, all presided over by a self-perpetuating National Society of Arts and Sciences modeled on the old Académie des Sciences, which would act to screen education from outside political pressures. Every district would have a free primary school open to all children of both sexes. Over a four-year course, students would learn to read, write, and do basic arithmetic and geometry. They would be introduced to geography and the principles of agriculture, and they would be given moral and political instruction through simple stories and fables. Secondary schools would be provided in every major district. While tuition would be free and state room and board scholarships would be available through competitive examinations, not all students would be able to attend these residential schools where emphasis would be on the pure and natural sciences, midwifery, political economy, foreign languages, the social sciences, and where instruction would involve the use of laboratories and botanical gardens as well as libraries. Scientific education, involving a special emphasis on proper experimental and logical method, was particularly important at this stage because it provided the best protection against error. In third-level *institutes* even greater emphasis would be placed on scientific and technical training, with the incorporation of practical apprenticeships in appropriate trades. Finally, at the fourth level, the major universities of France would offer preparation for all of the liberal professions— law, medicine, the sciences, history, literature—except for theology, which was limited to special seminaries. For those who lacked the talent or ambition to continue beyond the primary educational level, adult education focusing on new developments in law and science would be offered, so that even the common people would never again become "docile instruments in adroit hands," but would constitute a well-informed public.[137]

Though Condorcet's proposals were rejected in 1792, both because control by a National Society of Arts and Sciences seemed to revive too much of the corporate spirit of the old regime and because the Assembly was far too busy prosecuting a war to worry seriously about education, many aspects of his proposals were advocated by the so-called "Ideologues" who did push through plans for a new educational system in 1795. Once again a set of free primary schools were to be established in every canton. Above these an *Ecole centrale* would be provided in each department. In these institutions the six-year curriculum would begin with a two-year section in which three courses were offered—in natural history, in design, and in Latin (grounded in Condillac's analysis of general grammar). Following the first section there would be a two-year section devoted exclusively to the study of mathematics, physics, and chemistry, for not only are these sciences of critical practical importance, but their study trains the mind. Finally, in the third section, students would study the *grammaire générale* or *logique* of Condillac, the science of legislation (according to Helvetius and Condorcet), universal history, and belles lettres. Standing above these *Ecoles centrales* would be the two Grandes Ecoles in Paris: the *Ecole polytechnique,* devoted to mathematics, natural science, and engineering at the most advanced levels; and the *Ecole normale,* devoted to the subtleties of ideology and to training future educators. These schools would be subordinate only to the *Institut de France,* the reborn Académie des Sciences. Within five years, 97 of the 102 Departments of France had established their *Ecoles centrales,* and the *Ecole polytechnique* had become the world's premiere center of science and engineering education.

In spite of the fact that the new system had some success in offering practical preprofessional training in areas that served the needs of a commercializing and industrializing economy, the great plan for universal liberal civic education grounded in sensationalist psychology never really got off the ground. On the one hand, it faced tremendous practical difficulties, for the state failed to provide adequate funding for the all-important primary education that would have been necessary to realize the goals that the Ideologues had for the *Ecoles centrales.* Moreover, there was a critical shortage of acceptable teachers for the *Ecoles centrales;* so many, if not most, of the chairs of *grammaire generale,* legislation, modern history, and the physical and chemical sciences went unfilled. In time the *Ecole normale* might conceivably have produced a cadre of ideologically trained educators, but when Napoleon seized power in 1799, he suppressed the *Ecole normale* because the last thing he wanted to encourage was the spead of a critical, analytic spirit regarding government. Only among the educational leaders of the French Third Republic in the 1880s were the liberal aims of education associated with the debates of the Revolutionary period effectively revived and at least partially implemented in France.

If they did not achieve all of their goals, however, the ideological followers of Condorcet did have a substantial impact on education and politics. As a consequence of the new educational system, Saint-Simon was probably not far wrong when he claimed in 1812 that

> such is the difference in this respect between the state of . . . even thirty years ago and that of today that while in those not distant days, if one wanted to know whether a person had received a distinguished education, one asked, "Does he know his Greek and Latin authors well?"; today one asks: "Is he good at mathematics? Is he familiar with the achievements of physics, of chemistry, of natural history, in short, of the positive sciences and those of observation?"[138]

Furthermore, as we shall see in the next volume, the grand hopes of Helvetius, Condorcet, and the Ideologues for a new morality and politics grounded in the scientific study of human psychology and social institutions, did find important and enthusiastic supporters and developers among students and other intellectuals associated with the *Ecole polytechnique* who transformed them into a variety of programs for social engineering.

SUMMARY

To combat the conservative implications of Montesquieu's sociological analyses of social institutions and legal systems, a third tradition of social science emerged in France—a tradition grounded in the sensationalist/associationist psychology which had been advanced primarily by David Hartley in Britain. Like the Newtonian tradition in natural philosophy, to which its adherents constantly pointed as a model, the associationists sought to account for the phenomena of human behavior, morality, and law by deriving them from one or two basic characteristics of human nature. They had

little patience with the moral and political relativism of Montesquieu's analyses and even less with his willingness to use the existence of human prejudices and superstitions as basic explanatory
principles. That approach was tantamount to viewing the friction
in pulleys as central to an account of the operations of machines,
rather than as the source of minor perturbations of phenomena best
approached by ignoring such issues.

The associationists built upon John Locke's epistemology, revising it by viewing sensations as the sole source of our simple
ideas and the principle of association as the sole means by which
complex ideas were created and by which pleasure was annexed to
some ideas and acts while pain was annexed to others. The associationists concurred in defining man as a sensitive being who
seeks pleasures and shuns pains. All accepted the Hobbesian identification of evil with pain and good with pleasure. And all developed a secular morality that defined those acts as virtuous which
tended to the greater public good—i.e., toward the greatest aggregate pleasure or happiness. Though some associationists, including Hartley and Condillac, remained self-proclaimed Christians,
their doctrines certainly allowed for no concept like original sin;
and all saw man, not as a depraved and fallen being, but either as
a "blank slate," capable of being made vicious or virtuous through
education (Hartley, Helvetius, Hume) or as fundamentally good
(Condorcet).

The associationists did diverge substantially over the question
of whether men were motivated only by self-interests, as the political economists assumed. Hartley developed a theory in which benevolent and virtuous acts—acts which placed the public good
above mere private interests—were the source of men's greatest
pleasures. Sympathy, benevolence, and the other social virtues are
not innate; but they emerge and become dominant through associations made in the course of our social experiences. Thus a natural
confluence of private and public interests occurs without the need
for any special interference. Just as men can be expected to behave
to maximize public wealth if they act in an enlightened fashion to
increase their own, so, too, well-informed men can be expected to
act virtuously as they pursued their own greatest pleasures. Governments do not have to force men to act in the public interest; they
do so naturally if they are taught to recognize their true best interests and left free to do so.

A similar liberal perspective was developed by Condorcet in

France. Condorcet placed more emphasis on the human reason than most associationists, and he saw sympathy and benevolence as primitive rather than derivative social passions, but like Hartley and his English popularizer, Joseph Priestley, Condorcet emphasized the idea that the primary roles of the state were simply to remove the impediments to men's natural exercise of their benevolence and virtue and to educate men to recognize their true interests. Thus the protection of the citizens to act in any manner that does not interfere with the liberty of others becomes the first obligation of governments, and the provision of an education that teaches the critical use of our rational faculties becomes the second.

Claude Adrien Helvetius followed the implications of sensationalist /associationist psychology in a different and more radical direction. More rigorous in his analysis of men's passions, he could not accept the Hartlian notion that the social passions which grew out of self-interest could transcend the self-orientation from which they were constructed. Nor could he accept the notion that men possess some primitive sympathetic or benevolent sense like that posited by Adam Ferguson. Men must inevitably act only in their own perceived self-interest. Social passions certainly do exist, but they are always self-oriented. Love, for example, always directly or indirectly looks toward the gratification of sexual need; and benevolence always looks toward the need of the supposedly altruistic actor for social approval. Moreover, because of the historical circumstances in which human social passions develop, they are almost universally oriented toward special class interests rather than toward general public interests. There are real and objectively understandable conflicts that separate the interests of members of every privileged class from those of the public as a whole; accordingly, only the poor and powerless can be expected to act naturally in the public interest.

As a consequence of his more pessimistic and more historically sensitive approach to human motives—especially to the role of class interests—Helvetius argued that if human happiness was to be maximized, governments and legal systems needed to do far more than to leave men free to follow their natural inclinations. They had to create a set of circumstances which *forced* men to act in the public interest by making it so rewarding to serve the public or so painful to do otherwise that, in effect, there was an "artificial" identification of public and private goods.

In his early work, *De l'Esprit*, Helvetius argued that the

emergence of a better society—more egalitarian, more enlightened, more just—would involve a slow process guided by a "disinterested" class of scientific analysts of law and society not unlike the socially active elite of the Académie des Sciences or the Physiocrats, with whom he was closely associated. As the monarchy and the Church became increasingly repressive in his view, he changed his mind, and he argued in *De l'Homme,* that change must come, if at all, through violent insurrection.

Though not the most popular socialist theoretician of the eighteenth century, Helvetius presented shrewd psychological arguments that provided some of the most powerful supports for the radical activists of the French Revolution and played a key role in the passage of the so-called "agrarian law" of 1793, which demanded the equipartition of estates among heirs to order to reduce disparities of wealth. Moreover, those arguments were self-consciously developed by French and English Utopian socialists after the Revolution and by the Marxist scientific socialists of the mid-nineteenth century, giving Helvetius' works much the same kind of delayed impact in the growth of socialist ideology that those of Hobbes had in the growth of liberal ideology.

There is one final set of emphases and problems which unified those who took seriously the moral implications of associationist psychology, whether they followed them in the direction of liberal or socialist ideologies. The basic principles of associationist psychology implied the fundamental primitive equality of all men and denied all arguments that would justify the existence of inherited privileges. Thus it seemed that there were no grounds for denying political participation to any group or individual. This "democratic" implication of associationist psychology was almost universally acknowledged, becoming a central tenet of political liberalism. At the same time, as members of a scientific elite who saw the scientific analysis of society as a precondition for the formulation of effective governmental policy, all of the associationists shared the classical political theorists' fear of the political participation of the masses whom they viewed as ignorant, superstitious, and susceptible to the appeals of charlatans and demagogues.

Given their confidence in the basic equality of the intellectual abilities of all humanity and the power of education to shape ideas and values, however, the associationists did not retreat into a classical rejection of democracy in favor of rule by a leisured elite with inherited privilege. Instead, they emphasized the need for universal education to create an enlightened citizenry, capable of informed political consent to the policies formulated by a scientific, intellectual elite. As one of the chief educational leaders of the French Revolutionary period, Condorcet produced one of the fullest articulations of what has become the dominant attitude toward education within both modern liberal and socialist ideologies in the *Report of the General Organization of Public Instruction* of the National Assembly. State-sponsored secular education should be offered at the elementary level for all citizens of both sexes. The chief aim of this education should be the development of a critical intellect and the provision of those practical abilities needed to engage in ordinary economic and political life. Beyond this universal level, advanced education, accessible to all those who demonstrate sufficient talent, should be offered to provide society with an intellectual elite capable of generating the scientific, technological, and social knowledge that will ensure the continuing improvement of the human condition.

Science and Aesthetics: The Beautiful to the Sublime, ca. 1600–ca. 1800

7

The *esprit géométrique* and mechanical philosophy had a major impact during the seventeenth and eighteenth centuries on the theory and practice of architecture, landscape gardening, music, painting, and dance—all of which had become closely linked with science by the time of the Renaissance. The most dramatic impact of science upon early modern aesthetics, however, occurred in connection with poetry and drama, which had no significant connection with natural science until the early seventeenth century. In this chapter we will focus exclusively on the scientific stimulus to changing understandings of the nature and function of poetry.

As we suggested in Chapter 1, such philosophers as Descartes, Hobbes, and even Locke were obsessively concerned with the discovery of Truth. And whether they believed that ideas were ultimately traceable to sensation or that some might be innate, all concurred in arguing that it is human reason that allows us to compare and combine ideas to achieve true knowledge in all domains, including those of religion and morality. Moreover, because they identified reason with the techniques of mathematical demonstration, they often challenged the idea that *poetry,* in which the mind did not restrict its activities to the purely demonstrative, could be of any real value in the discovery of Truth. We will soon see that Hobbes offered alternative and critically important roles for poetry. But some of his followers, including John Locke, viewed the non-rational and fictive character of poetry as grounds for challenging its role in serious moral discourse. Locke warned parents to discourage their children from developing a taste and talent for poetry;[1] and at the hands of some of the less tolerant utilitarian

L'ART

DE DE'CRIRE

LA DANCE,

PAR CARACTERES ET FIGURES

DEMONSTRATIFS,

Avec lesquels on apprend facilement de soy-même toutes
sortes de Dances.

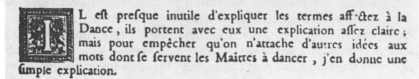

L est presque inutile d'expliquer les termes aff·êtez à la
Dance, ils portent avec eux une explication assez claire ;
mais pour empêcher qu'on n'attache d'autres idées aux
mots dont se servent les Maîtres à dancer , j'en donne une
simple explication.

Pl. 4. Title Page of Raoul Auger Feuillet's *Choréographie* (Paris, 1700),
an example of the trend to apply the *esprit géométrique* to such
arts as the dance.

developers of Locke's notions, the disrespect for poetic imagination led to the idea that poetry was the pre-rational, "mental rattle" of an infant humankind which both could and should be abandoned in its maturity. Isaac Newton, when asked his opinion of poetry, responded that he would repeat Isaac Barrow's opinion that it was "a kind of ingenuous nonsense."[2] Richard Bentley developed the distrust of poetry grounded in its lack of rationality and its disrespect for demonstrative *truth* into a series of angry and infamous commentaries on such poetic classics as the *Odes* of Horace and Milton's *Paradise Lost*. When Bentley glossed Milton's line, "Thither came Uriel gliding through the Even," for example, he complained: "I never heard but here, that the *Evening* was a place or space to glide through. Evening implies Time, and he might with equal propriety say, *Came gliding through six o'clock*."[3] And in response to Horace's pastoral description of the Italian countryside: "The festive village leopard idles with the ox, Lazy in the meadows," Bentley rages, "How did a leopard get into Italy? Their habitat is confined to Africa and Asia. . . ."[4]

Though the emphasis on calculative reason and truth sometimes shaded off into an overt distrust or denigration of the poetic art, more frequently it simply reinforced and extended the tendencies already present in aesthetic theories to emphasize the need for some kind of rational control over virtually limitless flights of imagination. As we have already seen in Chapter 1, both Hobbes and Dryden expressed this feeling by likening the imagination to a spaniel that needed to be kept under rigid control; and most aestheticians of the eighteenth century sought that control through some set of rules for poetry.

Though the search for aesthetic rules may resonate with Newton's search for rules for reasoning in philosophy and rules for methodizing the apocalypse, its most obvious historical precedent came from late Renaissance Aristotelianism. The Italian revivers of Aristotelian aesthetics viewed poetry as a special kind of logic that operates through examples rather than through formal syllogisms,[5] and they had emphasized the idea that poetry is subject to its own special laws. Initially there was a tendency simply to accept the classical—i.e., Aristotelian—rules for poetry on the authority of Aristotle. But the shrewder Renaissance critics argued that the Aristotelian rules were themselves drawn from a more fundamental source in nature. According to Menturno, whose *Arte Poetica* was published in 1564, for example, "Everything in nature is governed by some specific law which directs its operation; and as it is in nature, so it is in art, for art tries to imitate nature, and the nearer it approaches nature in her essential laws, the better it does its work."⁶ By the mid-seventeenth century, and under the influence of the *esprit géométrique,* major critics such as John Dennis in England insisted that it was a combination of nature and *reason* that provided the ultimate authority for the rules of art.

Dennis's discussion of the rules of poetry offers a particularly good illustration of how the emphasis on reason, with its focus on the order and symmetry associated with the *esprit géométrique* had come to shape at least one major attitude in the early eighteenth-century understanding of poetry:

> The work of every reasonable creature must derive its beauty from regularity; for reason *is* rule and order, and nothing can be irregular either in our conceptions or our actions any further than it swerves from rule, that is, from reason. . . . [T]he design of logic is to bring back order and rule and method to our conceptions, and want of which causes most of our ignorance and all our errors. The design of moral philosophy is to cure the disorder that is found in our passions, from which proceed all our unhappiness and all our vice; as from the due order that is seen in them come all our virtue and our pleasures. But how should these arts reestablish order unless they themselves were regular? Those arts that make the senses instrumental to the pleasures of the mind, as painting and music, do it by a great deal of rule and order: since therefore poetry comprehends the force of all these arts of logic, of ethics, of eloquence, of painting, of music, can anything be more ridiculous than to imagine that poetry itself should be without rule and order.⁷

We shall return in the next section to discuss some of the specific rules generally adopted by the poetic critics of the seventeenth and early eighteenth centuries. At this point it is enough to recognize that neoclassical poetic theory emphasized the importance of rule and regularity. That emphasis had a long-standing tradition in Aristotelian humanistic aesthetics, but the particular choices of rules and kinds of order to emphasize became closely associated during the late seventeenth century with Cartesian and Hobbesian emphases on mathematics. The mathematical side of late neoclassical aesthetics found its most extreme formulation in Francis Hutcheson's *Inquiry into the Original of Our Ideas of Beauty and Virtue* (1725). In this work, Hutcheson argues that we have an internal aesthetic "sense" that responds immediately to beautiful objects without any reflection about the causes of beauty. Analysis of the objects which this sense recognizes as beautiful leads to the discovery of a mathematical law of beauty:

> What we call beautiful in objects, to speak in the mathematical style, seems to be in compound ratio of uniformity and variety: so that where the uniformity of bodies is equal, the beauty is as the variety; and where the variety is equal the beauty is as the uniformity. . . . Thus squares are more beautiful than equilateral triangles because they have a greater variety of uniform sides, and regular solids are more beautiful than irregular ones with an equal number of faces because they have greater unity with the same number of faces. Theorems are more beautiful when a variety of consequences follows from them, and a work of poetry increases in beauty when there is a great uniformity in each character amid a variety of characters, as in the works of Homer.[8]

THE ARISTOTELIAN FOUNDATIONS OF NEOCLASSICAL AESTHETIC THEORY

Seventeenth- and eighteenth-century aesthetic theory fused elements of the *esprit géométrique,* mechanical philosophy, and associationist psychology with principles in the revived Aristotelian tradition. But all of the *new* perspectives served almost exclusively to modify an Aristotelian core. Thus the scope and central problems of subsequent "scientized" aesthetics continued to reflect the shaping force of Aristotle's works. Whether one took a rationalist or psychological–empiricist approach to questions of art, or whether one questioned the central *function* of poetry—i.e., to teach morality or to provide pleasure—or whether one was concerned primarily with the relative importance of "fancy" or "imagination" and "reason" in art, or with the nature of the emotional impact of art, or with the appropriate subject matter of artistic imitation, or with the "logic" of poetic figures, or with the taxonomy of artistic forms and the rules of art, or with the relationship among the various arts and between the arts and the sciences—including the critical relationship between artistic and philosophical "truth"—one's starting point was inevitably some doctrine or attitude found in or imputed to Aristotle's *Poetics.*

Virtually unknown to the Romans or to the Medieval Latin world, the *Poetics* was first translated into Latin and published at Venice in 1498. Within less than fifty years it became far and away the most important treatment of aesthetics because it provided a magnificent justification for the Humanists' fascination with poetry—a fascination that had found very little encouragement within the Platonic corpus or the Christian scholarly tradition. Plato's most extensive discussion of poetry, that in Book 10 of *The Republic,* was almost undeviatingly hostile; and the Christian tradition generally opposed poetry both because of the objections set forth by Plato and because of its "pagan" character. Thus Aristotle's justification of poetry found a grateful audience among literary humanists, for it offered a powerful complement to the one major classical justification of poetry that had remained known in the Latin West, Horace's *Ars Poetica.*

Horace emphasized the didactic function of poetry, acknowledging that poetry was intended to give pleasure, but insisting that the pleasure-giving aspect of poetry was a kind of sugar coating for the moral pill which was its essence. Aristotle, however, had seemed to raise the aesthetic, or pleasure-giving, function of poetry to centrality, though he admitted that the portrayal of immoral rather than moral actions would, no doubt, diminish the pleasure that poetry could give. Virtually all late Renaissance and early modern treatments of poetry and other art forms combined the perspective of Horace and Aristotle, admitting the didactic function of poetry, but drawing from Aristotle's *Poetics* for almost all details of critical analysis and for an emphasis on the purely aesthetic function of art.

We begin a consideration of the *Poetics* by exploring the notion of "imitation" in Aristotle, for this notion became the subject of extensive debate among seventeenth- and eighteenth-century aestheticians. Aristotle argues that poetic imitations portray men as better or worse than they are, not exactly as they are. This raises the question of the relationships among poetry, truth, and fact, a question which both Aristotle and his subsequent interpreters approached from several different perspectives throughout the *Poetics* and commentaries on this work. In general, Aristotle argues that while art should express truth, it may do so, unlike philosophy, by telling factual lies. Indeed, at one point Aristotle praises Homer for teaching other poets to lie skillfully,[9] and says that a poet may be perfectly justified in describing something impossible, "if the end of the art be thereby attained—i.e., if the effect of this or any other part of the poem is rendered more striking."[10] Of the three kinds of things that might be imitated—things as they are, things as they are said or thought to be, and things as they ought to be—the latter two are no less important for the poet than the first. In fact, that which ought to be is even more worthy of imitation than that which actually is, "for the ideal type must surpass the reality."[11] At another point Aristotle reiterates the notion that art may serve some truth that is higher than mere fact by comparing poetry with history:

> It is not the function of the poet to relate what has happened,
> but what may happen—what is possible according to the law
> of probability or necessity. . . . Poetry therefore is a *higher* thing
> than history: for poetry tends to express the universal, history the
> particular.[12]

Each of the short passages cited above demonstrates both that the artist need not be limited to portraying facts and that he is nonetheless subject to certain constraints. If he exercises his legitimate "poetic license" to speak falsely, he must always do so in order to advance some higher truth or to increase the emotional impact of his work, and he must remain true to what is either probable or consistent with received opinion. Precisely how far the poetic license to speak falsely extends and under what conditions it might be exercised became a central problem for the scientific/psychological critics that we will be discussing soon; for though such mathematically oriented theorists as Hobbes could accept the proposition that ideal truths were somehow higher and more fundamental than facts, such empiricists as Locke, Bentley, and the utilitarians could not. This is perhaps the best way to understand the seemingly perverse literalism which we noted in Bentley's approach to Horace and Milton.

Aristotle argues that men have a powerful instinct for imitation—through which most learning occurs—and that they feel a related and no less universal pleasure in experiencing "things imitated." As evidence for the latter claim, he offers the signs of pleasure that humans express in looking at carefully done paintings, even when those paintings are of objects which, in their originals, may be the source of pain. Why we should get pleasure from experiencing likenesses or representations of objects that inspire fear or pain is a question that obviously disturbed Aristotle, as it did his seventeenth-century readers. His explanation—that we get pleasure from learning and that when we reflect on representations of things we become conscious of them as a source of knowledge without an accompanying sense of threat[13]—provided the starting point of at least one line of argument later on; but the general problem raised by Aristotle led in a number of different directions for those with a psychological bent.

There is one odd feature of Aristotle's notion of poetic imitation as a source of pleasure that needs at least to be mentioned here, though we will return to it again. Aristotle reverses the places of learning and pleasure which were part of the notion of poetry expressed by Aristophanes for whom, as for Horace, the pleasure associated with poetry is important because it serves the didactic end of the poetic art. For Aristotle, however, the roles are inverted, and it is because mimetic learning *pleases* us that learning is important. As we shall see again and again, for Aristotle, poetry— and art in general—addresses itself to the emotions. Whenever the intellect or reason is mentioned, it is in terms of heightening the emotional impact of art. This feature of Aristotelian aesthetic theory distinguished it from all other theories known to the Renaissance and made it the most important starting point for the psychological treatment of art that was to emerge in the writings of Hobbes and Descartes.

Because all elements of poetry are included in tragedy, Aristotle offers a detailed analysis of the nature of tragic poetry as that form which will best illuminate the important characteristics of all poetry, and indeed, of all art. He begins with a complex definition of tragic poetry that provides a starting point both for his subsequent analyses and for all of the neoclassical discussions of the rules governing dramatic and epic poetry:

> Tragedy is an imitation of an action that is worthy, complete, and of a certain grandeur; it is in speech embellished with appropriate artistic conventions; it is in the form of action rather than narrative; and it operates through pity and fear to effect the proper catharsis [purification or purging?] of these emotions.[14]

Once more Aristotle emphasizes the emotional end of art. Tragedy is particularly aimed at inspiring fear or pity in order to effect a "catharsis" of these emotions. Most Renaissance and twentieth-century commentators agree in interpreting the term *catharsis* as one borrowed from medical usage and as best translated by the term *purging*. The general idea is that the pity and fear that we feel in daily life create a "morbid and disturbing element"[15]—a

pathological condition—which demands relief. Just as many ancient medical treatments involved the idea that an illness should
be treated with something similar to its symptoms—i.e., a sour
stomach might be treated with curdled milk—Aristotle's theory of
tragedy argued that ordinary pity and fear could be purged, or elimi-
nated, by exposing men to a heightened and similar kind of pity and
fear evoked by dramatic art. Thus, for the audience of tragedy the
final end is a release from painful emotions and the attainment of
tranquility. As we shall see, when Hobbes and his psychologizing
followers got hold of the *Poetics,* they agreed with Aristotle's emphasis on the critical importance of creating heightened emotions;
but they totally rejected the Renaissance idea, derived from Aristotle, that the end of tragic art was a state of tranquility.

Through the remainder of the *Poetics,* Aristotle analyzes tragic
poetry into six different "elements": plot, character, thought, diction,
song, and spectacle, and suggests how each element should be
handled in order to achieve the definitional character of tragedy. I
want to consider just one element of his comments on poetic diction
or language, for this was a topic that Hobbes, Locke, and their followers found particularly interesting. According to Aristotle, the
most perfect poetic style should be clear without being common or
vulgar.[16] Maximum clarity, on the one hand, could be achieved by
using only current, ordinary words in their literal meanings; but
such usage could not lift our thoughts above the commonplace. The
loftiest style, on the other hand, would employ unusual words—
esoteric, archaic, newly coined, lengthened or contracted words—
and would often use them metaphorically; but such usage would
create a riddle rather than a work of poetry, hiding the meaning
from the audience. The key to stylistic perfection is thus to mix
clear and ordinary language with rare usages so as to heighten the
impact of the language without hiding its meaning.

Best of all techniques for heightening this impact is the use
of metaphor. As Aristotle argues, "this alone cannot be imparted
by another; it is the mark of *genius,* for to make good metaphors
implies an eye for resemblances."[17] For the seventeenth- and eighteenth-century psychological aestheticians, the "eye for resemblances" embodied in metaphor became not just the symbol for art,
but the very essence of the faculty or characteristic, variously
labeled "wit," "fancy," "imagination," or "genius" that constituted
artistic creativity.

HOBBESIAN PSYCHOLOGY AND THE
SHAPING OF ENGLISH LITERARY CRITICISM

I have insisted that there was at least one theme developed by Hobbes—the identification of Reason with calculative and demonstrative rationality and the subsequent claim that poetry could lead only to absurdity in the search for truth—which led through Locke to a general distrust of poetic expression as a legitimate source of knowledge during the late seventeenth and early eighteenth centuries. There was another Hobbesian tradition, however, derived to some extent from the *Leviathan* and *Elements of Law* but more fully developed in two short essays, *The Answer to Davenant* (1650) and *The Virtues of an Heroic Poem* (1675), which not only defended the importance of poetry, but which formed the foundation of a new kind of poetic criticism in England. Ever since the beginning of the twentieth century there has been a general acknowledgment that Hobbes initiated a new seventeenth-century concern with artistic creativity. J. E. Spingarn wrote in 1911:

> The Seventeenth Century first attempted to deal accurately with the relation between the creative mind and the work of art; it began to analyze the content of such terms as "wit," "fancy," and taste. Hobbes is here a pioneer; he left an impress on critical terminology, and his psychology became the groundwork of Restoration criticism.[18]

In recent years interest in Hobbesian literary theory has continued to grow, and Hobbes is seen as centrally important in the whole British trend to view literature increasingly in terms of the two problems which ultimately obsessed the Romantic poets and theorists: the pleasures peculiar to art and the mental processes through which art is created.[19]

Like Horace and the most important of English Renaissance critics, Sir Philip Sidney, Hobbes accepted the notion that the end of poetry was to stimulate men to virtuous action. Moreover, he agreed with Sidney that in order to accomplish this end, poetry needed both to delight or please and to teach or inform. It needed to delight in order "to move men to take that goodness in hand which, without delight, they would fly from as a stranger," and it needed to teach "to make them know that goodness whereunto they are moved."[20] Passionless pedantry and pointless passion were equally undesirable from Hobbes's perspective and, as we shall see, he emphasized a balance between the didactic and emotional elements of poetry. What differentiated Hobbes from Sidney was that Hobbes clearly joined with Aristotle in shifting the balance point away from pedagogy toward passion.

Since most of Hobbes's comments focus on the emotional effects of art and the means to achieve those effects, we will emphasize such issues; but first it is important to understand how Hobbes could allow a teaching function to an enterprise that effectively abandoned the kind of calculative reason that alone guaranteed the truth of its claims. We begin by considering a series of terms discussed by Hobbes in the *Leviathan*—i.e., "Wit," "Appetite," "Reason," "Science," "Fancy," "Judgment," and "Prudence." Though Hobbes is not absolutely consistent in his use of all of these terms, the general thrust of his comments is very clear. For Hobbes the term *wit* is used to indicate the quality of a person's thought processes, of which "reason," "fancy," and "judgment" are among the most important. All thinking depends ultimately on experience, memory (which is retained experience), and imagination (which is recalled or recovered experience). And all thought also depends at some fundamental level on "appetite," "passion," or "desire"; for as Hobbes says: "The thoughts are to the desires as Scouts and spies, to range abroad and find the way to things desired. All steadiness of the mind's motion, and all quickness of the same, proceeding from hence."[21] The important point is that the mind is not passive in thought, but active, and the person who has the greater appetite or desire for knowledge—i.e., the greater curiosity—will have the more penetrating wit. According to Hobbes, wit is of basically two kinds: natural wit, which depends only on use and experience "without method, culture, or instruction," and acquired wit, which "is grounded on the right use of speech and produceth Science."[22]

Artificial wit alone proceeds primarily through calculative reason and is capable of producing that certain knowledge which is rightly called Truth. It is this kind of wit that Hobbes is most concerned with in his strictly philosophical writings. He does acknowledge, however, the utility of natural wit, synonymous with "prudence," which operates through fancy and judgment alone rather than through these faculties supplemented by reason. "Fancy" is that capacity of the mind which considers all of the sensations and memories or "phantasms" present in the mind and discovers similarities among them. Judgment is that capacity of the mind which recognizes differences among the phantasms. If someone has an unusual ability to discover similarities seldom noticed by others, he is said to have a good fancy; if he is able to observe differences, "in case such discerning be not easy," he is said to have good judgment.[23]

Fancy and judgment, working on the phantasms provided by experience, constitute a kind of prudential knowledge which forms the ground for wise human action. When someone has some end in mind, according to Hobbes, he or she remembers a whole range of past actions. The fancy collects those that produced results similar to those desired, whereas the judgment helps discard those that would not work in present circumstances because of salient differences. In this way a course of action is selected without rigorous reasoning from cause and effect. In most cases demanding human action, such a prudential decision-making procedure is the only one available; and for this reason Hobbes has deep admiration for the prudent man, the man with a large fund of experience and with good fancy and judgment.

Fancy in particular plays a key role for Hobbes in all creative activities, including the practical arts and philosophy or science, for it is fancy which draws forth from the storehouse of experience those elements that can be manipulated effectively by the reason:

> So far forth as the fancy of man has traced the ways of true Philosophy, so far it hath produced very marvelous effects to the benefit of mankind. All that is beautiful or defensible in building, or marvelous in Engines and Instruments of motion, whatsoever commodity men receive from the observations of the heavens, from the description of the Earth, from the account of time, from walking on the seas, and whatsoever distinguishes the civility of *Europe* from the barbarity of the *American* savages, is the workmanship of fancy but guided by the principles of true philosophy.[24]

After this paean to fancy supplemented by reason, or philosophy, Hobbes goes on to reaffirm his confidence in fancy in places where philosophy has not yet gone; and he specifically ties this consideration to the teaching function of poetry:

> But where these precepts [i.e., those of science or true philosophy] fail, as they have hitherto failed in the doctrine of moral virtue, there the architect, *Fancy*, must take the philosopher's part upon herself. He, therefore, that undertakes an heroic poem, which is to exhibit a venerable and amiable image of heroic virtue, must not only be the Poet to place and connect, but also the philosopher, to furnish and square his matter, that is, to make both body and soul, color and shadow of his Poem out of his own store. . . .[25]

In this passage Hobbes clearly indicates that the fancy must create not only the aesthetic form but also the moral substance of the poem. It must draw forth from the mind a set of images which will not only delight, but provide a concept of virtue that is worthy of emulation. Nowhere is Hobbes's notion of the critical importance of fancy more clearly distinguishable from that of Locke. Certainly, Hobbes would always insist that fancy without judgment may be dangerous; nonetheless, it is at the heart of all constructive and creative intellectual activity. For Locke, on the contrary, all "fantastical" ideas "are for nothing else but to insinuate wrong ideas . . . and thereby mislead the judgment, and so indeed are perfect cheats."[26] "There are so many ways of fallacy [offered by] the fancy," he argues, "that he who is not wary to admit nothing but the truth itself . . . cannot but be caught."[27]

Given Hobbes's acknowledgment that fancy, always controlled by judgment, plays an essential role in the production of prudential knowledge and in the creation of suitable models of virtue which justify the didactic function of poetry, we must turn to the peculiarly aesthetic or "moving" power of fancy. In his early *Elements of Law,* Hobbes suggested that the recognition of "unexpected similitude" expressed through such devices as similes, metaphors, and other tropes is what gives poetry its special power to please and move.[28] In his late *The Virtues of an Heroic Poem,* he continued to argue that an elevated fancy gives poetry its unique character:

> . . . men more generally affect and admire fancy than they do either judgment or reason, or memory, or any other intellectual virtue; and for the pleasantness of it, give to it alone the name of Wit, accounting reason and judgment but for a dull entertainment. For in fancy consists the sublimity of a poet, which is that poetical fury which the readers for the most part call for. It flies abroad swiftly to fetch in both matter and words.[29]

Of course, Hobbes admitted that the "delight and grace" of fanciful constructions would be lost without the application of judgment to decide what images and words "are fit to be used, and what not," but he agreed with the generality of men in granting fancy first place in the creation of poetry.

It was in *The Whole Art of Rhetoric,* in the course of a discussion which addresses Aristotle's claims that strange words and metaphors heighten the impact of poetic discourse, that Hobbes provided his clearest account of *why* "unexpected similitude" is so pleasing:

>   Forasmuch as there is nothing more delightful to a man, than
> to bid that he apprehends and learns easily; it necessarily follows
> that those *words* are most *grateful* to the ear, that make a man seem
> to see before his eyes things signified.
>   And therefore *foreign* words are unpleasant, because *obscure*;
> and *plain* words, because *too manifest,* making us learn nothing
> new. But *metaphors* please; for they beget in us, by the *genus,* or by
> some common thing to that with another, a kind of *science.* As
> when an *old man* is called stubble; a man suddenly learns that he
> grows up, flourishes, and withers like grass, being put in mind of it
> by the qualities common to *stubble* and *to old men.*[30]

In this passage Hobbes draws from at least two basic notions expressed in Aristotle's *Poetics,* amplifying and modifying them in a
fascinating way. He begins by accepting the notion, appealed to in
Aristotle's very definition of poetry as an art, that learning is intrinsically the source of pleasure. Then he goes on to use this principle
to explain that metaphor pleases on the small scale in much the
same way that artistic imitation pleases on the large, i.e., by providing a kind of unexpected learning that depends on the recognition
of likeness. Moreover, Hobbes uses this insight both to account for
Aristotle's recognition that ordinary language fails to provide special pleasure and to support Aristotle's claim that archaic or strange
language should not, by itself, produce pleasure. In the first case
nothing *new* is learned, while in the second, the audience fails to
get the special feeling of satisfaction that derives from illuminating
metaphor.

    This analysis of the pleasure derived from metaphor and other
expressions of similitude is not yet traced back to the mechanistic
foundations of Hobbesian psychology; for it still leaves unanswered
the question of why learning pleases. Hobbes does complete the job
in *An Answer to Davenant,* however. Here he is particularly interested in emphasizing the notion that it is the *novelty* of "apt metaphors," and the "rarity of their expression" that produces pleasure,
"by excitation of the mind; for novelty causeth admiration, and admiration curiosity, which is a delightful appetite of knowledge."[31]
One must recall that for Hobbes pleasure is always connected with
some increase in vital motion. Moreover, our sensations always depend upon motions impressed upon our sense organs by some
change of configuration in the outside world. Since, with rare exceptions, the impression of motion associated with sensation leads
to increases in vital motions, our sensations, which increase our
knowledge, also provide pleasure. Hobbes draws directly from this

theory to account for the pleasure we derive from the novel use of language in poetry:

> . . . as the sense we have of bodies consists in change and variety of impression, so also does the sense of language in the variety and changeable use of words. I mean not in the affection of words newly brought home from travail, but in new and with all significant translation to our purposes of those that be already received and in far fetched but withal apt, instructive and comely similitudes.[32]

John Dennis, among the least successful playwrights, but perhaps the greatest of eighteenth-century British practical literary critics, closely followed Hobbes in his simultaneous insistence that poetry pleases by literally producing an excitation or "agitation" of the mind and that such an agitation depends on constant novelty and surprise. In *The Advancement and Reformation of Modern Poetry* (1701) he wrote:

> It is impossible that any pleasure can be very great that is not at the same time surprising. . . . [T]he mind does not care for dwelling too long upon an Object, but loves to pass from one thing to another; because such a transition keeps it from languishing, and gives it more agitation. Now agitation only can give delight. For agitation not only keeps it from mortifying reflections, which it naturally has when it is not shaken, but gives it a force which it had not before, and the consciousness of its own force delights it.[33]

It may seem odd that Hobbes, who was so insistent on limiting the use of words to their unique, unambiguous, and denotative meanings in connection with science, should so strongly advocate using the variability and connotative elements of language in poetry; but this position is perfectly consistent with his arguments regarding the distinction between artificial and natural wit and between science and prudence. Hobbes has no desire at all to turn poetry or other works of art into scientific works. Indeed, he offers one of the clearest distinctions between science and art to be discovered in the early modern period.

What makes Hobbes's scientific psychology so important for modern aesthetics is its new way of understanding both artistic creativity and the emotional effects produced by works of art, simultaneously elevating those effects to a position of prominence they had not enjoyed earlier. Whereas Aristotle unquestionably insisted that poetry and art in general should be judged primarily in terms of its emotional impact, he said relatively little about *why* emotion is important in the general scheme of things. In fact, given his notion of catharsis, Aristotle often reflects Plato's fundamental distrust of all emotion, and he seems to want to intensify temporarily emotions only in order to destroy them in the long run. Within Hobbes's mechanistic psychology, however, passions become both the sources and symptoms of all happiness and vitality. This idea found one of its most powerful seventeenth-century expressions in the writings of John Dennis, who turned it into an apparent attack on reason.

I present an extensive passage from Dennis's *The Usefulness of the Stage* (1698) because it is so important in the transition between neoclassical aesthetic theory, with its high regard for reason, and sentimental aesthetics, which assigned calculative reason a distinctly secondary—often even a negative—role in human life:

> . . . reason may often afflict us and make us miserable by setting our impotence or our guilt before us, but that which it generally does is the maintaining us in a languishing state of indifference, which is perhaps more removed from pleasure than it is from affliction, and which may be said to be the ordinary state of men.
>
> It is plain then that reason, by maintaining us in that state, is an *impediment* to our pleasure, which is our happiness; for to be pleased a man must come out of his ordinary state. Now nothing in this life can bring him out of it but passion alone, which reason pretends to combat. . . . Since nothing but pleasure can make us happy, it follows that to be very happy we must be much pleased; and since nothing but passion can please us, it follows that to be very much pleased we must be very much moved.

At this point Dennis goes beyond any kind of Hobbesian idea to assert that even in the afterlife our pleasures will "proceed from passion, or something which resembles passion":

> Reason shall then be no more. We shall then have no more occasion from premises to draw conclusions and long trains of consequences. . . . We shall lead the glorious life of angels, a life exhausted of all Reason, a life consisting of ecstasy and intelligence.

Thus, he concludes, "it is plain that the happiness both of this life and the other is owing to passion and not to reason."[34]

Dennis was not as anti-rationalist as this passage, taken out of context, might suggest. Just as Hobbes emphasized the constructive role of fancy in creative acts, but balanced it with an insistence that judgment must be present to reign in that undisciplined faculty, so too did both he and Dennis admit that even if the force of passion alone was able to make us happy it could never do so *against* reason. In Dennis's words, "If reason is quite overcome, the pleasure is neither long, nor sincere, nor sage."[35] On this issue Dennis clearly borrowed from arguments developed in connection with the defense of religious revelation. Secular poetry, like the religious "poetry" of the Bible, might possibly go beyond reason, but it could not contravene reason. Indeed, as we saw a bit earlier, Dennis could at times forget Hobbes and write almost like a neoclassical critic.

In time, the tendency to elevate the creative power of fancy over the merely corrective powers of judgment which had been initiated by Hobbes and Dennis *was* transformed into an aesthetic doctrine that viewed philosophy and reason not just as different from but coexistent and compatible with poetry and *imagination* (a term that replaced *fancy* in the mid-eighteenth century), but as fundamentally opposed to them. In his youthful speech on "whether philosophy be of use to poetry," delivered in 1747, for example, Edmund Burke defended the following propositions:

> That the provinces of Phil[osophy] and poetry are so different that they can never coincide, that Phil[osophy] to gain its end addresses to the understanding, poetry to the imagination, which by pleasing, it finds a nearer *way* to the heart, that the coldness of philosophy hurts the imagination and taking away as much of its power must consequently lessen its effect, and so prejudice it. That such is the consequence of putting a rider on Pegasus that will prune his wings and incapacitate him from rising from the ground.[36]

Twenty-two years later, as the most important British aesthetician of the century, Burke continued to hold virtually the same position. "The judgment," he wrote, "is for the greater part employed in throwing stumbling blocks in the way of the imagination, in dissipating the scenes of its enchantment, and in tying us down to the disagreeable yoke of our own reason."[37]

On one particular issue the Hobbesian psychological insistence on heightened emotions as the source of pleasure and happiness forced an extremely interesting modification of Aristotelian and neoclassical doctrines. Aristotle and his neoclassical interpreters adamantly insisted that tragedy operates through a catharsis of fear and pity that leads to a state of emotional quiescence. However, to be passionless or tranquil was, for Hobbes, quite literally to be *dead,* and to have weak passions was to be *dull;*[38] so if tragedy was to have a positive function, it had to be to intensify emotions rather than to tranquilize them. "Felicity," he argued, "consisteth not in the *repose* of a mind satisfied," but rather in an *"agitation of the mind"* associated with increased vital motions. Furthermore, since *fear* and *pity* are both emotions associated with *pain,* and therefore with the destruction of vital motions rather than with their enhancement, tragedy could not achieve its positive goals primarily by increasing fear and pity.

Hobbes did not argue that fear and pity are irrelevant to tragedy, but rather that these emotions are mixed in those who view a tragedy with positive emotions that overcome them. Furthermore, these positive emotions spring both from the general pleasure that comes from the novel use of language which characterizes all poetry and from a special pleasure that comes to those in the audience from reflecting on the fact that they do not really share the pain and suffering of the characters on the stage. In Hobbes's terms, part of our pleasure in tragedy occurs in the "remembrance of our own security present, which is delight."[39] John Dennis once again followed Hobbes on this issue, arguing that "no passion is attended with greater joy than enthusiastic *terror,* which proceeds from our reflecting that we are out of danger at the very time that we see it before us."[40] This doctrine becomes the foundation of many later Romantic interpretations of the "thrilling terror" experienced by the reader of a Gothic romance, a "terror" modified into "delight" or "joy" by our relief at being a spectator rather than a participant in the action. Of course, few subsequent interpreters accepted the Hobbesian mechanistic underpinnings of this notion; but most did come to accept the general claim that poetry should heighten and intensify our emotions rather than pacify them, and many accepted the notion that our emotional responses to all works of art are strongly influenced by our awareness of the differences between the condition being depicted and that which we are in.

THE SENSATIONALIST AESTHETICS OF EDMUND BURKE

Though Thomas Hobbes initiated a British psychological tradition in theories of art to which such authors as John Dennis contributed, no systematic attempt to develop a full-scale psychology of aesthetic experience appeared until the eighteenth century. By the time such attempts were begun, the more empirically oriented sensationalist psychology of John Locke had replaced the theories of Hobbes as a foundation for aesthetic theory.

We have already seen that there was at least one line of Lockean thought, authorized by Locke himself and found in such writers as Richard Bentley and Isaac Newton, that was at best suspicious of and more often overtly hostile to poetic emphases on fancy and emotionality. A second line, initiated by Joseph Addison in a series of seventeen essays on "Wit" and "The Pleasures of Imagination" published in the *Spectator* during May 1711 and June and July of 1712, eventually outdid Hobbes and Dennis in its enthusiasm for the creative imagination and for the moving power of art. To this very popular tradition, the Abbé du Bos's *Reflexions et Critiques sur la Poésie et sur la Peinture* (1719), Francis Hutcheson's *Inquiry into the Original of Our Ideas of Beauty and Virtue* (1725), John Ballie's *Essay on the Sublime* (1747), David Hartley's *Observations on Man* (1749), David Hume's essays "Of Tragedy" and "Of the Standard of Taste" (1757), Alexander Gerard's *Essay on Taste* (1757), and the first six chapters of "Essay IV" of Claude Adrien Helvetius's *De l'Esprit* (1757) added a great deal. But the tradition culminated in Edmund Burke's *A Philosophical Enquiry into the Origin of Our Ideas of the Sublime and Beautiful* (1757, with major revisions in 1759). Because of its long-term impact on both British and German aesthetics, Burke's version of sensationalist aesthetics as it appeared in 1759 will occupy our attention.

Burke's *Enquiry* was as much *the* manifesto of the pre-Romantic movement as Smith's *Wealth of Nations* was the manifesto of economic liberalism. It explained and intensified the growing taste "for ruins and melancholy terror, for graveyard poetry, for wild and desolate scenery, for . . . infinity, vastness, power, magnificence and obscurity."[41] Furthermore, it systematized a change in aesthetic values by providing a new and well-defined meaning for the critical term *sublime* and by generally popularizing and legitimizing a whole set of ideas which became central to the subsequent Romantic movement. Yet Burke was completely non-Romantic in the central scientific approach to aesthetic issues that informed the *Enquiry*; and William Blake was absolutely correct when he argued that "Burke's Treatise on the Sublime and Beautiful is founded on the Opinions of Locke and Newton."[42] Ironically, just as the great rationalist Hobbes provided the single strongest impetus to a nearly exclusive focus on the emotional factor in aesthetics, the great Newtonian aesthetician, Burke, ended up creating an aesthetic theory that portrayed scientific thought as fundamentally destructive.

As a fifteen-year-old student at Trinity College, Dublin, in 1744, Burke centered his interests on literary studies. Before he began his legal studies in 1750, he had been editor and chief author of *The Reformer,* a short-lived but high-quality weekly drama review notable for its unstinting praise of Shakespeare's tragedies. Burke had read widely in the available classical and contemporary literature on aesthetics and was frustrated by the confused and inconsistent use of key terms like *beauty, the sublime,* and *taste.* Thus, as early as 1747, he began work on a treatise that was intended to do for problems associated with judgments of the works of imagination and the elegant arts what John Locke and David Hume had attempted for our "understanding." It was this work that finally appeared in 1757 as *An Enquiry into the Origin of Our Ideas of the Sublime and Beautiful.*

Burke began by lamenting the confusion of ideas that makes virtually all artistic criticism uncertain and inconclusive; then he goes on to explain the method by which he hoped to remedy the situation:

> Could this admit of any remedy, I imagined it could only be from a diligent examination of our passions in our own breasts; from a careful survey of the properties of things which we find by experience to influence those passions; and from a sober and attentive investigation of the laws of nature, by which those properties are capable of affecting the body, and of thus exciting our passions. If this could be done, it was imagined that the rules deducible from such an enquiry might be applied to the imitative arts, and to whatever else they concerned, without much difficulty.[43]

Note from the very outset that Burke implicitly accepted the idea that the appeal of art is to the passions rather than to the reason or understanding.

All passions, according to the tradition of Hobbes and Locke, are derived from pleasure and pain. This premise guarantees to Burke that there must be some universal aesthetic principles. Explicitly appealing to Newton's Rule II of Reasoning in philosophy, Burke argued:

> It must necessarily be allowed that the pleasures and pains which every object excites in one man, it must raise in all mankind, whilst

it operates naturally, simply, and by its proper powers only; for if we deny this, we must imagine that the same cause operating in the same manner, and on subjects of the same kind, will produce different effects, which would be highly absurd.[44]

Since he is concerned with pleasures of the imagination rather than with those of the senses, Burke's argument from the uniform nature of sensory pleasures would be irrelevant except for the fact that he fully accepts Locke's notion that "the power of the imagination . . . can only vary the disposition of those ideas it has received from the senses." The combined images provided by the imagination will then produce pleasure or pain in the same way as the original ideas of sense, and will produce "natural" responses which are no less determinate than those to ideas of sense.[45] It is true that an additional pleasure may arise when imagination "imitates" an original; for we get pleasure from the resemblance which the imagined object has to its original. But once again there is every reason to think that this kind of pleasure is the same in all persons.

Just as Burke appealed to a Newtonian formulation of the principle that like causes must produce like effects in order to justify his expectation that a "science" of aesthetics must be possible, he also drew explicitly from Newton to set those limits beyond which legitimate analyses of the causes of sublimity and beauty could not reach. "When Newton first discovered the property of attraction and settled its laws," Burke wrote, "he found it served very well to explain several of the most remarkable phenomena in nature; but yet with reference to the general system of things, he could consider attraction but as an effect whose cause . . . he did not attempt to trace." By the same token, Burke admitted that he did not expect to be able to "explain *why* certain affections of the body produce . . . a distinct emotion of the mind and no other; or *why* the body is at all affected by the mind or the mind by the body." Instead he would seek only to explore "*what* distinct feelings and qualities of body shall produce certain determinate passions in the mind, and no others." If he could achieve this much, Burke argued, "a great step will be accomplished towards a distinct knowledge of our own passions. . . ."[46]

Burke was perfectly aware that aesthetic tastes do vary considerably from one individual to another, from place to place, and from time to time; but he made the same argument regarding aesthetic issues that we have seen Helvetius and Condorcet make about morality. Habits, customs, or different associations may modify our natural responses to things; but we must begin any attempt to understand aesthetic responses by exploring the "natural and simple" cases before we can understand the complicating effects produced by "the habits, the prejudices or the distempers" of particular men.[47] Thus Burke insisted that although "many things affect us after a certain manner, not by any natural powers they have for that purpose, but by association," nonetheless:

> . . . it would be absurd . . . to say that all things affect us by association only; since some things must have been naturally agreeable or disagreeable, from which the others derive their associated powers; and it would be . . . to little purpose to look for the cause of our passions in association, until we fail of it in the natural properties of things.[48]

Two additional major issues concerning the foundations of taste concern Burke in the "Introduction on Taste" which he added to the beginning of his *Enquiry* in 1759. Burke agreed with Hobbes and others that a significant part of our appreciation for poetry arises out of the unusual resemblances which the artist calls to our attention through a variety of devices, including metaphoric language. Burke was aware, however, that the impact of metaphoric language on a reader depends on the prior experience which he or she has of the things or relationships being compared. Though the principle by which metaphor operates is the same in all humans, the *degree* of pleasure derived depends heavily upon knowledge, which in turn depends upon prior "experience and observation."[49] Burke thus admitted a "natural" difference of taste which produces a superior taste in some persons—a taste based not in any intrinsic superiority of the person but in a range of knowledge derived from more extensive experience and reflection. In this way reason reenters consideration of taste in at least a small way, for it plays a role in generating the background knowledge of objects and their relationships which makes possible a refined and superior taste.[50]

Finally, Burke ends his introduction on taste with a methodological critique of those prior sensationalist aesthetic theories like Hutcheson's (which assumed that taste was a single internal sense different from the traditional five "external" ones) or Gerard's (which posited the existence of several new internal senses such as "novelty, sublimity, beauty, imitation, harmony, ridicule, and virtue"). Again drawing from Newton's rules of reasoning, Burke insisted that taste could be fully accounted for in terms of the normal set of external senses accepted by Hobbes and Locke. Positing additional "internal" senses was thus unacceptable, for "to multiply principles for every different appearance is useless and unphilosophical. . . ."[51]

After a very short set of comments on the importance of novelty for effective art, Burke launched into the central theme of his *Enquiry* by exploring the ideas of pleasure and pain in a way that deviated from that of any prior sensationalist psychologists but which built upon the distinction between self-interested and social passions which had been explored by Hartley. Burke rejected the Hobbesian theory that pain accompanies a diminution of vital motions whereas pleasure accompanies their increase. Instead he agreed with Hartley that both pleasure and pain are associated with *positive* states. Moreover, like Hartley, he was absolutely certain that painful emotions are more powerful and intense than mere pleasure. But Burke did *not* see pain as produced by a simple quantitative increase of the same sensations or ideas that produce pleasure in their more moderate states; so pleasure does not arise in connection with the removal or moderation of painful ideas. Indeed, Burke used the terms *pleasure* and *pain* to deal with passions that have very different sources; and this assignment of pleasure and pain to different classes of experience is the key to his whole aesthetic theory.

All ideas capable of arousing our passions are related either to self-preservation or to society. Those passions connected with self-preservation "turn most frequently on *pain* and *danger* and they are the most powerful of all passions."[52] Those passions connected with other persons—our *social* passions—whether connected with the most basic sexual interactions or with our more general social interactions linked to the notion of sympathy, "have their origins in gratifications and *pleasures*." These passions, associated with the term *love,* are much more tender and gentle than those associated with pain and fear.[53]

The basic difference between the sublime and the beautiful arises directly out of this distinction between the more powerful, painful passions connected with self-preservation and the more gentle, pleasurable passions connected with social intercourse. Burke reserved the term *sublime* for our response to objects or ideas that produce the former, and the term *beautiful* for our response to those that produce the latter:

> Whatever is fitted in any sort to excite the ideas of pain and danger, that is to say, whatever is in any sort terrible or is conversant about

terrible objects or operates in a manner analogous to terror, is a source of the *sublime*; that is, it is productive of the strongest emotion which the mind is capable of feeling.[54]

Beauty, however, is

a social quality; for where women and men, and not only they, but when other animals give us a sense of joy and pleasure in beholding them (and there are many that do so) they inspire us with sentiments of tenderness and affection towards their persons. . . .[55]

Leaving aside for the moment all considerations of beauty, we shall consider Burke's treatment of the sublime, especially as it relates to art. If art works operated on us by producing only terror, pain, and fear, no human beings would subject themselves to artistic experiences because humans naturally flee from pain. But, said Burke, under certain circumstances the terror and pain associated with the sublime can be transformed into *delight*, which is an even more powerfully positive and attractive emotion than *pleasure*.[56] The problem, of course, is to figure out how the most aversive experiences we have can be transformed into the most attractive. Burke attempts to solve this problem on two different levels: the purely emotional and the psycho-physical.

On the purely emotional level, Burke began from the empirical observation that poetry, painting, and the other affecting arts are capable of grafting a delight on "wretchedness, misery, and death itself."[57] The usual Hobbesian explanation for this phenomenon— i.e., that our delight depends on our recognition that we are free "from the evils we see represented"—Burke dismissed on the grounds that it calls for a kind of *reasoning* about our situation; and he was quite certain that an explanation can be found which depends only on "the mechanical structure of our bodies, or the natural frame and constitution of our minds," without any appeal to the influence of reason. In place of any kind of reasoning, Burke appealed first to the natural passion of *sympathy* which had been strongly emphasized by David Hume, and which was made the foundation of virtue by Hutcheson and most other members of the Scottish school, including Ferguson:

Whenever we are formed by nature to any active purpose, the passion which animates us to it is attended with delight or a pleasure. . . . And as our Creator has designed we should be united by the bond of sympathy, he has strengthened that bond by a proportionable delight; and there most where our sympathy is most

> wanted, in the distress of others. If this passion were simply pain-
> ful, we would shun with the greatest care of all persons and places
> that could excite such a passion. . . . But the case is widely different
> with the greater part of mankind; there is no spectacle so eagerly
> pursued, as that of some uncommon and grievous calamity; so that
> whether the misfortune is before our eyes, or whether they are
> turned back to it in history, it always touches with delight . . . by
> an instinct that works to its own purposes, without our [rational]
> concurrence.[58]

This is why tragedy produces delight in the audience rather than
mere aversion and why tragedy is the more powerful, "the nearer it
approaches the reality, and the further it removes us from all idea
of fiction," rather than when we remain aware of the difference
between the representation and that which it represents.[59]

Burke was aware that some art forms, such as landscape
painting, seem to create sublime effects without reference to human
misfortune and therefore without evoking any immediate sympathetic response to transform the objects of terror into objects of
delight. How do these art forms work? In these cases it is our natural pleasure "in imitating, and in whatever belongs to imitation
merely as it is such, without any intervention of the reasoning faculty" that produces our delighted responses.[60]

On the purely descriptive emotional level it would seem that
Burke argued that the terror and fear associated with the sublime
can be transformed to delight in two very different ways. On the
psycho-physical level these two different methods of transforming
fear, terror, and aversion into delight and attraction find a unified
explanation in a theory that was almost certainly suggested by
Hartley's vibrational theory, even though it differs from it in certain
critical features. According to Burke, terror produces "an unnatural
tension and certain violent [e]motions in the nerves."[61] In their most
extreme forms these tensions and motions become dysfunctional,
producing a kind of paralysis connected with those things which
we call "horrible." In a slightly less extreme form, they give us that
extra degree of energy that allows us to flee more rapidly from a
dangerous situation than we could manage under ordinary circumstances; moderated a bit more, they become the physiological
correlates of delight. Thus, wrote Burke:

> . . . if the pain and terror are so modified as not to be actually
> noxious, if the pain is not carried to violence, and the terror is not
> conversant about the present destruction of the person . . . they are

capable of producing *delight*; not pleasure, but a sort of delightful horror, a sort of tranquility tinged with terror; which, as it belongs to self-preservation, is one of the strongest of the passions.[62]

Both sympathy and the imitative passion act by lessening the tensions or violent motions associated with objects of terror just enough to make them delightful rather than painful.

One can recognize in this theory a close resemblance to Hartley's vibrational theory of pleasure and pain according to which increasing pleasure accompanies nervous vibrations of increasing magnitude up to a point beyond which the vibrations become so violent that they literally begin to produce permanent deformations which are painful. Beyond the threshold of pain, yet greater vibrations create yet greater pains. In connection with Burke's theory of sublime effects that are *like* terror but not identified with objects of terror, the Hartlian physiological theory leaves even more obvious traces. Burke, for example, argued that infinite objects and what he called "artificial infinity" are capable of producing sublime responses. Extremely large and uniform objects—i.e., "vast" objects—produce a sublime effect by producing a tension in the muscles of the eyes and in the optic nerves because "the eye must traverse the vast space of such bodies with great quickness, and consequently the fine nerves and muscles destined to the motion of that part must be very much strained."[63] No such strain, however, is produced when the eye scans the same visual field interrupted by a variety of objects because in the process of stopping and starting at the various objects, tensions are relieved. There is, however, a way for artists to create an "artificially" vast or infinite effect through the repetition of a series of similar objects or sensations. Suppose, for example, that an architect constructs or an artist depicts a colonnade of identical pillars:

> It is plain that the rays from the first round pillar will cause in the eye a vibration of that species . . . the pillar immediately succeeding increases it; that which follows renews and enforces the impression; each in its order as it succeeds, repeats impulse after impulse, and stroke after stroke, until the eye long exercised in one particular way cannot lose that object immediately; and being violently roused by this continued agitation, it presents the mind with a grand or sublime conception.[64]

In much the same way the rhythmic beat of a military drum creates a sublime effect, "the tension of the [eardrum] increasing at every blow . . . it is worked up to such a pitch as to be capable of the sublime; it is brought just to the verge of pain."[65] In both of these examples, the resonance with Hartley's vibrational theory is clear.

It would carry us far beyond our present aims to explore precisely how Burke accounted for the sublime effect produced by a wide variety of sensations and ideas, but it is important to give some sense of the ideas Burke considered most important in producing sublime effects in poetry and the other arts. Burke's inventory of sublime ideas clearly influenced subsequent practicing poets and artists—including such painters as Joshua Reynolds, George Barret, David Cox, John Martin, J. H. Mortimer, Henri Fuseli, and James Barry (one of William Blake's favorites),[66] and the great theorist of the picturesque English garden, Uvalde Price.[67] Of course, whatever is associated with terror or danger may produce sublime effects; thus, rugged mountains and the tremendous ocean are paradigmatic sublime objects. Anything connected with the idea of immense and awesome power, like a strong current of water, violent winds, or—above all—the idea of God, is sublime. Vast objects, solitude, darkness, irregularity, loud sounds, sudden changes, or anything which is difficult or obscure produces sublime effects.

On no other single issue was Burke more at odds with his neoclassical predecessors than he was in his treatment of obscurity. For Aristotle, and even more critically for this neoclassical followers in the seventeenth century, clarity of expression, with just enough ornamentation to make it striking, was the chief sign of a great poet or visual artist. For Burke, however, to be clear was to be "little" or trivial. "A great clearness," he wrote, "helps but little towards affecting the passions, as it is in some sort an enemy to all enthusiasms whatsoever."[68] Two specific considerations make obscurity or confusion critical in creating sublime emotional effects. First, "it is our ignorance of things that causes our admiration, and chiefly excites our passions." That is, we are not in awe of or fearful of that which we can understand and control; so that which is clear to us, while it may be beautiful, cannot be sublime.[69] Similarly, Burke argued that our most striking ideas "approach toward infinity, which nothing can do whilst we are able to perceive its bounds; but to see an object distinctly, and to perceive its bounds, is one and the same thing. A clear idea is therefore another name for a little idea."[70]

The power of the obscure to heighten emotions associated

with the sublime leads Burke to insist that poetry must be more emotionally powerful than the visual arts:

> The most lively and spirited verbal description I can give, raises a very obscure and imperfect idea of . . . objects; but then it is in my power to raise a stronger emotion by the description than I could do by the best painting . . . so that poetry, with all its obscurity, has a more general as well as a more powerful dominion over the passions than the other art[s].[71]

Though subsequent Romantic poets, like Blake, often overtly denied Burke's argument on behalf of the emotional force of obscurity and mystery, there is no question that they frequently exploited his insight into the emotional power of the evanescent and not quite graspable. Wordsworth's reflections, for example, are constantly punctuated with a yearning for lost and not quite recoverable insights. In *The Prelude,* he writes:

> . . . The days gone by
> Return upon me almost from the dawn
> Of life: the hiding places of man's power
> Open; I would approach them, but they close.
> I see by glimpses now; when age comes on,
> May scarcely see at all. . . .

And in Book XII of *The Prelude,* Wordsworth exclaims:

> Oh! mystery of man, from what a depth
> Proceed thy honours. I am lost, but see
> In simple childhood *something* of the base
> On which thy greatness stands. . . .[72]

Burke's *Enquiry* played a major role in codifying and popularizing most of those attitudes toward the sublime which characterized the so-called "gothic revival" of the late eighteenth and early nineteenth centuries. It was much less important in its treatment of beauty, but it did embody a powerful and at least partially successful attack on the neoclassical notion that beauty consists in "certain proportions of parts" or in a "fitness to answer some end";[73] so we will briefly consider this portion of the *Enquiry.*

In attacking traditional notions of beauty, Burke attempted to be even more systematically scientific than in his treatment of the sublime; for he knew that on this topic he was fighting a set of well-entrenched doctrines. Thus he adopted a strategy that seems to have been borrowed at least in part from the beginning of Book III of Newton's *Principia*. He began by laying out a set of four "rules" to be followed in his enquiry:

1. If two bodies produce the same or a similar effect on the mind, and on examination they are found to agree in some of their properties, and to differ in others; the common effect is to be attributed to the properties in which they agree, and not to those in which they differ.
2. Not to account for the effect of a natural object from the effect of an artificial object.
3. Not to account for the effect of any natural object from a conclusion of our reason concerning uses, if any natural cause may be assigned.
4. Not to admit any determinate quantity, or any relation of quantity, as the cause of a certain effect, if the effect is produced by different or opposite measures and relations; or if the measures and relations may exist, and yet the effect may not be produced.[74]

Next, Burke considered three different classes of things that humans almost universally agree contain beautiful elements—flowers, animals, and human beings. We find almost every kind of flower beautiful, said Burke, "but flowers are almost of every sort of shape. . . . [I]t is in vain that we search here for any [regular] proportion between the height, the breadth, or any thing else concerning the relation of the particular parts to each other."[75] It is true that each kind of flower has a regular figure and arrangement of leaves, but there are flowers of equal beauty with vastly different proportions. Among animals, too, we find variations of shape with no corresponding differences in beauty. The peacock, with its short neck and long tail vies with the long-necked, short-tailed swan in beauty, and among dogs and cats there is no agreement about what ratio of limb length to body length is most beautiful.[76] Even among human beings, from whose proportions architects like Vitruvius and artists like Leonardo Da Vinci and Albrecht Dürer drew their very principles of beautiful proportions, it is often the case that the same ratios of various parts to one another are found in handsome or beautiful as well as ugly bodies.

From these phenomena and rule 4, it follows that "you must lay by the scale and compass, and look for some other cause of beauty. For if beauty be attached to certain measures which operate from a principle in nature, why should similar parts with different measures of proportion be found to have beauty."[77] Indeed, it seems clear to Burke that architects in particular have really worked backwards. Finding that certain proportions in buildings produce a pleasing effect, they have tried to devise ways of giving credit to their works by linking them to the form of the "noblest work of nature," the human body. In this way they have attempted—wrongly—to impose an artificial principle of beauty on a natural object, in violation of rule 2. By the same token they have mistakenly transferred their ideas to their gardens:

> They turned their trees into pillars, pyramids, and obelisks; they formed their hedges into so many green walls, and fashioned walks into squares, triangles, and other mathematical figures with exactness and symmetry.[78]

But now, in the picturesque English garden movement to which Burke's disciple, Uvalde Price, gave a major impetus, "nature has at last escaped from their discipline and their fetters."

If proportion has little or nothing to do with beauty in nature, how did so many come to believe that it does? Burke is certain—and probably correctly so—that the tendency arose from the reasonings of the Pythagorean number theorists and "from the Platonic theory of fitness and aptitude," in violation of rule 3.[79]

This brings us to the major neoclassical alternative to proportion as a source of beauty; i.e., the idea that "a part's being well adapted to answer its end is the cause of beauty, or indeed beauty itself."[80] To this theory Burke offers a series of dramatic counterexamples. Almost nothing is better adapted to its functions than "the wedge-like snout of the swine, with its tough cartilage at the end . . . so well adapted to its offices of digging and rooting," yet no one would offer the snout of the swine as an example of great beauty, nor would one choose the marvelously useful bag under the bill of a pelican or the porcupine with its protective coat of quills. In human beings we find little beauty in such well-adapted parts as the stomach, lungs, and liver. There is no question that we approve of and admire the fitness of things to their uses, but it often takes long study and sophisticated reasoning to recognize and appreciate the relationship between form and function. Such relationships do not operate naturally, as any primary source of pleasure must. Excellent as fitness for use is, it cannot be a source of beauty except through some train of associations; and even then, many admirably adapted objects will never seem beautiful.

Having undermined the major traditional claims regarding the sources of beauty by using rules 2, 3, and 4, Burke turns to rule 1 in order to isolate those characteristics of objects and ideas that *do* seem found constantly correlated with our experiences of beauty. Of mental characteristics he finds that the "softer virtues," including compassion, kindness, liberality, and easiness of temper, "engage our hearts [and] . . . impress us with a sense of loveliness" which makes for a feeling of beauty.[81] Among sensible objects, he discovers that smallness, smoothness, gradual variation, delicacy, and soft colors like "light greens; soft blues, weak whites; pink reds; and violets" produce that "sinking, that melting, that languor, which is the characteristic effect of the beautiful."[82]

We will not discuss the psycho-physical arguments by which Burke links all beautiful experiences to a relaxation of the nerves and muscles from their normal state—as sublime experiences heighten tensions—but one consequence of that theory does deserve mention because of its subsequent development in connection with the term *synaesthesia*. Strictly speaking, according to Lockean sensationalist psychology, the experiences connected with each sense are independent of one another. In one famous experiment, William Mollyneaux, a mathematician and close friend of Locke, seemed to demonstrate this independence by showing that a man who had been blind and was easily able to distinguish between square and round objects by touch, was unable to tell which was which by sight alone when his sight was restored. Only after simultaneous tactile manipulation and visual observation did the two kinds of sensory experience become associated in the ideas of round and square.

Given the independence of different kinds of sensory experiences, terms like "soft" colors or "sharp" sounds make no natural sense, for they mix terms proper to quite different modes of experience. If they are to make sense at all, it must be as metaphors that depend upon their simultaneous connection with certain objects; but it is frequently difficult if not impossible to construct any rational notion of the "similitudes" that such metaphoric uses must imply. Within Burke's theory, however, there is a natural connection among those sensory experiences associated with beauty; for they all reduce nervous and muscular tensions. Likewise, there are natural connections among those sensations associated with the sublime, for they all tend to produce tension (up to a point). Thus, Burke argued that "all the senses bear an analogy to and illustrate one another," even in the absence of particular habitual associations.[83] This notion was used extensively by Romantic poets, especially Keats, who constantly used synaesthetic couplings like "delicious moan" and "velvet summersong" to intensify the emotional impact of his work.[84]

Before leaving Burke, we must once again address a recurrent issue in this work. Was Burke's *Enquiry* merely a reflection of emergent aesthetic ideas whose scientific character was irrelevant to long-term trends, or did it serve as a shaping force whose scientific character was essential to the paths it opened up? As usual the answer is not straightforward. There is no doubt that many of the general attitudes expressed by Burke reflected earlier seventeenth- and eighteenth-century artistic practices and that his ostentatiously scientific analysis of aesthetic issues sometimes served merely to provide an excuse for his biases. At the same time it seems clear that by his codifying and popularizing activities, he accelerated the growing interest in "the sublime" and established some of the special stock techniques of artists and poets for creating "sublime" effects. Moreover, in at least a few cases—for example, in his discussion of the artificially infinite and in his development of the idea of the mutual reinforcement of the different senses which came to be known as synaesthesia—he produced doctrines, adopted by subsequent artists, which depended critically on the "scientific" psycho-physical theory that he explored.

Science, Technology, and the Industrial Revolution: The Conflation of Science and Technics

8

To this point I have carefully skirted a central issue connected with the impact of science on Western culture—i.e., the relationship between science and technology. After Western science had become virtually identified with Platonic and Aristotelian intellectualism, its institutions and personnel had very little to do with the means by which men and women produced and distributed the goods and services that shaped the material environment in which they lived—i.e., with their technology—until at least well into the early modern period. Even then, most claims that technological advances were dependent on scientific knowledge were more prophetic than reflective of actual conditions. Outside of the medical tradition, which never saw a rigid divorce between theory and praxis, there was only a handful of unusual pre-modern scientists—Archimedes comes to mind as the most outstanding—who clearly combined what we would now call theoretical and applied or scientific and technological concerns. Neither the spectacular engineering accomplishments of the Romans nor the exploitation of water and wind power to replace human and animal power for productive activities in twelfth- and thirteenth-century Europe depended on contemporary scientific knowledge or institutions.

If the monastic institutions of the medieval world served both to preserve elements of the scientific tradition of antiquity and to initiate aspects of the technological dynamism of the late Middle Ages,[1] it was not because the monks recognized any intrinsic connections between science and technology. Instead, they viewed the contemplation of God's works—manifested in scientific attempts to understand the universe—and labor in the world—often manifested in economic activities—as quite different and unequal but nonetheless valid forms of worship.

A few pedagogues did try to incorporate technological topics into the arts curriculum of the medieval monastic schools as early as the twelfth century, but even they made no attempt to link directly scientific with technological concerns. Hugh of St. Victor, for example, accepted the long-standing Christian/Platonic notion that philosophy aims ultimately at "restoring the likeness of man's soul to divine and supernatural substances," whereas the mechanic arts, including fabric making, armament, commerce, agriculture, hunting, medicine, and theatrics, are aimed merely at "relieving the weakness of man's body which belongs to the lowest, or temporal category of things."[2] Nonetheless, Hugh insisted that these arts are deserving of attention, for it is in the construction of the useful arts that man's reason shines forth most impressively:

> . . . it is not without reason that while each living thing is born equipped with its own natural armor, man alone is brought forth naked and unarmed. For it is fitting that nature should provide a plan for those beings which do not know how to care for themselves, but that from nature's example, a better chance for trying things should be provided to man when he comes to devise for himself by his own reasoning those things given to all other animals. Indeed, man's reason shines forth much more brilliantly in inventing these very things than ever it would had man naturally possessed them; nor is it without cause that the proverb says: "Ingenious want hath mothered all the arts." Want it is which has devised all that you see most excellent in the occupation of men. From this the infinite varieties of painting, weaving, carving, and founding have arisen, so we look with wonder not at nature but at the artificer as well.[3]

For our purpose what is most important about this passage is that while it demonstrates a serious interest in topics which fall under our category of technology, it does not suggest that the mechanic arts are tied to scientific knowledge. Indeed it is part of a discussion that is concerned with clearly *distinguishing* between the more important goal of salvation—to which scientific, or philosophical, knowledge contributes—and the less important, merely temporal concerns of daily life to which the mechanic arts are relevant. The idea that the improvement of material conditions of life might depend upon "applied" science or philosophy, which became part of the Renaissance Humanist heritage, was far from Hugh's mind.

Hugh's argument that students in monastery schools should

learn about technological topics made great sense, for monasteries were self-sufficient communities in which economic activities were necessary for survival; however, the intellectual elitist tradition derived from Greco-Roman learning was very strong. Few responded to the practicability of Hugh's didactic plans, and in most monastic communities there was a continuing division of labor in which most economic and artisanal activity was turned over to lay brothers, whereas the monks "labored" largely in scriptoria, copying texts, or in other appropriate intellectual positions; thus, no regular pattern connecting scientific activities with temporal affairs developed. In a few cases the unavoidable contact of intellectuals with technological artifacts led medieval scientists to draw illustrative materials from technology. But there is almost no notion—except in the optical writings of Roger Bacon—that scientific knowledge might be valuable for producing objects of daily use. Indeed, most of the medieval thinkers who concerned themselves seriously about the relationships between science and technology at all were much more likely to be interested in technological artifacts and processes as providing a means "to assist the development of science" rather than vice versa.[4]

Even during the sixteenth and early seventeenth centuries, after Humanists like Juan Luis Vives, Giordano Bruno, Johann Andreae, and Francis Bacon insisted that material progress must rest on scientific knowledge and that science should be judged in terms of its practical utilitarian applications, science was slow to make any major impact on economic life. Accordingly, A. R. Hall, who tried to estimate the practical consequences of the science of ballistics to seventeenth-century military activities, found that science virtually worthless and concluded that in spite of its promise, science had little technological and economic impact until well into the nineteenth century:

> The seventeenth-century revolution in thought and method had molded a science which was potentially capable of effecting profound changes in the means of production, and in fact many writers on science at the time found an important justification for the study of science in the fuller exploitation of natural resources, with the consequent enrichment of human life and alleviation of daily toil which it promised. But this promise was only fulfilled through the industrial and agrarian revolutions of the nineteenth century and the changes in the organization of economic activity which they brought about.[5]

Almost precisely the same judgment has been recently stated by Richard Westfall. In spite of vigorous seventeenth-century claims regarding the scientific foundations of numerous technological innovations, Westfall finds no evidence that "anything to do with the word *science*, as I understand it, properly applies."[6]

The prevailing current view among economic historians and historians of technology seems to be that at least until the mid-nineteenth century, scientific knowledge had relatively little bearing on technological change in general and industrial processes in particular.[7] Since the early 1950s, however, there has been an increasing minority of scholars who see science as central to the "first industrial revolution" of the late eighteenth century. We will follow this minority view, pushing back the economic importance of science-based technology well into the eighteenth century, and to some extent even into the late seventeenth century.

During the seventeenth and early eighteenth centuries, most non-scientists still responded to science in terms of its perceived religious implications—i.e., in terms of the natural theology which it stimulated. Since the mid-eighteenth century, however, most non-scientists have responded to science in terms of its perceived technological implications. It is this changed foundation for widespread responses to science that we seek to understand in this chapter.

WHAT MAKES A TECHNOLOGY "SCIENTIFIC"?

Some of the major disagreements regarding the role of science in transforming the means of production and distribution in modern society hinge upon whether science is understood to be a special body of knowledge or whether it is more fruitfully understood as a set of activities and habits of mind, as I have been arguing. In the former case, in order for a technology to be said to depend upon science, it must incorporate knowledge that had its first historical appearance as an explicit part of an accepted science rather than as an implicit element of some practical activity. Such, for example, is the case in connection with electrical power generation and transmission technologies in the late nineteenth century. These technologies built self-consciously on Michael Faraday's discovery of the laws of electromagnetic induction, laws articulated in the course of a series of scientific investigations undertaken with no immediate concern for economic activities. Even in this case it is perhaps more proper to call the technology science-dependent rather than scientific, according to C. C. Gillispie; for those technologists who appropriated Faraday's discoveries for commercial gain are more properly understood as exploiting scientific knowledge than engaging in the scientific enterprise, which can by definition have no proper end but the creation of abstract theoretical knowledge. From this point of view the "spirit" of science and that of technology are totally independent:

> The scientist works for love of science and to increase his own reputation. When he makes a discovery, he is eager to publish it, and his only concern is to secure his intellectual property for the achievement. The artisan, on the other hand, whether in his own research or in using the research of others is always thinking of his economic advantage. He publicizes only what he cannot keep secret and tells only what he cannot hide. Society benefits from both the disinterested investigation of the scientist and the interested speculation of the artisan. Confound the two, however, and both will lose the spirit distinctive to them.[8]

In many times and places the attitudes associated with science and technology *did* diverge as much as those attributed to scientists and artisans by Gillispie; but the eighteenth century in Western Europe saw a much greater overlap between the values and habits of mind associated with science and those associated with productive activities. Science and technology are not identical; but neither have they been totally isolated from one another in the Western world at any time since the late Renaissance.

The Baconian adage that "knowledge is power" expressed a profound belief of many eighteenth-century scientists and economic entrepreneurs, bridging the intellectual gap which often separates the two groups. Especially in Britain and North America, the utilitarian vision of David Hume and the Scottish common sense philosophers virtually identified the aims of science and technology. Hume insisted that the chief goal of science "is to teach us how to control and regulate future events by their causes."[9] This vision shaped the way in which large numbers of scientists understood their own principal functions. William Cullen, one of the great Scottish chemists of the eighteenth century, explained the aims of chemistry to his students very much in terms of its potential practical applications:

Does the joiner want a particular glue? Does the mason want a cement? Does the dyer want the means of tinting a cloth of a particular color? Or does the bleacher want the means of discharging all colors? It is the chemical philosopher who must supply these. In short, I need but say this one thing: Wherever any art requires a matter endued with any particular physical properties, it is the chemical philosophy which either informs us of the natural bodies possessed of these properties or induces such in bodies which had them not before, or, lastly, produces new bodies endued with the necessary qualities.[10]

Cullen's most distinguished pupil, Joseph Black, insisted on maintaining a distinction between the scientist, or philosopher, and the technologist, or artisan. Like his mentor, he viewed the scientist as the source of improvements in technology. In fact, Black viewed science and technology as mutually supportive; but he was inclined to think that science was more critical for technology than vice versa. Speaking of chemistry in particular, he argued:

Chemistry is not an Art but a science. The Artist is he who puts in practice what the philosopher conceives. . . . But he who studies deeply and thinks on rational methods . . . is a philosopher. . . . The study of the mechanical arts has no doubt done a good deal of service to this science; but those arts are more obliged to chemistry than it to them.[11]

As long as we define science as a systematized body of knowledge, innovations in eighteenth-century agricultural technology, for example, owed little if anything to science; for the spread of new crops (including turnips, clover, and rye), the use of new fertilizers, crop rotations, and irrigation techniques, and the improvement of animal stocks through breeding experiments generally preceded the systematic genetic and chemical theories which can, in retrospect, account for their successes. Even where contemporaries openly claimed a theoretical scientific basis for some innovation, such claims are often dismissed by modern historians who view science as a body of knowledge; because such historians frequently have a bias that blocks them from accepting as part of science any theory that has since been discarded. Thus, for example, G. N. von Tunzelmann, who has many illuminating things to say about agricultural technology and steam power technology in the eighteenth century, cannot assign a scientific component to the development of Watt's steam engine. This is so because in his view the caloric theory of heat to which Watt was committed was part of a "wrong-headed," "so-called science" that must have "positively obstructed study of the theory of the steam engine." Similarly, he denies science a role in the development of Jethro Tull's extremely useful seed drill on the grounds that Tull developed his drill because of a "fallacious" belief that air was an excellent manure—a belief that depended, not on true or real science, but on "bogus" science.[12] By the same token, C. C. Gillispie denies that theoretical science had an important bearing on Nicholas Leblanc's process for making soda on the grounds that the chemical reactions involved were not clearly understood and that Leblanc was guided by a "fallacious analogy with the smelting of Iron ore."[13] Gillispie argues across the board for eighteenth-century chemistry that "it proves extraordinarily difficult to trace the course of any significant theoretical concept from abstract formulation to actual use in industrial operations." Moreover, he complains that when manufacturing chemists did refer to theory, it was to the theory of affinities, which was "devoid of abstract interest."[14]

For Gillispie in particular, this conclusion is oddly disturbing, because as one deeply knowledgeable about late eighteenth- and early nineteenth-century French academic chemists, he is aware that almost without exception the men he studies *believed* that their scientific work was transforming industrial practice. J. A. C. Chaptal even published a three-volume work on this theme, *Chimie appliquée aux arts,* in 1807. In order to square his own perception of the theory-independence of most eighteenth- and early nineteenth-century industrial processes with the undeniable perception among most scientists of the period that science *was* relevant to utilitarian ends, Gillispie has developed an interpretation of the relations between eighteenth-century science and technology which, though it is perversely stated, contains many extremely important insights.

According to Gillispie, eighteenth-century French chemists served industry not through their primary function as scientists— i.e., by developing new theories that could generate new applications—but in their secondary capacity as promoters of Enlightenment and scientific method, as formulated in terms of Condillac's associationist psychology. As educators, scientists tried to impose the orderly, logical, and analytic approach of science on technology largely through the publication of mammoth descriptive projects like that by the Académie des Sciences, *Description des arts et métiers* and that most famous of Enlightenment projects, the *En-*

cyclopédie of Diderot and d'Alembert. These works, Gillispie argues, were "attempts to lift the arts and trades out of the slough of ignorant tradition and, by rational description and classification to find them their rightful place within the great unity of human knowledge."[15] Just as Voltaire and the other *philosophes* detested the hold of superstition and tradition in religion and politics, the scientists detested the hold of tradition on the way in which artisans produced the material objects of the world. So the scientists' aim was not so much to produce technical innovations themselves as to rescue artisans from a bondage to tradition, freeing them to adapt the best techniques available.

Gillispie is certainly correct in viewing the educational aims and activities of d'Alembert, Diderot, and other scientists as playing a key role in spreading knowledge of effective industrial practices; but he both understates the innovative activities of elite scientists—activities such as Berthollet's use of chlorine for bleaching and Gay-Lussac's development of vacuum-sealed vessels for the preservation of cooked foods, for example—and the scientifically influenced activities of technologists like the Montgolfier brothers in France, and Watt, Smeaton, Wedgwood, and others in Britain.

If we move away from the tendency to view science almost exclusively as a body of knowledge and to suppose that the only way in which science might contribute to technological innovation is through the direct application of advanced abstract scientific theory to productive processes, and if we focus instead on the activities and habits of mind that constitute science as a social enterprise, then a scientific technology or science-based technology may develop in one of many ways. It may still be the consequence of applying some undeniable scientific theory or proposition, but it might also be the consequence of adopting (1) a particular pattern of activities modeled on those of the scientists, (2) certain scientific habits of mind in approaching productive activities, or (3) new information about the properties of materials or the effects of processes developed in connection with scientific activities (whether or not that information was adequately explained by contemporary scientific theory). We might then readily admit that scientific *theories* were relatively unimportant in connection with technological innovation until well into the nineteenth century at the same time that we insist that *science* was nonetheless a major contributor to innovations that greatly increased agricultural and industrial productivity in eighteenth-century Europe.

The eighteenth century *did* see the spread of an *approach* which incorporated self-conscious, controlled experimentation, the careful measurement and quantification of relevant features of phenomena, the testing of provisional hypotheses, and the systematic collection and dissemination of information—all features drawn from the sciences—to a broad range of technological problems. From this perspective the encyclopedists were acting *scientifically* as they surveyed current technological practices, rationally analyzed industrial procedures, experimentally evaluated the relative efficiency of different processes, and urged practitioners to abandon traditional methods in favor of more effective and profitable ones.

SCIENTIFIC ATTITUDES AND THE DIFFUSION OF TECHNOLOGY: THE AGRICULTURAL REVOLUTION IN EIGHTEENTH-CENTURY ENGLAND

If we consider the agricultural revolution that established the necessary preconditions for the British Industrial Revolution, we see a pattern of developments that spread through almost every sector of British economic life by the beginning of the nineteenth century. Between 1720 and 1820, the yields of both plant and animal products roughly trebled in Britain, and large tracts of previously unproductive land were put into production through the use of new crops and new techniques of irrigation and fertilization, making Britain a frequent exporter of foodstuffs during the second half of the eighteenth century in spite of a rapid rise in population. Some of this increase in productivity came about because of changes in the organization of farming and the commitment of additional capital and labor resources to traditional agricultural technologies; but the vast majority was a consequence of technological innovations— the spread of new crops, new implements, new irrigation techniques, the "setting" of seeds in the soil, the use of more and new manures, and the improvement of breeds of cattle, sheep, swine, and fowl.[16] Thomas Coke, for example, increased his farm income tenfold during a forty-year period during the late eighteenth century by forcing his tenants to use the latest in manuring techniques and new crops. The efforts of stock breeders such as Thomas Bakewell led to the startling increases in animal size indicated in the following table:[17]

Average Weight of Animals Sold at Smithfield (in lbs.)		
	1710	1795
Oxen	370	800
Calves	50	150
Sheep	38	80

Though it is true that none of the technical innovations in eighteenth-century agriculture can be traced directly and unambiguously to any specific lasting advance in chemical or biological theory, they did follow in large part from the adaptation of scientific activities and habits of mind to agricultural concerns. Moreover, scientists—i.e., publishing members of organizations like the Royal Society of London, the Académie des Sciences in Paris, and the Royal Society of Edinburgh, as well as holders of academic chairs in chemistry and medicine—were frequently to be found among the chief promoters of the application of experimental methods to agricultural problems and the chief proponents for the spread of new techniques. In England, for example, Robert Boyle and Stephen Hales were among the strongest advocates of agricultural experimentation. In Scotland, William Cullen, the chemist; Francis Home, Professor of Medicine at Edinburgh; and Henry Home (Lord Kames), amateur scientist and founding member of the Royal Society of Edinburgh were key members of or advisors to "The Honorable The Society of Improvers in the Knowledge of Agriculture in Scotland" (founded in 1723) which stimulated agricultural improvement not only in Scotland, but also in northern England. In France, both Claude Adrien Helvetius and Antoine Lavoisier established model experimental farms that were widely admired and emulated, self-consciously exploiting behavior patterns which they understood as scientific, for purposes of increasing agricultural productivity.[18]

A typical example of those innovations whose cumulative effect so dramatically increased eighteenth-century British agricultural productivity was the introduction of smaller, lighter plows that required fewer draught animals, but still turned the earth over effectively. The Rotherham plough, based on Dutch designs, was introduced into northeastern England around 1750. Almost immediately, the appearance of the new plough inspired several scientists interested in agricultural improvement to promote both the use of the Rotherham plow and the design of increasingly efficient plows. During the early 1750s, for example, William Cullen offered a public lecture at Glasgow, "On the Construction and Operation of the Plough."[19] More importantly, the mathematicians John Arbuthnot and James Small designed new plows and published works on plow design which included plans and detailed descriptions that allowed local blacksmiths to make the new plows, thus increasing their use.[20]

Another device that gradually penetrated British agricultural practice during the mid-eighteenth century, largely due to the promotional efforts of the scientific members of the Society of Improvers in the Knowledge of Agriculture in Scotland, was Jethro Tull's seed drill. Designed in 1701, and described in Tull's book, *The New Horse Houghing Husbandry* in 1731, the chief benefit of the seed drill was that it set seeds in parallel rows, allowing for efficient mechanical hoeing techniques to replace labor intensive hand hoeing. What fascinated the Scots most, however, was Tull's incorrect insistence that frequent plowing and hoeing brought air, which was an excellent manure, into contact with the crop roots. This claim intrigued Cullen and the two Homes particularly because it appealed to their theoretical and experimental interest in developing new "manures" or fertilizers. During the early 1750s their interest is evidenced in the correspondence between Cullen and Henry Home. In 1752, Home writes to Cullen:

> Having been entertained with no new theory now for a long while, I am sinking into a mere practical farmer. I have not a single new thing at present, except one experiment I am making to convert moss into dung, by endeavouring to rot it in a dunghill, by mixing it with fresh horse-dung. I shall let you know the result. If I succeed I shall be able to multiply my manure greatly.[21]

Francis Home explored the subjects of manuring and tillage in his *Principles of Agriculture and Vegetation* (Edinburgh, 1757) which set out to discover "how far chemistry could go in settling the principles of Agriculture," and to promote "a spirit of experimental farming over the country."[22] Henry Home followed, nearly twenty years later, with his immensely popular *The Gentleman Farmer; Being an Attempt to Improve Agriculture by Subjecting It to the Test of Rational Principles* (Edinburgh, 1776). Both works encouraged the use of a variety of valuable soil supplements as well as the use of seed drills and horse-hoeing.

The agricultural writings and lectures of scientists such as William Cullen, John Arbuthnot, James Small, Francis Home, and Henry Home, with their apparent influence on the diffusion of new implements and fertilizing techniques, illustrate the centrality of spreading scientific attitudes for the diffusion of new technologies during the eighteenth century. Peter Mathias has nicely summarized a mass of evidence on this issue. He writes:

> Coupled with false premises about chemical reactions were urgent pleas for experimentation, shrewd observation and recording, the comparative method, seeking alternative ways of doing things which could be measured and tested to see if they were superior to the old. This was a programme for rejecting traditional methods justifiable only because they had always been done in that way. . . . Scientific procedures and attitudes encouraged by the scientists may have been more influential than the scientific knowledge they dispensed . . . [and] the publicity given to new methods, new crops, rotations, and implements by these same groups may also have increased the pace of diffusion of innovations in agriculture.[23]

SCIENTIFIC ACTIVITIES AND TECHNOLOGICAL INNOVATION: SMEATON'S IMPROVEMENT OF WATER WHEELS AND WEDGWOOD'S INNOVATIONS IN POTTERY TECHNOLOGY

The British Industrial Revolution depended heavily on the work of a class of men who began to designate themselves as civil engineers during the mid-eighteenth century. These men designed and constructed the canals, bridges, roadways, and later, the railroads, which produced the transportation revolution that was a key to industrialization. At least until the third decade of the nineteenth century they were primarily responsible for creating the prime movers—the water wheels and column-of-water engines—which powered the factories and forges until steam engines finally overtook them for many productive purposes. Largely self-educated, the civil engineers were nonetheless well read in both mathematics and natural philosophy.[24] Many of the most distinguished viewed themselves as men of science (about half of the initial membership of the Society of Civil Engineers were also publishing members of the Royal Societies of London and Edinburgh), they viewed engineering as heavily dependent on the sciences, and they frequently based their innovations on scientific studies.

On all of these counts, John Smeaton (1724–1794) typifies the elite among eighteenth-century British civil engineers. Like James Watt, Smeaton was apprenticed to a scientific instrument maker and spent much of his early career making air pumps, electrical machines, and working models of various devices for scientific lecturers.[25] In connection with his model construction he developed an interest in empirical studies of engine efficiencies, much as Watt did. In fact, during the 1760s and 1770s, just before Watt's separate condenser engine came into use, Smeaton was experimenting with a Newcomen-type steam engine, carrying out some 130 experiments over a four-year period.[26] Building on his results, Smeaton was able to construct the most efficient commercial steam engines then operating, nearly tripling the efficiency of Newcomen's engines.[27]

Smeaton sought out the company of scientists, becoming a member of the Royal Society of London and a frequent visitor at the meetings of the Lunar Society of Birmingham, whose members included Joseph Priestley; the zoologist, Erasmus Darwin; and the potter, Josiah Wedgwood, as well as both James Watt and Matthew Boulton. By any reasonable measure, Smeaton was himself a scientist, publishing a number of papers in the *Philosophical Transactions of the Royal Society of London* on mechanics[28] and astronomical instruments[29] in addition to the major paper on water and wind power which will be discussed below. Finally, when Smeaton joined with a group of like-minded friends to establish the first British Society of Civil Engineers, he emphasized the dependence of engineering on the sciences, advocating in particular, "superior knowledge in practical chemistry, mechanics, [and] natural philosophy."[30] To emphasize the close links between the sciences and civil engineering, the Society created a special class of honorary members to include "men of science . . . who had applied their minds to subjects of Civil Engineering, and who might, for talents and knowledge, have been real engineers."

Smeaton's greatest claim to fame as a scientific engineer and as a dominant figure of the early Industrial Revolution came from his innovations in water wheel design, which were among the most important engineering achievements for the conduct of eighteenth-century industry. They allowed for an effective fifty percent increase in power extracted from England's rivers and streams at a time when there was a developing crisis of power needs. The development of steam power was even substantially delayed because of the dramatic success of Smeaton's new water wheel designs, which were widely emulated throughout Britain between 1759 and 1800.[31]

Few engineering developments of the century were a more direct consequence of scientific activities. When Smeaton considered the power transferred to water wheel blades by flowing water, it occurred to him that water wheels could be made much more efficient if less power were wasted in creating turbulence. In order to test and quantify his suggestion, Smeaton constructed model water wheels to measure the amount of work done in a given time by determinate volumes of water falling through known heights under different configurations. The results were reported to the Royal Society of London in May of 1759 and published in "An Experimental Enquiry Concerning the Natural Powers of Water and Wind to Turn Mills and Other Machines." Assuming that in his devices (fig. 10) the total power available to drive the wheels was given by the rate of flow of the mass of water times its height of fall, or "head," Smeaton discovered that the greatest fraction of input power was converted to usable power by overshot wheels in which water entered buckets near the top of the wheel moving very slowly, driving the wheel by the gravitational effect of the falling water. In the most effective case, sixty-three percent of the water's power was converted to usable power. At the other end of the scale, an undershot wheel struck by water emerging rapidly from a restricted opening converted only about twenty-two percent of the water's power to usable power. In general, overshot wheels were found to be about twice as efficient in utilizing water power "under the same circumstances of quantity and fall" as undershot wheels, and overshot wheels were the more efficient, "the higher the wheel is in proportion to the whole descent."[32]

Though Smeaton's experiments had not included breast wheels (fig. 11), he concluded that the total power of a breast wheel would be given by the sum of the powers of an undershot wheel with a head given by the head of water minus the height at which the water struck the wheel plus that of an overshot wheel with a head equal to the difference between the tail water and the height at which the water struck the wheel.[33] In many cases, since water was easier to channel for breast wheels and since large wheels with their long lever arms could be used with low heads of water, Smeaton argued that breast wheels would usually be the most practical wheels to use. He immediately began constructing large breast and

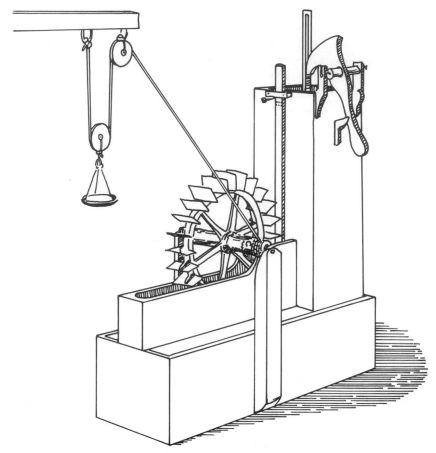

Fig. 10. One of Smeaton's models for measuring the power generated by
different water wheel designs.

overshot water wheels in place of the almost universal undershot
installations. During the next fifty years almost all engineers followed his patterns in creating new water wheels, and many old
installations were converted. Very few undershot wheels had generated as much as 10 horsepower; but by the early nineteenth
century, high breast wheels using iron axles and spokes (another
innovation pioneered by Smeaton because of links with the Carron
iron works in Scotland) were generating up to 200 hp. while the
largest commercial steam engines were producing about 60 hp.[34]

Smeaton's experiments received the Royal Society's Copley

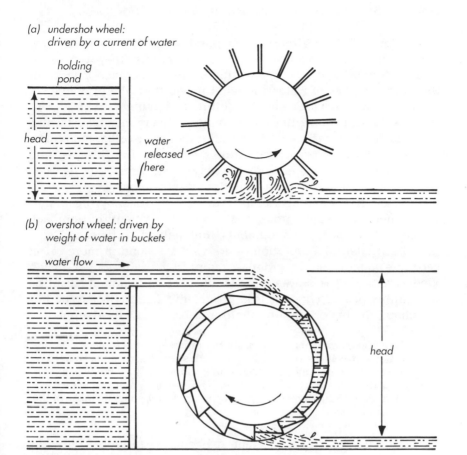

(a) *undershot wheel:*
driven by a current of water

holding
pond

head

water
released
here

(b) *overshot wheel: driven by*
weight of water in buckets

water flow

head

(c) *breast-wheel: combines*
both the above principles

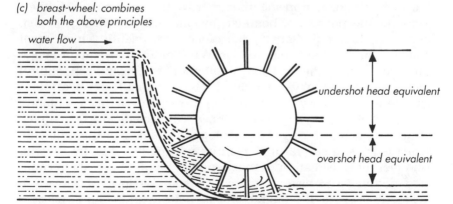

water flow

undershot head equivalent

overshot head equivalent

Fig. 11. The different water wheel configurations discussed by Smeaton.

Medal for 1759, were reprinted separately in 1794, and became classics in applied science. Nearly a century after they were done, George Stephenson, the pioneering railroad engineer, wrote of Smeaton's studies of water and wind power that "the principles of mechanics were never so clearly exhibited as in his writings. . . . To this day there are no writings so valuable as his in the highest walks of scientific engineering. . . ."[35] Whether there were other more valuable contributions or not, Smeaton's writings, his designs, and his completed projects embody the very close connection between eighteenth-century scientific activity and technology.

If the civil engineers of Britain—exemplified by Smeaton—constituted one major group that brought a scientific approach to the technological developments that helped initiate and sustain the Industrial Revolution, there was a second, a more select yet more heterogeneous group—exemplified by the potter Josiah Wedgwood—that was even more famous for its fusion of scientific with industrial activity. Of this group, the Lunar Society of Birmingham, the historians Archibald and Nan Clow have argued:

> Taken *en masse* this group probably represents the highest concentrations of Fellows of the Royal Societies that has been associated at one time with any industrial undertaking. . . . [T]here was not an individual, institution, or industry with pretention of contact with advancing technology throughout the land, but some member of the Lunar Society group had connexions with it.[36]

Over its roughly thirty-year existence (ca. 1765–1795), the Lunar Society had no more than fourteen members, beginning with the buckle maker, iron manufacturer, and entrepreneur Matthew Boulton; the physician, botanist, mechanical inventor, geologist, and poet, Erasmus Darwin; the clock and scientific instrument maker George Whitehurst; and Dr. William Small, friend of Benjamin Franklin, teacher of Thomas Jefferson, mathematician, natural philosopher, and physician. By 1765 this small group of close friends was regularly corresponding or meeting to discuss literary or scientific topics, especially those connected with electricity, steam engines and carriages, geology, and heat. In the following year the first of several new recruits was gathered in by Darwin, who met the new man in connection with shared interests in canal construction. The new recruit was Josiah Wedgwood, already a modestly successful manufacturer of pottery and soon to become one of the most successful entrepreneurs of the century. Wedgwood introduced the architect and geologist John Whitehurst to the group, and soon after, the early membership was completed with the addition of James Kier, an army officer turned industrial chemist; Joseph Priestley, one of the outstanding chemists of the century; James Watt, the instrument maker and inventor of the separate condenser steam engine; Richard Lowell Edgeworth, the gentleman inventor; and Thomas Day, the eccentric poet and generous financial backer (we might now say venture capitalist).

Though the careers of several of this group could be used to illuminate the connections between scientific activities and technological innovation, we will focus on that of Josiah Wedgwood, because his work best embodies the curious mixture of almost blind empirical emphases with experimentation guided by clever— if sometimes false—scientific theories and hypotheses, typical of the amateur scientist/inventory/entrepreneur of the period.

Wedgwood was born in 1730 into a family of artisan class potters at Burselm, Staffordshire. He had very little formal schooling, and by age nine he was working full days in the family pottery. At fourteen he was apprenticed to his brother to learn the craft; but when he was about twenty, Wedgwood contracted a severe case of smallpox, whose complications left him unable to continue his work for a period of nearly two years. During this period, he undertook a program of self-education, guided by a brother-in-law who was a Unitarian minister with a passionate interest in science. Under his brother-in-law's tutelage, Wedgwood began both his scientific reading and a program of experimentation on clays, glazes, and methods of controlling the materials and processes for producing pottery which eventually led to the creation of Jasperware, Queensware, and the factory production system at Etruria, all of which were keystones of the mass production and marketing program that made Wedgwood's the greatest financial success among Lunar Society members.

One of the claims frequently made by those who deny or minimize the scientific element within technological innovation during the eighteenth century is that the widespread empirical or "experimental" emphasis of technological innovators was the kind of crude empiricism of unlettered artisans that has always been part of craft traditions.[37] From their point of view, scientific empiricism differs from mere craft empiricism in being more systematic, more highly quantified, and—according to the hard-liners—being informed by "pre-existent theoretical knowledge of a non-trivial character."[38]

Even by these stringent standards, much of Wedgwood's industrially oriented empirical work was undoubtedly scientific. Consider, for example, his work on pyrometry (the measurement of high temperatures). Wedgwood was concerned with the determination of temperatures well above those measured by contemporary alcohol or mercury thermometers because regulating the temperatures of pottery kilns is crucial for practical purposes (the colors of glazes and of fired clays depend on the temperature at which they are fixed; and cooling kilns too slowly is inefficient, while cooling them too rapidly may produce cracked and defective products). Thus temperature was an obvious variable to be considered in experiments on clays and glazes. During the 1740s, two of Wedgwood's relatives had invented "pyrometric beads," small pieces of various material that changed color at different points as the temperature of a kiln increased.[39] But there were serious drawbacks to this means of measuring oven temperature. In most cases the color changes were irreversible, so while the "beads" might indicate temperature on the way up, they were useless in controlling cooling. Furthermore, the beads were not calibrated to any standardized temperature scale. They simply indicated a hierarchy of degrees of heat. By 1780 Wedgwood was seriously disturbed by the lack of effective ways of measuring high temperatures; so he set out to develop some kind of new pyrometer. He began by searching systematically through some fifteen *Philosophical Transactions* articles spanning the previous fifty years on thermometry. (In a typical note he wonders how a Dr. Fordyce determined by a mercury thermometer that bodies begin to be luminous in the dark at about 700° Fahrenheit since mercury boils at about 600°.)[40]

In his first experimental approach to that problem, Wedgwood extended the pyrometric bead concept, suggesting a way to relate the points at which pieces of Earth of Alum, iron calx, and clay reached certain colors to the Fahrenheit temperature scale. His method depended on the theory, explored by Joseph Black and first published by Adair Crawford in 1779, that materials had virtually constant specific heat capacities over large temperature ranges, so that it took almost the same amount of heat to raise one unit of mass or volume of water, for instance, one degree Fahrenheit from 32° to 33° as it did to raise it from 210° to 211°. Assuming "Crawford's principle" and ignoring the difference between the specific heat capacities per unit volume of water and that of his ceramic thermometer pieces, Wedgwood determined the temperature of each sample at a certain color by dropping the hot piece into a large known volume of water and multiplying the temperature rise of the water by the ratio of water to ceramic volume. The resulting number was the excess of the initial ceramic piece temperature over the initial water temperature.[41]

Though his failure to take into account the differences in specific heat capacity between water and his thermometer pieces vitiated the validity of his calibration, Wedgwood's proposed method was clever; and almost immediately, William Playfair, who managed Boulton's Soho Iron Works, wrote to Wedgwood suggesting how to straighten out the problem.[42]

Within little more than a year, Wedgwood had given up on the thermometer pieces in favor of a ceramic pyrometer whose color varied continuously from a dull red to nearly white as the temperature rose. At each of several critical points—e.g., the melting point of iron—the color was matched to a portion of a colored spectrum printed on paper, and the Fahrenheit equivalent was established by the calorimetric technique suggested in the first paper but corrected by Playfair. This new device was described in "An Attempt to Make a Thermometer for Measuring the Higher Degrees of Heat, from a Red Heat up to the Strongest that Vessels made of Clay can Support" in 1782. It was immediately adopted by potters to regulate their kiln temperatures and remained the preferred industrial pyrometer for several generations. Wedgwood also developed a ceramic pyrometer that depended on shrinkage, and he was asked by the outstanding chemists of the time, including Priestley, Antoine Lavoisier, and Armand Seguin, and by the geologist Faujas de Saint Ford, to provide pyrometers for their use; moreover, one of his pyrometers was used by Spallanzani to determine the temperature at which volcanic lavas begin to flow.[43] In this case there seems to be no doubt that a new device of substantial commercial value had emerged out of a set of experiments that were scientific in every sense of the word. That is, they were consciously set within a scientific tradition of studies of thermometry; they were guided by nontrivial theoretical insights that justified the calorimetric techniques used to relate fixed points on the color scale or degrees of shrinkages to the Fahrenheit scale; and the resulting techniques for measuring high temperatures were in turn used by scientists studying other phenomena as well as by industrialists who used them for controlling the production of ceramics.

A similar story can be told of the work Wedgwood undertook

in connection with problems faced by Lunar Society member James Kier in his glass manufacturing. Society members frequently helped one another, so when Kier complained in 1779 that it was virtually impossible to create flint glass without waves and streaks in it, Wedgwood decided to try to find out how the waves were produced and how to eliminate them. This time, Wedgwood had his secretary, Alexander Chisholm, survey the scientific literature and copy out relevant passages.[44] After discussion with Kier, Wedgwood recorded two hypotheses regarding the imperfections in flint glass: first, that the veins were somehow produced because "it [flint glass] is composed of materials of . . . different densities"; second, that the wave lines were produced by the motion of heavy particles through the cooling glass, "leaving an indelible tract [*sic*] behind them."[45] Wedgwood showed that the many lines in flint glass could be perfectly modeled by mixing "gum waters" of different densities; so the supposed motion of heavy particles was unlikely. Then he had experiments done by two independent glass makers—one in Liverpool, and one in London—as well as by his own staff at Etruria. These experiments showed that under normal manufacturing conditions, the specific gravity of the molten glass increased substantially from the top of the melting pot to the bottom; and tests of samples from different strata showed significantly different indices of refraction. When the melt was poured, different strata were incompletely mixed, leaving apparent "waves" and "veins" because of local variations in refractive index. As a consequence, Wedgwood recommended either striving to ensure homogeneity or making lenses or plates from material taken from a single stratum of glass in the melt.[46] The paper reporting Wedgwood's results, "An Attempt to Discover the Causes of Colors and Waviness in Flint Glass, and the Most Probable Means of Removing Them," was completed around 1783, but it was never read to a public society even though it had been intended for delivery at the Royal Society. It is at least possible that Kier wanted the information for proprietary use. In any event, within about fifteen years after Wedgwood's manuscript was completed, his recommendations were adopted throughout the glass-making industry.

The most important of Wedgwood's experiments were those done on various clays and glazes in the search for aesthetically pleasing, durable, and easily produced pottery and specialty items. Sometimes these experiments were clearly guided by theoretical knowledge, as when Wedgwood sought to use clay with less "fixed air" to reduce the blistering in mortars that he was producing for the chemical trade. Wedgwood's commonplace book shows that as a result of reading Priestley's theoretical discussions, he concluded that blistering was caused by the release of fixed air. Even when the "theoretical" element was minimal in his experiments, they very frequently depended on following up effects reported in scientific journals. Chisholm was directed to pore over virtually anything available on the behavior of clays; and he filled page after page of Wedgwood's commonplace book with the discussions of chemists, geologists, mineralogists, and other naturalists for his employer's consideration. Finally, when he began a series of experiments, Wedgwood was systematic in varying one factor (the amount of one ingredient, the temperature of firing, etc.) at a time; he provided a careful quantitative record of all relevant factors; and he was almost unbelievably persistent and patient. Over ten thousand samples of clay mixes survive from the series of experiments which Wedgwood undertook to improve his Jasperware, although experiment 3681 ultimately provided the best jasper, and 3839 produced a desirable yellow jasper.[48] Neil McKendrick, one of the shrewdest recent students of Wedgwood comments:

> Experimental inquiry carried to this level of persistence, backed by scientific advice, sometimes made possible by scientific method, and rendered industrially reliable by accurate measurement, not surprisingly produced a high yield in successful innovations.[49]

Some scholars have argued that men like Josiah Wedgwood and John Smeaton were so rare in eighteenth-century England that even if it can be shown that their work was significantly influenced by science, this does not mean that such influence was widespread and of major industrial significance. It is true that Smeaton and Wedgwood were unusually successful in their trades and more science-oriented than many eighteenth-century Englishmen; but the degree of scientific literacy among English artisans and businessmen was substantial;[50] and scientific efforts to achieve innovations were constantly being promoted in provincial societies, masonic lodges, and the pages of popular journals like *Gentlemen's Magazine* as well as through premiums offered by the Royal Society of Arts, Manufactures, and Commerce (founded in 1754). Thus it seems quite wrong to accept the claim of A. R. Hall that during the eighteenth century "the impact of any question of abstract science upon a human brain . . . could only happen to, say, one individual in a hundred thousand."[51] In England at least one in a hundred would seem a much more reasonable guess. The extent of scientific literacy and enthusiasm is irrelevant at one level, for the works of Smeaton, Wedgwood, and their like were clearly imitated even by those who were shrewd businessmen almost totally innocent of scientific—or any other—education. So even if Wedgwood had been the only potter whose works were *directly* influenced by science, the whole industry would have been rapidly transformed by a raft of Wedgwood imitators capable of no more than bribing one of his workmen to get a little inside information.

SCIENTIFIC HABITS OF MIND AND THE ORGANIZATION OF WORK: WEDGWOOD'S ETRURIA POTTERY

It should hardly be surprising that industrialists who directed their attention to the development of new products and processes should also have attempted to order, quantify, analyze, and improve the organization of work in their manufactures. Foremost among the rationalizers of eighteenth-century industry stood Josiah Wedgwood, whose pottery became a model of efficient manufacturing. With a passion for precision and detail worthy of a William Petty before him, but with greater shrewdness and authority that came from owning what he managed, Wedgwood sought to achieve the greatest possible division of labor, the most efficient flow of materials through his shops, and the most waste-free use of raw materials. As for his workers, his prime goal was to control their every move so as "to make such *machines* of the men as can not err."[52]

At Etruria the workship for each class of objects was kept separate from the others so that neither the materials nor the workers involved with coarse and useful works intermingled with those for fine and ornamental ones. Within each assembly line, workers did not fabricate an entire object, going from one procedure to another, as in other potteries. Each perfected a single task. In fact, in the painting rooms, tasks were divided so that the border painters, the flower painters, and the fine figure painters were separated, with differential wage rates for different levels of skill. By the early 1790s, of 278 employees at Etruria, only five were general laborers. All the rest were highly specialized workers whose skills had been honed so as to produce goods of a quality that no competitor could match at double the price.

For the quality and quantity of Wedgwood's wares, specialization was an unalloyed blessing; for the workers involved, however, the situation was not so clear-cut. In the first place, reduced demand for certain kinds of goods put highly specialized workers in danger of unemployment; thus, when gilded dishes went out of fashion, Wedgwood's gilders lost their jobs. Wedgwood sympathized with their plight, but as he wrote, "gold, the most precious of metals, is absolutely kicked out of doors, and our poor Gilders I believe must follow it."[53] More important, in such highly specialized work, the ability of all workers to get on with their tasks often depended on the productivity of each. So the typically relaxed atmosphere and independent pace of traditional craft manufacture had to be replaced with a schedule in which everyone began and broke from work at the same time and in which the pace of work was coordinated and regulated.

Wedgwood was among the first to use a system of bells to signal times to begin work or take breaks, and he expended great care and thought to encouraging and enforcing habits of punctuality and reliability. A student of Hartley's utilitarian psychology, he well knew the power of the carrot as well as the stick, so he engaged a "clerk of the manufacturing," whose major function was "to encourage those who come regularly to their time, letting them know that their regularity is properly noticed, and distinguishing them by repeated marks of appreciation, from the less orderly part of the work people by presents or other marks suitable to their age, etc."[54] Of course, those who were repeatedly late had their wages docked in proportion to the time they missed. In order to determine the arrival and departure times of his employees efficiently, Wedgwood experimented with a number of schemes, finally settling on a method devised by his Lunar Society and clockmaker friend, George Whitehurst. Whitehurst invented the first punch-in time clock, and Wedgwood installed the system at Etruria.

Finally, Wedgwood gave the same attention to controlling materials in his potteries as to controlling the workers. Initially he simply tried to teach his workers to cut down on clay waste; next he purchased scales for the workers so they could weigh out just the amount of clay needed for each job; and finally he installed a "clerk of weight and measure," whose sole job was to weigh out appropriate amounts to be given to each potter. In proposing this job Wedgwood wrote to a friend that "he shall save me three times his wages in clay—the first clever fellow I can spare shall certainly be set down to this business."[55]

It is certainly true that the division of labor was something that had been developing unconsciously for a long time without scientific attention. It is also true that scientific attitudes had little to do with manufacturers' desires to get a penny's worth of work for a penny's wage. But the systematic concern with organization and efficiency shown by Wedgwood, the extreme conscious emphases on division of labor, the development of reward and penalty systems borrowing from contemporary psychological theory, and the tendency to experiment constantly with new techniques of control, all seemed to owe something to scientific habits of mind.

THE PRODUCTS OF SCIENCE AND TECHNOLOGICAL INNOVATION: THE CASE OF CHLORINE BLEACHING

Scientific activities often lead to the discovery of new processes for producing materials, the recognition of new properties of traditional materials, or to the creation of new substances or objects, any of which might be exploited for technological ends. In some cases, as when the chemist Peter Woulfe discovered a new inexpensive method for producing pure cobalt (a key material for producing blue glazes) in the course of a series of experiments on the distillation of acids and alkali salts, the new techniques could be incorporated without further modification into industrial practice. Woulfe sent his recipe for purifying cobalt to Josiah Wedgwood who used it immediately in producing his blue jasperware.[56] More frequently, scientific discoveries become the bases for industrial research efforts that modify them into commercial products or processes. Such was the case in connection with chlorine bleaching, one of the most crucial innovations in textile manufacture during the late eighteenth century. By the late 1760s a serious bottleneck had developed in textile production at the bleaching and dyeing stage because mechanized techniques for spinning and weaving had become so successful.[57]

Traditional methods of bleaching involved dipping the fabric in water, boiling it in weak lye water, exposing it to sunlight for several days or weeks in bleach fields, and finally "souring" the cloth by soaking in sour milk (or vinegar, where it was available). This process was both exceedingly slow and—in terms of the total production costs of textiles—exceedingly costly. Some of the raw materials, especially sour milk and vinegar, were always in short supply because of priority uses given to the raw materials (milk and wine) from which they were produced. And huge investments in both land and barrels were required for large-scale bleaching.

Because of the availability of cheap, flat land in the early eighteenth century and because of the access to lye water produced from burned kelp, the region around the Firth of Forth in Scotland saw the growth of large bleaching establishments serving textile manufacturers from the English Midlands and as far away as London. Bleaching thus became a key element in the Scottish economy. The Board of Trustees for Fisheries, Manufactures, and Improvements in Scotland (founded in 1717) subsidized the laying down of bleachfields and promoted improvements in bleaching;[58] and Scottish chemists took a special interest in problems related to bleaching. William Cullen, Francis Home, Joseph Black, and even James Watt devoted substantial efforts to the problems.[59] As a consequence of Home's suggestions in *Experiments on Bleaching* (1754), diluted sulfuric acid replaced sour milk for souring with a ninety-five percent reduction in souring time.[60] Numerous attempts were made to find a way to replace potash (made from scarce wood ashes or limited seaweed supplies) with lime water prepared from limestone as a source of alkali; but no commercial and officially acceptable techniques were discovered. (Most cheap ways of producing lime water left traces of caustic alkali suspended in it, with destructive long-term consequences to the fabric or its user.)

In 1774, the Swedish chemist Karl Scheele made a dramatic new discovery that led to a complete transformation of the bleaching industry. Involved with Priestley, Lavoisier, and other chemists in the search for new gasses, Scheele produced a new chemical entity which he called "dephlogisticated marine acid" (i.e., Chlorine, Cl_2) by reacting marine acid (HCl) with black calx of manganese (MnO_4).[61] During his attempts to test his new "air," Scheele discovered that it rapidly removed the color from plants, flowers, and paper. When the French chemist C. L. Berthollet learned of the new gas, he not only reinterpreted it as oxymuriatic acid (not until 1809 was the new chemical recognized as the element chlorine) but also recognized its potential as a textile bleaching agent and developed a cheaper method of production starting from the plentiful and inexpensive materials, salt (NaCl) and "oil of vitriol" or sulfuric acid (H_2SO_4). Berthollet reported his discoveries in several articles between 1785 and 1790,[62] and during the fall of 1786 he demonstrated to James Watt, who was visiting in Paris, his new technique for bleaching cloth in a liquid made of chlorine dissolved in plain water.

Watt almost immediately communicated information about Berthollet's process and sent a bottle of chlorine water to James McGrigor, his father-in-law, who happened to be a Glasgow bleacher; and after a series of trial experiments on various types of cloth, McGrigor was bleaching cloth in substantial amounts—500 yards at a time—by early in 1787.[63] Though Watt attempted to get a patent for using the new process, the secret was soon out, and a whole series of bleachers were demonstrating the speed and quality of the new bleaching process. In March of 1788, for example, the Manchester bleacher Joseph Barker took in a 28-yard piece of Grey calico one afternoon, bleached it that evening, had it printed the next day, and made it available to the public on the third day. Two weeks later Thomas Henry of Manchester showed that a variety of different cloths could be whitened with chlorine water in times ranging from three hours to five minutes.[64]

In spite of the speed of early chlorine water bleaching, it was not an immediate commercial success for several reasons: the costs of producing and transporting the bleaching liquor (which was so volatile and corrosive that special production and handling techniques had to be developed) were very high; various impurities in cloth—especially in linens—reacted with the chlorine, leaving an unacceptable yellow color; and chlorine tended to weaken fabrics as it bleached. These hurdles were gradually overcome, but the additional expenses involved in producing and handling chlorine water left the new process barely competitive with traditional methods rather than economically superior for nearly a decade.

The economic stumbling block to the superiority of chlorine bleaching was removed when Charles Tenant combined the long-standing Scottish chemist's interest in limewater with that in chlorine bleaching. He discovered that if chlorine was passed into lime water, the gas reacted with the lime to form a highly soluble compound (in modern terms, calcium hypochlorite) and that the resulting solution was more reliable than regular chlorine water. In January 1798 Tenant patented his process for making what he called "oxygenated muriate of lime."[65] His patent, however, was overturned because a number of bleachers were beginning to do the same thing at the same time. In 1799 he successfully patented a process for producing bleaching powder (calcium hypochlorite) in its dry undissolved form, solving the key problem of transporting bleach. From that time, hypochlorite bleaching gradually replaced all other methods, with major financial benefits to Britain. In the mid-nineteenth century, Justus Liebig estimated the importance of the creation of the new method of chemical bleaching:

> But for this new bleaching process it would scarcely have been possible for cotton manufacture in Great Britain to have attained the enormous extent which it did. Finding capital to purchase land for the old methods of bleaching would have presented a considerable problem especially when it is realized that by 1840 a single establishment near Glasgow was bleaching 1,400 pieces of cotton daily throughout the year.[66]

SUMMARY

By the eighteenth century, the promises made by humanistic and Baconian scientists that scientific knowledge would provide the foundation for material progress were being realized in a major way. Other factors were almost certainly as important as technological innovation in shaping the early Industrial Revolution. The availability of investment capital, the rising demand for consumer goods among the middle classes, and the availability of cheap labor were all critically important. Even in connection with technological innovation, scientific influences were not universally important. Increasingly, as the eighteenth century progressed, however, science and technological change became closely linked both in the popular imagination and among industrial innovators themselves.

Very seldom, if ever, was some new abstract scientific theory used directly for productive purposes. Instead, science began to have its major impact on technological development first through the popularization and spread of scientific attitudes and habits of mind. In connection with agricultural developments, for example, scientists urged experimenting with new crops, manures, implements, and breeds of animals. In response, advanced farmers began to vary practices and to keep systematic yield and weight-gain records so that improvements could be recognized and preserved. As a consequence, British agricultural productivity almost tripled during the period between 1720 and 1800. Attitudes and habits of mind popularized in connection with science became associated with the structuring of work in the new factory setting, leading to increasing division of labor and more thorough control of productivity and working conditions.

First in connection with engineering, which had been closely linked to mathematical learning, and soon after in those industries to which chemical knowledge was relevant, the eighteenth century began to see a more pervasive and systematic linkage between scientific activities and technological innovation. Experimentation, which had probably always been part of craft traditions, took on a more systematic and precise quantitative cast. In some cases, as in Smeaton's experiments on the efficiency of water wheels and Wedgwood's on pyrometry, industrially useful experiments were undertaken in order to test and extend insights derived from scientific theory. In many others, as in Wedgwood's glass-making and ceramic experiments, results depended on observations reported in scientific literature, though contemporary theory had very little to offer. In just a small number of cases—of which chlorine bleaching was the most economically important—some new material or process discovered in the course of a purely scientific investigation was adapted to commercial uses. Overall, there was a real and perceived acceleration of ties between science and technology as the eighteenth century proceeded.

Romantic Reactions to
Scientized Politics
and Production

9

The early decades of the eighteenth century saw an unprecedented series of attempts to apply scientific understanding to improving social, political, economic, and technological processes and institutions. Yet there were small numbers of intellectuals who renounced the dominant trends or whose adherence was at least highly qualified. Samuel Johnson in England and Jean-Jacques Rousseau in France, for example, were both concerned about the spread of scientific mentality. Each purported to admire natural scientists of true genius—Bacon, Descartes, Galileo, and above all, Newton—but each felt that in the hands of lesser lights and applied outside the realm of nature, scientific knowledge brewed a dangerous pride of intellect with a consequent disdain for the great collections of wisdom embodied in traditional institutions. There were also some elements within traditional institutions, especially conservative elements within the Catholic and the Anglican Church hierarchies and groups within the traditional aristocracies of Europe, who rightly feared that the new intellectual trends were threatening to their authority and who sought to suppress them. In spite of these pockets of opposition, during the period between 1730 and 1790, the vast majority of middle- and upper-class intellectuals throughout Europe came to view scientific learning as an unalloyed blessing and as the engine that might bring humankind into an unprecedented era of peace and plenty.

At the turn of the century, however, both the innovative political tradition and the innovative technological tradition which derived so much of their impetus from scientific developments showed ugly and disturbing features. The French Revolution

345

Pl. 5. Joseph Wright of Derby's "The Air Pump" catches the range of eighteenth-century attitudes toward science. While most of the figures watch in fascination as a pigeon is deprived of air, the girl on the right covers her eyes in horror, and the woman on the left shows her total disinterest. (Tate Gallery, London)

spawned first regicide and then the Terror, whereas the Industrial Revolution spawned pollution, cyclical unemployment, and a demoralized working class. Three distinctive critical responses to the negative characteristics of industrialization and the French Revolution emerged between 1790 and 1812: one response shaped European Socialism; one grew within the powerful Liberal tradition; and the third virtually created the modern Conservative tradition. Each of these traditions was characterized by a unique interpretation of the responsibility of science for the social and political turmoil of the 1780s and 1790s.

We will defer consideration of socialist and liberal attitudes until the beginning of the next volume and will focus here on conservative responses to the French and Industrial Revolutions. These responses briefly dominated European intellectual life during the nineteenth century, and though the liberal and socialist traditions ultimately became dominant, the conservative arguments of the early nineteenth century have remained alive in almost all anti-modernist social criticisms both of the right *and* left. They thus form key links between such otherwise divergent groupings as the European fascists of the 1930s, the German Frankfurt School of social critics which emerged during the height of the Cold War, and the American "New Left" in the 1960s and 1970s.

JOSEPH DE MAISTRE AND THE RELIGIOUS THEME OF CONTINENTAL CONSERVATISM

The most impassioned and articulate spokesperson for continental opponents of the French Revolution and the intellectual trends which nurtured it was Joseph de Maistre. Born into a noble Savoyard family in 1753, de Maistre was a committed Catholic, a lawyer, and a public prosecutor. French Republican armies entered Savoy in October 1792, and de Maistre fled to become an active counter-revolutionary.

No reactionary thinker was more self-consciously anti-scientific than de Maistre. As long as philosophical thought was restricted to the domain of natural philosophy, de Maistre had no problem with it; but once the attempt was made to apply it outside that domain, de Maistre saw it as corrosive, as he wrote in the *Study of Sovereignty*:

> [Philosophy] is useful when it does not leave its own domain, that
> is the sphere of the natural sciences. Here all its efforts are useful
> and merit our gratitude. But once it sets its foot inside the moral
> sphere, it should remember that it is no longer on its own ground.
> It is the general mind that holds sway in this sphere and philoso-
> phy, that is to say the individual intellect, becomes noxious and
> thus culpable if it dares to contradict *or bring into question* the
> sacred laws of the sovereign, that is to say, the national dog-
> mas. . . . When it enters the domain of the sovereign, its duty is to
> act in concert with it. This distinction . . . shows us the confines of
> philosophy. It is good when it remains in its own domain or when
> it enters a sphere higher than its own only as an ally or even as a
> subject; it is *hateful* when it enters as a rival or an enemy.[1]

Many conservatives viewed themselves as defending an empirical—
as opposed to an overly rationalistic—vision of science. Even de
Maistre frequently viewed uncontrolled *Reason* as the evil element
within the scientific tradition and spoke favorably of history as "experimental politics."[2] But unlike many of his colleagues, de Maistre
even devoted an entire work, the *Examen de la philosophie de
Bacon,* to a condemnation of scientific *empiricism* as he understood
its moral applications.

Before we explore his overt opposition to the scientific thought
of the eighteenth century, it is worth pointing out that even this most
harsh critic of the extension of scientific patterns of thought into
the realms of the human and the divine was deeply indebted to the
tradition he was so eager to discredit. Evidence of his debt lies in
his frequent use of metaphors and analogies based on the sciences.
De Maistre, for example, viewed the French Revolution as divine
punishment for evil human acts; and because he also viewed it as
a uniquely terrible punishment, he felt that the sins of the French
must have been uniquely horrible. To express his feeling that God
surely fitted the punishment to suit the crime, he draws explicitly
on Newton's third law of motion in the following passage from *Considerations on France:*

> From the fact that the action and reaction of opposing powers is
> always equal, the greatest efforts of the Goddess of Reason against
> Christianity were made in France.[3]

Talking about why it is unnecessary to *devise* ways of limiting
monarchical power, de Maistre argues that God has willed that all
power is *self-destructive* when it goes beyond its natural bounds,
and he illustrates this idea from the physical sciences. During the

1780s William Hershel had failed to make a successful reflecting telescope with a 48-inch-diameter parabolic mirror because the massive mirror sagged under its own weight. De Maistre derived from this failure an important general principle: "Look at this telescope; up to a certain point, the bigger you make it, the more powerful it will be; but go beyond that, and invincible nature will turn all of your efforts to perfect the instrument against you. This is a crude image of power. To conserve itself, it must restrain itself."[4]

In yet another example, de Maistre cleverly appeals to the recently developed theory of probability to reject Enlightenment arguments for developing France as a "great Republic." If a die were thrown millions of times and 1, 2, 3, 4, or 5 always turned up, we would be convinced that there was no face containing a 6. By the same token, history has repeatedly thrown up monarchies, aristocracies, small republics, even small democracies. But it has never given rise to a great republic:

> The comparison with the dice [*sic*] is therefore perfectly exact: the same numbers having always been thrown from the dice box of fortune, we are allowed by the theory of probabilities to maintain that there are no others—and a great republic is impossible since there has never been a great republic.[5]

Similarly, when de Maistre talks of God's ordering of the universe, he refers to the "eternal geometry,"[6] and he borrows the stock watchmaker image of the natural theologians.

Even as he uses these borrowings, de Maistre often twists them to emphasize the incomprehensible character of God's works. The watch of de Maistre's watchmaker, unlike that of the natural theologians from Boyle to Payley, is one, "all of whose springs continually vary in power, weight, dimension, form, and position, and which nevertheless, invariably shows the right time."[7]

More significant than de Maistre's use and occasional perversion of scientific images is the fact that many of his basic assumptions and commitments owe more than he would admit to the scientific traditions of social theory and aesthetic theory. For example, not only does he accept without acknowledgment or criticism the four-stage stadial theory of social development built up by Montesquieu, Turgot, Smith, and Ferguson but he also repeatedly appeals to the "law of unintended consequences" or the "invisible hand" notion that was central to that theory. Speaking of the unintended benefits which derive from war, for example, de Maistre argues that "humanity can be considered as a tree that an *invisible hand* is continually pruning, often to its benefit."[8] Talking of the origins of representation and the balance of powers in the British governmental system, he argues:

> It was not an invention, or the product of deliberation, or the result
> of the action of the people making use of its ancient rights;
> but . . . in reality, an ambitious soldier, to satisfy his own designs,
> created the balance of the three powers after the Battle of Lewes,
> without knowing what he was doing, *as always happens.*[9]

Additionally, in speaking of the course of the French Revolution itself, he insists that "those who established the Republic did so without wishing it, and without realizing what they were creating; they have been led by events: no plan has achieved its intended end."[10]

From the sociological tradition, de Maistre and the rest of the conservative tradition also developed their focus on the importance of special local circumstances and their denial of the relevance of any abstract notion of humanity. "A constitution that is made for all nations," he writes, "is made for none." Furthermore, he asks,

> is not a constitution a solution to the following problem: *Given the
> population, customs, religion, geographical situation, political rela-
> tions, wealth, good and bad qualities of a particular nation, to find
> the laws which suit it*? Yet this problem is not even approached in
> the 1795 constitution [for the French Republic] which was aimed
> solely at *man.*[11]

Finally, de Maistre and most other conservatives followed the sociological tradition in denying that humans were to be found outside of society. In a close paraphrase of Adam Ferguson's words, de Maistre argues that "history continually shows us men joined together in more or less numerous societies; [so] to talk of a state of *nature* in opposition to the social state is to talk *nonsense* voluntarily."[12] It is certainly true that de Maistre and his conservative friends drew different implications than their eighteenth-century sociological predecessors did from the ineluctable social circumstances in which humans are discovered. They sought to use the social context argument to justify rejecting the social utility of individual human reason—something which neither Montesquieu nor Ferguson would have countenanced. Nevertheless, in a typical example of unintended consequences, eighteenth-century scientific attempts to understand social institutions provided the very foundation for arguments that not only rejected the utility of scientific attempts to understand society but saw such attempts as basically destructive.

Even the hated utilitarian tradition of Helvetius shaped de Maistre's political notions to some degree; for de Maistre's understanding of the criteria for judging governments is an only slightly modified paraphrase of the utilitarian principle:

> The best government for each nation is that which, in the territory occupied by this nation, is capable of producing *the greatest possible sum of happiness and strength, for the greatest possible number of men, during the longest possible time.*[13]

Of course, deviations are as important as similarities. The utilitarians would have thought it unnecessary to add a temporal consideration, for they were confident of the stability of any happiness-procuring government. But the conservatives had a much greater sense of the temporal and historical character of institutions, and they were particularly convinced that "democratic" institutions were inevitably short-lived and therefore undesirable, even if they were sometimes glorious while they lasted.[14] The utilitarians would also have thought the addition of *strength* to happiness quite unnecessary; but most conservatives, like de Maistre, retained a special admiration for strength and power in its own right and not merely for its role in achieving happiness.

Their admiration for power was but one way in which conservative attitudes paralleled the new aesthetic which emerged out of Burke's emphasis on the sublime. There were two additional critical ways in which conservative ideas found important resonances with the Burkean aesthetic. Both renounced the moral power of the prosaic written word—especially in the clear, distinct, and unambiguous form advocated by Hobbes, Locke, and Condillac. Instead, they praised language which carried implicit rather than explicit meaning, which was ambiguous, and which even said different things to different audiences—language characterized as "poetic" by the aestheticians and as "spoken" by de Maistre. In *The Generative Principle of Political Constitutions,* de Maistre argued that any association with writing indicated some weakness in an institution: "The greater the role of deliberation, science, and *above all writing*—the more fragile will be the institution."[15] In particular, he argued that the attempt to formulate a written constitution for a state was an exercise in futility. He decried Tom Paine's claim that "a constitution does not exist if one cannot put it in one's pocket."[16] Furthermore, he enumerated the fallacies involved in such an assumption, claiming that the written word "does not know what to say to one man or to hide from another" and, citing a passage from Plato's *Phaedrus,* argued that it cannot defend itself, "for its father is never there to protect it."[17] In addition, he who sees written laws and constitutions as superior because they provide "the proper clarity and stability—disgraces himself—for he has thereby shown that he is equally ignorant of *inspiration* and madness, *justice* and injustice, *good* and evil."[18]

This mention of inspiration raises another key parallel between political conservatism and the Romantic aesthetic. Just as Romantic poets insisted that poetic insight, genius, or inspiration transcended mere reason and was essential to the creation of a poetry that will arouse the emotions, de Maistre insisted that political genius transcended reason and that it alone could provide what was needed to be a law-giver to a nation. In the *Study of Sovereignty,* de Maistre describes the ideal political founder who

> divines those hidden powers and qualities which shape a nation's character, the means of bringing them to life, putting them into action, and making the greatest possible use of them. . . . [H]is mode of acting derives from inspiration, and if sometimes he takes up a pen, it is not to argue but to command.[19]

God did not usually employ an individual human agency in forming governments, allowing them to simply grow like plants. When he did want to speed things up, however, he always granted some person these unusual powers.

Starting from a belief in the divine and superrational origins of all human communities and institutions, de Maistre insisted that these institutions depend for their continued existence on the renunciation of individual human reason and on submission to divine and communal authority:

> To conduct himself well, man needs beliefs, not problems. His cradle should be surrounded by dogmas . . . . Nothing is more vital to him than *prejudices* . . . [i.e.,] opinions adopted without examination. Now these kinds of opinions are essential to man; they are the real basis of his happiness. . . . Without them there can be neither religion, morality, nor government. There should be a state religion just as there is a state political system; or rather religion and politi-

cal dogmas, mingled and merged together, should form a *general* or *national* spirit sufficiently strong to repress the liberation of individual reason which is of its nature, the mortal enemy of any association whatever because it gives birth only to divergent opinions.[20]

The first great evil associated with the scientific tradition is its critical and antiauthoritarian character. Speaking of all who had been tainted with scientific attitudes, de Maistre wrote, "It is not this or that authority they detest, but authority itself . . . ." For that reason, although they do not recognize it, "they do not want any government, since there is no government which does not lay claim to obedience."[21] There is a hard kernel of truth in de Maistre's claim that scientific attitudes involve a deep-seated suspicion of anything recognized as authority and that the scientist's resulting unwillingness to submit to collective decisions, though understandable, can sometimes be both demoralizing and destructive. Even more critical than the philosophers' questioning of governmental authority was the fact that they extended their renunciation of unquestioned *authority* even to that of God, imparting a *Satanic* element to their views. In fact, from de Maistre's point of view, such scientistic social philosophers as Helvetius and Condorcet had moved beyond skepticism into a vicious frontal attack on Christianity itself.[22]

Perhaps nothing symbolized the philosophers' rejection of God more clearly than Condorcet's plan for a totally secular and nondogmatic education. In a mock paraphrase of Condorcet's views, de Maistre wrote:

> Must we forever tremble before the priests, and receive from them whatever instruction they care to give us?—Truth is hidden by the smoke of incense; it is time that it emerged from this fatal cloud. We shall no longer speak of you [God] to our children; it is for them, when they become men, to decide if you are and what you are and what you ask of them. . . . We wish to destroy everything and to re-create it without your help. Depart from our councils of state, our schools, our homes; we shall be better off alone, reason will be a sufficient guide.[23]

De Maistre's own position was diametrically opposed to what he viewed as the anti-Christian and antiauthoritarian thrust of the secular, scientistic philosophers: "If one does not rely on the ancient maxims," he wrote, "if education is not given back to the priests; and if science is not put everywhere in second place, the evils that await us are incalculable; we shall be brutalized by the sciences, and that constitutes the worst kind of brutalization."[24]

In the face of the prideful and evil rebellion of the "cult of philosophy" against Christianity, according to de Maistre, God decided to punish all for the sins of the few and chose the vehicle of the French Revolution for his retribution. De Maistre certainly reviled the violent activists of the Revolution—Robespierre in particular—but he sees Bacon, Spinoza, Locke, Hobbes, Hume, Montesquieu, Condorcet, Voltaire, and Helvetius as the real villains. The crimes of the Revolution "are their work, since the criminals are their disciples. . . . The tiger that rips men open is following his nature; the real criminal is the man who unmuzzles him and launches him on society."[25] Once God took his revenge, de Maistre was convinced that traditional monarchical government, which is natural to the spirit and geography of France, would be reestablished in its rightful place; and each wise citizen would be prepared to please God by subordinating himself to the restored monarchy. In the meantime he should welcome the benefits, arranged by God, that unintentionally flow from revolutionary fanaticism—the purification of the clergy, the expansion of territory, and, above all, the Terror, through which ever more radical elements are literally killing off those who initiated the Revolution, eliminating the leadership of any potential antimonarchical movement.

EDMUND BURKE AND THE WISDOM OF OUR ANCESTORS

Virtually all of the major themes of de Maistre's works had been raised in Edmund Burke's *Reflections on the Revolution in France,* written in 1790, well before the bloody character of the Revolution had manifested itself. Radical English theorists such as Richard Price and Joseph Priestley were sympathetic to the principles of the Revolution and found in them support for their desire to disestablish the English National Church. Burke was appalled.

In spite of the fact that most of the extreme reactionary principles of de Maistre can be dug out of Burke's *Reflections*—Burke stated that "religion is the basis of civil society,"[26] that prejudices are often to be admired, that scientistic antiauthoritarianism and "atheistical fanaticism" are at the root of the entire revolution,[27] and that through the vehicle of the Revolution, God might be punishing the French for "some great offenses"[28]—the tone and major thrust of Burke's work are vastly different. Burke's religious references are those of the eighteenth-century anthropologist and psychologist, not those of an unquestioning believer, for example. In direct contrast with de Maistre, he explicitly rejected the divine source of kingly authority.[29] His opposition to the confiscation of Church lands arose not because he had some special reverence for the Church, but because he had special reverence for property, and if the property of the Church is not safe, no citizen's property is safe.[30] At one point, Burke even insisted that the theater is often "a better school of moral sentiments than churches."[31] When he discussed the root *causes* of misery in the world they lie in the *vices* (not sins) of pride, ambition, avarice, revenge, lust, sedition, hypocrisy, ungoverned zeal, "and all the train of disorderly appetites."[32] Religion is just one of many "pretexts" used to excuse our vices.

Burke was willing to admit many of the philosophers' criticisms of traditional religion and traditional political institutions. It is true that the fiscal policies of the French monarchy were improvident and that the institutions of the Gallican Catholic Church "savor of superstition in their very principle and they nourish it by a permanent and standing influence."[33] But just because they are imperfect in comparison with some theoretical ideal, they do not deserve to be totally abandoned. The reformation and improvement of existing structures is to be preferred over revolutionary rejection for several reasons, all of which are closely related to the limitations of human reason.

Since, in Burke's words, "the science of constructing a commonwealth, or renovating it, or reforming it, is, like every other experimental science, not to be taught a *priori*,"[34] we must look to past political experience to discover any principles upon which to build. When we do so, we recognize that the "rights" that belong to human beings in any particular society are defined by *convention* or habitual behavior within that society.[35] Governments develop over time in response to the wants of humans in particular societies: "They admit of infinite modifications, they cannot be settled upon by any abstract rule; and nothing is so foolish as to discuss them upon that principle."[36] In the absence of any clear external criteria against which to test present institutions, the presumption must be in favor of their continued existence. Burke was perfectly aware that historical conditions change, bringing changed demands on government; so he did not argue for total stasis, as de Maistre seemed to want. Instead he argued that men should build upon the strengths of existing institutions as they reform them.[37]

Burke's conviction that morality and society were created by convention did not mean that he thought they were arbitrary. Like Ferguson and Hartley, Burke was convinced that human institutions emerge out of our natural social *passions*. Thus, repeatedly he emphasized the priority of passions, emotions, and sentiments over *rationality* in the arrangements of society and government. We "feel inwardly" the importance of religion for society, for example.[38] It is in our "breasts" rather than in our "speculations" or "inventions" that we are to find "the great conservatories and magazines of our rights and privileges."[39] And the marvelous character of the English constitutional monarchy derives from "following nature, which is *wisdom without reflection, and above it*."[40] Even our unease over revolutionary political events is the result of a natural reaction:

> In events like these our passions instruct our reason . . . when kings are hurled from their thrones . . . and become the objects of insult to the base, and of pity to the good . . . we are alarmed into reflection . . . upon the unstable condition of mortal prosperity, and the tremendous uncertainty of human greatness.[41]

Given the central role of emotional attachments in our social and political lives, any social theory that grounds human action in purely rational understandings of human nature is bound to be not merely erroneous but fundamentally destructive because it undermines our natural emotional reactions.[42] Materialism of the kind expressed by Helvetius destroys all the "pleasing illusions" that help us cope in a harsh world. From within the mechanic philosophy, which spawned utilitarianism, Burke argued, "our institutions can never be embodied . . . in persons; so as to create in us love, veneration, admiration or attachment."[43] Without these affections, which are banished by reason, societies and governments lack that cement that binds humans together for long-term survival.

To find a period when institutions were self-consciously built on the kind of personal, emotional ties that are most immediate and natural, we must look back to the Middle Ages, said Burke, linking conservative political theory to the medievalism of Romantic attitudes. But now, in the new revolutionary age, "the age of chivalry is gone.—That of sophisters, economists, and calculators has succeeded; and the glory of Europe is extinguished forever."[44]

Like Hobbes and Petty, Burke saw political reasoning as a form of computing, but for Burke, it was a peculiar form of *moral* "adding, subtracting, multiplying, and dividing," not a form that was metaphysical or mathematical.[45] Burke's greatest objection to the new scientistic political philosophy and the acts of the National Assembly which it stimulated was its replacement of moral reason by geometric and arithmetic reason. Looking at the new geometrical/geographical redistributing of France into eighty-three departments, each broken into Communes, and each Commune into Cantons, making 6,400 political territories in all, Burke remarked first that "in this new pavement of square within square, and this organization and semi-organization made on the system of Empedocles and Buffon, and not upon any *politic* principle, it is impossible that innumerable local inconveniences, to which men are not habituated, not arise."[46] What overshadows the possible inconveniences is the fact that local and regional emotional ties will be unraveled. One goal of the new geometric system was, of course, to break down the local commitments of Gascons, Bretons, Normans, and Savoyards, so that all would view themselves as *French*, with a single interest:

> But, instead of being all Frenchmen, the greater likelihood is, that the inhabitants of that region will shortly have no country. No man was ever attached by a sense of pride, partiality, or real affection, to a description of square measurement. He will never glory in belonging to the chequer number 71. . . . Such divisions of our country as have been formed by habit, and not by a sudden jerk of authority, were created so many little images of the great country in which the heart found something with which it could fill.[47]

When Burke turned to express his fears of what might happen if the French Revolutionary spirit were imported to Britain, he stressed above all his fear of giving up traditional laws and institutions "in favor of a geometrical and arithmetical constitution."[48] In addition, he pointed to France as a real-life imitation of Jonathan Swift's literary nightmare. From recent events, he insisted that "one would conclude that it had for some time past been under the special direction of the learned academicians of Laputa and Balnibarbi."[49]

Finally, Burke argued that for all of their recent emphasis on constitutional reform, many of the theoreticians of the Revolution (the Physiocrats and their friend Condorcet stand as among the clearest examples) seemed almost indifferent to the form of government as long as it allowed them to institute their favorite economic reforms. We have seen that up until the mid-1770s, most of the scientists in the economic tradition—including Turgot and, to some degree, Condorcet—expected enlightened monarchy to be the governmental form best suited to their reforming spirit. Only after the fall of the Turgot ministry did they turn toward republican and democratic principles out of disgust with the weakness of the monarchy. Burke correctly saw this changed perspective and viewed those who were involved as a group bent on their own private goals regardless of the consequences. To them, he wrote that "it was indifferent whether these changes were to be accomplished by the thunderbolt of despotism, or by the earthquake of popular commotion."[50] For those to whom institutional forms are vastly more significant than particular policies—as was true for Burke, de Maistre, and their conservative followers—this willingness to sacrifice institutional stability to advance a particular policy seemed to be an act of bad faith and added to the revulsion that they had toward a science-based politics.

Burke either could not or did not wish to acknowledge that there *was* a scientific tradition—which we have associated with the term *psychology,* and for which Condillac and Helvetius stand as the prime French exemplars—a tradition that consistently concerned itself with fundamental institutional issues. Unfortunately, members of this tradition had at least two critical characteristics that infuriated Burke. First, unlike the economists, many of whom had at least some practical ministerial experience in the government, the psychological theorists were not generally practical men of affairs but were rather self-styled "savants." Burke had little patience for ivory-tower intellectuals who presumed to offer advice without having faced the complex problems of balancing relative goods and evils that was an essential part of governing. "Practical wisdom," he argued, must always take precedence over "theoretic science."[51] Theoreticians tend to "despise experience as the wisdom of unlettered men"; they constantly oversimplify conditions, and they tend to act "rashly" and "inconsiderately" because they never have to face the consequences of their beliefs.[52] In addition, Burke saw most members of the psychological tradition as associated with the "spirit of atheistical fanaticism" that hardened the hearts of men, shrinking them "from their natural dimensions by a degrading and sordid philosophy . . . fitted for low and vulgar deceptions."[53] Thus Burke, like de Maistre and the entire conservative tradition, placed the major blame for the French Revolution on the intellectual traditions of economics and psychology which had developed during the eighteenth century, while many of his arguments were grounded in the more empirically oriented sociological tradition.

WILLIAM BLAKE: EARLY ROMANTIC POETRY ON THE FRENCH AND INDUSTRIAL REVOLUTIONS

Burke's *Reflections on the Revolution in France* is fascinating in part because it saw latent in the theoretical background to that Revolution the fanaticism and violence which manifested themselves only in the years after Burke's attack. In much the same way, the early poetic reactions to industrialization on the part of writers like William Wordsworth and William Blake are fascinating because they saw implicit in the early stages of industrialization many of the social and environmental problems that only became manifest during the course of the nineteenth century. Moreover, the Romantic poets tended to place blame for many of the negative consequences of industrialization on the same scientistic tendencies that the conservatives blamed for the French Revolution.

William Blake was born in 1757 at London to the hosier James Blake and his wife Catherine. He had minimal formal schooling, though he began to study drawing at age nine. When he was about fourteen, he was apprenticed to an engraver; and after his apprenticeship he made a regular but meager living as an engraver for several London booksellers. He began writing poetry by the time he was about twelve years old; and though he subsequently printed his poetry by a special process that he developed, there is no indication that he even expected to make any money by selling it. Though a handful of important younger poets were deeply influenced by Blake's work, it was not until almost half a century after his death (in 1827) that Blake was widely recognized as among the greatest of English poets—in a class with Chaucer, Shakespeare, and Milton.

In 1788 Blake first introduced his new method of illuminated printing and began to make explicit the grounds of his intense opposition to the scientistic mentality, which he always associated with Baconian utility, Newtonian physics, and—above all—

Lockean psychology. In *There is no Natural Religion,* Blake first rehearsed the central arguments of Locke's psychology:

    I. Man cannot naturally perceive but through his natural or bodily organs.
    II. Man by his reasoning power can only compare and judge of what he has already perceived.
    III. The desires and perceptions of man, untaught by any things but the organs of sense, must be limited to objects of sense.[54]

But then, instead of accepting these claims and deciding, like Hume, that one should reject as illusory all that is not "fact" or calculative reason, Blake takes the path of John Wesley and concludes that Locke's psychology must be incomplete:

    I. Man's perceptions are *not* bounded by organs of perception; he perceives more than the senses (tho' ever so acute) can discover.
    II. Reason, or the ratio of all we have already known, is *not* the same as it shall be when we know more. . . .
    Conclusion: If it were not for the Poetic or the Prophetic character the Philosophic and Experimental would soon be at the ratio of all things and stand still, unable to do other than repeat the same dull round over again.[55]

Nothing angered Blake more than placing limits on what he viewed as an infinite human appetite for knowledge and experience. In a clearly autobiographical vein he argued that "the bounded is loathed by its possessor. The same dull round, even of a Universe would soon become a mill with complicated wheels."[56] Even more important, Blake had a belief very similar to that of the Renaissance Hermetics that man should seek to make himself as like God as possible. And because God is infinite, any human admission of finitude was an admission of defeat: "He who sees the Infinite in all things sees God. He who sees the Ratio only, sees himself only."[57]

In 1793 Blake returned to these themes in his early prophetic works, *Visions of the Daughters of Albion* and *The Marriage of Heaven and Hell.* Against the Lockean notion that we receive all of our information through our five senses, Blake cried out:

They told me that I had five senses to inclose me up,
And they inclos'd my infinite brain into a narrow circle,
and sunk my heart into the Abyss, a red, round globe, hot burning,
Till all from life I was obliterated and erased, . . .
With what sense does the bee form cells? Have not the mouse and frog

Eyes and ears and sense of touch? Yet are their habitations
And their pursuits as different as their forms and as their joys.
Ask the Wild ass why he refuses burdens, and the meek camel
Why he loves man: is it because of eye, ear, mouth, or skin,
Or breathing nostrils? No, for these the wolf and tiger have.
Ask . . . the wing'd eagle why he loves, the sun;
And then tell me the thoughts of man, that have been hid of old.[58]

With respect to the calculative reason Blake became increasingly fearful; for he saw it as increasingly intrusive. Not only is the domain of reason limited, as he insisted in 1788, but reason is expanding its hold on men and closing down the range of human aspirations. Beginning from the older Platonic notion of reason as the rightful governor of the passions, Blake turned this notion on its head, replacing Plato's fear of ungoverned passion with a fear that desire might be totally deadened by rational control:

Those who restrain desire, do so because theirs is weak enough to be
   restrained,
And the restrainer or reason usurps its place and governs
The unwilling.
And being restrained, it by degrees becomes passive,
Till it is only the shadow of desire.[59]

Until the late 1790s Blake did not closely associate his opposition to eighteenth-century notions of sensation and reason with any hostility to either the French Revoluton or the Industrial Revolution. In fact, Blake was a vehement advocate of freedoms of all kinds and a supporter of lower-class democratic ideals; so he was an initial supporter of the French radicals. Moreover, Blake viewed the arts and sciences—as vehicles for human creativity—as among the great achievements of mankind; thus, he had no opposition to technological innovation *per se*. But as the French Revolution became bogged down in abstract discussions of human rights, theories of the balance of powers, schemes for secularizing education, and military invasions of other countries, Blake increasingly saw it as betrayed by repressive reason. As the Industrial Revolution began to produce increasingly regimented jobs, cycles of unemployment, and the breakdown of family structures, he began to identify technological developments in the arts as the outgrowth of Baconian utilitarianism and a mechanistic vision symbolized for him by Newton's physics.

In 1804, Blake finished *Jerusalem,* the last of his majestic, mythic, prophetic poems condemning the limits to spirituality and imagination which were imposed by the scientistic mentality. This time, however, he also cried out against the political and industrial implications of this mentality:

> . . . [T]he Spectre, like a hoar frost and a Mildew, rose over Albion,
> Saying, "I am God, Oh Son's of Men! I am your Rational power!
> Am I not Bacon and Newton and Locke who teach humility to Man,
> Who teach Doubt and experiment? And my two Wings, Voltaire,
>   Rousseau?
> Where is that Friend of Sinners? that rebel against my Laws
> Who teaches Belief to the Nations and an unknown eternal Life?
> Come hither into the Desert and turn these stones to bread.
> Vain Foolish Man! Wilt thou believe without Experiment
> And build a World of Phantasy upon my great Abyss,
> A world of Shapes in craving lust and devouring appetite?
>   So spoke the hard cold constrictive Spectre.[60]

Now Blake explicitly turned this argument on the scientistic element in the French Revolution. The Sons of Albion [England] are mocking the torment of some undefined being named Luvah who is suffering from some undefined anguish. Against their own wills, they begin to become more like Luvah, whom they mock. Then comes the following passage, in which the reader is told the significance of Luvah, who is France, suffering under the revolutionary attempts to rationalize her government and constitution:

> [The Sons of Albion] build a stupendous Building on the Plain of Salis-
>   bury, with chains
> Of rocks round London Stone, of Reasonings, of unhewn
> Demonstrations
> In labyrinthine arches (mighty Urizen [your reason] the Architect) thro'
>   which
> The Heavens might revolve and Eternity be bound in their chain.
> Labour unparallell'd! a wondrous rocky world of cruel destiny,
> Rocks piled on rocks reaching the stars, stretching from pole to pole.
> The Building is Natural Religion and its Altars Natural Morality,
> A building of eternal death, whose proportions are eternal despair.
> Here Vala stood turning the iron Spindle of destruction.
> From Heaven to Earth, howling, invisible; but not invisible,
> Her two Covering Cherubs, afterwards named Voltaire and Rousseau,
> Two frowning Rocks on each side of the Cove and Stone of Torture,
> Frozen Sons of the feminine Tabernacle of Bacon, Newton and Locke;
> *For Luvah is France, the Victim of the Specters of Albion.*[61]

What sense and calculative reason have done to punish France through the French Revolution, they are now conspiring to do to

England through the Industrial Revolution. Amplifying Smith's and Ferguson's fears that industrialization must produce an increasingly insensitive, ignorant, and alienated class of workers, doomed to jobs which involve hour after hour of mindless drudgery, Blake wrote:

> And all the arts of life they chang'd into the arts of Death in Albion.
> The hour glass contemn'd because its simple workmanship
> Was like the workmanship of the plowman, and the water wheel
> That raises water into cisterns, broken and burn'd with fire
> Because its workmanship was like the workmanship of the shepherd;
> And in their stead, intricate wheels invented, wheel without wheel,
> To perplex youth in their outgoings and to bind to labors in Albion
> Of night and day the myriads of eternity: that they may grind
> And polish brass and iron hour after hour, laborious task,
> Kept ignorant of its use: that they might spend the days of wisdom
> In sorrowful drudgery to obtain a scanty pittance of bread,
> In ignorance to view a small portion and think that *All*,
> And call it Demonstration, blind to all the simple rules of life.[62]

One final passage from *Jerusalem* contains perhaps the most powerful of all Romantic cries against the stultifying and hardening effect of the eighteenth-century emphases on sense and reason combined into a materialistic and determinist philosophy. The son of a hosier, Blake fused once again in his mind the tyranny of ideas and the tyranny of regimented textile production:

> ... Oh Divine Spirit, sustain me on thy wings,
> That I may wake Albion from his long and cold repose;
> For Bacon and Newton, sheath'd in dismal steel, their terrors hang
> Like iron scourges over Albion. Reasonings like vast Serpents
> Infold around my limbs, bruising my minute articulations.
> I turn my eyes to the Schools and Universities of Europe
> And there behold the Loom of Locke, whose Woof rages dire,
> Washed by the Water-wheels of Newton: black the cloth
> In heavy wreaths folds over every Nation: cruel Works
> Of many Wheels I view, wheel without wheel, with cogs tyrannic
> Moving by compulsion each other, not as those in Eden, which
> Wheel within wheel, in freedom revolve in harmony and peace![63]

WILLIAM WORDSWORTH: PASSIONATE POETRY VS. ANALYTIC SCIENCE

When we turn from the intensity of William Blake to the reflective calm of William Wordsworth, we seem at first in the presence of a

totally different temperament. For Blake what is "Natural" is fundamentally evil; whereas for Wordsworth, natural objects and natural,
innocent humanity provide the source of virtually all inspiration.
For Blake, to discover only one's finite self was a horrible falling
short in the search for the divine infinite. Wordsworth, however,
virtually transformed the focus of modern poetry into an inward-
turning search for self. Even in his youthful writings Blake expressed a deep suspicion of the sciences; but in 1784, at age fourteen, Wordsworth still expressed a totally uncritical faith in science,
religion, and their fundamental unity:

> Science with joy saw superstition fly
> Before the lustre of Religion's eye;
> With rapture she beheld Britannia smile,
> Clapp'd her strong wings, and sought the cheerful isle.[64]

Blake remained a political and religious radical throughout his life,
whereas Wordsworth traveled from active support for the French
Revolution as late as 1792 to active support for Tory and High
Church positions in the 1820s. Yet in the works of Wordsworth's
decade of greatest creativity, 1797–1807, there are very important
similarities which led Wordsworth, like Blake, to see both the
French and Industrial Revolutions as directed into unfortunate and
stultifying paths by the scientistic mentality. Though he expressed
his beliefs in a very different way, Wordsworth was almost as concerned with the active, imaginative role of the mind in seeking
knowledge; he was equally concerned with the centrality of passion.
Though he was more inclined than Blake to accept the traditional
association of passion and desire with the "heart," he is as concerned as Blake to make the senses serve the human will rather than
vice versa.

In the *Preface to the Lyrical Ballads,* revised in 1802, Wordsworth offers us a vision of the relationship between poetry and science which, while more generous to science than that of Blake, is
careful to stress the fundamental social and emotional necessity of
poetry and the limited and private character of science:

> The knowledge both of the poet and the man of science is pleasure;
> but the knowledge of the one [poetry!] cleaves to us as a necessary
> part of our existence, our natural and unalienable inheritance; the
> other [science] is a personal and individual acquisition, slow to
> come to us, and by no habitual and direct sympathy connecting us
> with our fellow-beings. The man of science seeks truth as a remote

and unknowing benefactor; . . . the poet, singing a song in which all human beings join with him, rejoices in the presence of truth as our visible friend and hourly companion. Poetry is the breath and finer spirit of all knowledge; it is the *impassioned expression* . . . [and] if the time should ever come when what is now called science, thus familiarized to men, shall be ready to put on, as it were, a form of flesh and blood, the poet will lend his divine spirit to aid the trans-figuration, and will welcome the being thus produced, as a clear and genuine inmate of the household of man.[65]

Until such time as science is transfigured by acquiring a soul, Wordsworth is not inclined to admit it into "the household of man."

Wordsworth was particularly distressed by the "cold" reason-ing of the utilitarian political economists; so he opposed French developments in the 1790s and Whig politics in the early nineteenth century on the grounds that power should never be transferred to "mere financiers and *political economists.*"[66] In poems like *The Old Cumberland Beggar,* Wordsworth attacked the liberal political economists' notions that virtue consists merely in not interfering with the rights of others and that a person's economic productivity is the true measure of his usefulness. Considering the beggar, Wordsworth wrote:

> But deem not this Man useless—Statesmen! ye
> Who are so restless in your wisdom, ye
> Who have a broom still ready in your hands
> To rid the world of nuisances . . . .
> But of the poor man ask, the abject poor;
> Go, and demand of him, if there be here
> In this cold abstinence from evil deeds,
> And these inevitable charities,
> Wherewith to satisfy the human soul?
> No—man is dear to man; the poorest poor
> Long for some moments in a weary life
> When they can know and *feel* that they have been,
> Themselves, the fathers and the dealers-out
> Of some small blessings; have been kind to such
> As needed kindness, for this single cause,
> That we have all of us one human heart.[67]

Summing up his attitudes to utilitarianism in 1833, Wordsworth expressed a fear of the domination of fact and reason over inspira-tion and passion almost as intense as Blake's:

> Avaunt this economic rage!
> What would it bring?—an iron age,

When Fact with heartless search explored
   Shall be Imagination's Lord,
And sway with absolute control
The god-like functions of the Soul.
Not *thus* can knowledge elevate
Our Nature from her fallen state.[68]

Turning to the effects of the Industrial Revolution, Wordsworth was bothered by several trends. First, he saw that industrialization was destroying the economic viability of the traditional agrarian economy in which small amounts of "home-manufacture" undertaken by women and children—and even by men during slack agricultural periods—supplemented farm income enough to keep families going. Under this old economic regime, Wordsworth felt that a perfect republic of self-sufficient families sustained by their emotional ties to the land had existed. With the establishment of factory production, home manufacture nearly disappeared, marginal farms became untenable, and a proud and productive segment of humanity migrated into cities to increase the class of urban poor. Once there, they faced a series of circumstances brought on by a combination of industrialization and misguided—because heartless—utilitarian policies to relieve the poor, policies that separated children from their parents in the name of efficient care. In a letter to Charles Fox, the Whig leader, Wordsworth expressed his feelings with greatest clarity:

> . . . by the spreading of Manufactures through every part of the country—by workhouses, Houses of Industry, and the invention of Soup-shops, etc., etc. superadded to the increasing disproportion between the price of labor and that of the necessaries of life, the bonds of domestic feeling among the poor, as far as the influence of these things has extended, have been weakened and, in innumerable instances entirely destroyed. The evil would be less to be regretted if these institutions were regarded only as Palliative to a disease; but the vanity and pride of their promoters are so subtly interwoven with them that they are deemed great discoveries and blessings to humanity. In the mean time parents are separated from their children and children from their parents.[69]

Just as important to Wordsworth as the transformation of a self-sufficient, rural lower class into a dependent and displaced urban poor was the soul-wrenching regimentation of labor that came about from considering human beings as objects and tools (in Wedgwood's terms, as machines) rather than as fellow humans with active and important desires. In *The Excursion,* published in 1814, Wordsworth describes a gas-lit factory at night, referring explicitly to the system of bells introduced by Wedgwood:

> . . . an unnatural light
> Prepared for never-resting Labour's eyes
> Breaks from a many-windowed fabric huge;
> And at the appointed hour a bell is heard—
> Of harsher import than the curfew-knoll
> That spake the Norman Conqueror's stern behest—
> A local summons to unceasing toil!
> Disgorged are now the ministers of day;
> And, as they issue from the illumined pile,
> A fresh band meets them, at the crowded door—
> . . . Men, maidens, youths,
> Mother and little children, boys and girls,
> Enter, and each the wonted task resumes
> Within this temple, where is offered up
> To Gain, the master-idol of the realm,
> Perpetual sacrifice.[70]

Just as Wordsworth saw science as potentially beneficial but corrupted by its lack of social conscience and its extension into domains where a moral and poetic element belonged, so too he saw productive machinery as intrinsically admirable but as destructive when its use was extended into systems in which people became transformed into mere means, rather than ends-in-themselves.

Wordsworth hinted at a negative aspect of industrialization that would become a major issue for Romantic critics of the next generation. The physical environment was, in his mind, being made almost as ugly as the moral. Describing the transformation of the beautiful countryside into a smoke-blanketed urban chaos, he wrote:

> Where not a habitation stood before,
> Abodes of men irregularly massed
> Like trees in forests,—spread through spacious tracts,
> O'er which the smoke of unremitting fires
> Hangs permanent . . . .[71]

With the growing use of coal, both in iron production and as fuel for steam engines, soot increasingly covered the landscape, and as manufacturing towns grew far more rapidly than clean water supplies and sanitation systems, there was a definite deterioration in air and water quality. By the 1850s John Ruskin was crying out against this trend with a combination of Wordsworth's sensitivity to nature and Blake's prophetic intensity:

> You might have the rivers of England as pure as the crystal of the rock; so beautiful in falls, in lakes, in living pools; so full of fish that you might take them out with your hands instead of nets. Or you may do always as you have done now, turn every river of England into a common sewer, so that you cannot so much as baptize an English baby but with filth. . . . Are they not what your machine gods have produced for you?[72]

SUMMARY

Because the French Revolution and the Industrial Revolution had become closely linked with scientific thought, it was natural that their negative features were blamed by some on the extension of scientific thinking into inappropriate domains. In fact, conservative reactions against the French Revolution and the Industrial Revolution usually placed scientific ways of thinking at the very heart of what had gone wrong. Conservative critics attacked the scientific emphasis on an analytic approach to social phenomena, and they denied the basic ability of human reason to recognize the intricacies of human interactions. Against the authority of reason, the obsession with quantification, the worship of universal truth, the assumption of analyticity, and the optimism with respect to individual's abilities to formulate intelligent rational courses of action, the new intellectual leaders of reaction offered the authority either of tradition or of the poetic imagination. They almost uniformly denied that only what can be counted counts. They insisted that wholes are more than the aggregates of their parts. They valued the unique over the universal, the distinctive over the regular, the divine over the natural. With few exceptions, they argued that only in collective wisdom or in some transrational illumination was effective guidance for human action to be found.

For all of their differences, the two greatest early Romantic poets, Blake and Wordsworth, had remarkably similar responses to both the French and Industrial Revolutions and to the scientific liberalism which they saw at the bottom of what had gone wrong with each of them. Both started out and remained sympathetic to lower-class interests and to the democratic tendencies of the French Revolution.[73] But each was repelled by the deistic and atheistic tendencies of revolutionary ideology; and they were appalled by the coldly rational, quantitative, and managerial mentality that infused both the early French Revolution and the Industrial Revolution. Finally, both admired science as a creative activity, opposing only its imposition of limits on human activity. At the root of both Blake's and Wordsworth's common attitudes was their mutual insistence that human moral truth emerges fully only out of a *passionate, active,* and imaginative mental engagement with the world, and not through the *passive* acceptance of sensory impressions combined through *reason* which lay at the foundation of almost all eighteenth-century notions of science.

Political conservatism and Romantic poetry flourished only briefly during the early nineteenth century in the face of the dominant products of scientistic thought. They were, like Henry Clerval, the socially conscious poet created by one of Wordsworth's greatest admirers, Mary W. Shelley, killed off by the product of what Blake, Wordsworth, and Mrs. Shelley all saw as an egotistical and antisocial scientific enterprise. But just as Mary Shelley's allegory, *Frankenstein, or The Modern Prometheus,* continues to inspire new generations, early Romantic ideas continue to live—especially within our contemporary literary subculture—rising up occasionally to inform social ideals.

Notes

INTRODUCTION

1. Sandra Harding, *The Science Question in Feminism* (Ithaca: Cornell University Press, 1986), p. 16.

2. Richard Olson, *Scottish Philosophy and British Physics* (Princeton: Princeton University Press, 1975).

3. John Herman Randall, *The Making of the Modern Mind* (Boston: Houghton Mifflin Co., 1926).

4. B. F. Skinner, *Beyond Freedom and Dignity* (New York: Vintage Books, 1971), p. 206.

5. Almost all social psychology textbooks discuss the self-fulfilling prophecy in connection with experimenter expectations, teacher expectations, and stock market behavior. See, for example, David G. Myers, *Social Psychology*, 2d ed. (New York: McGraw-Hill, 1987), pp. 27–42, 37–46, 140–150.

6. George Sarton, *The Study of the History of Science* (Cambridge: Harvard University Press, 1936), p. 5.

7. For an excellent example of the implicit argument that only "mistaken" science can lead to undesirable applications, see Stephen Jay Gould, *The Mismeasure of Man* (New York: W. W. Norton, 1981).

8. John Heilbron, *Electricity in the 17th and 18th Centuries: A Study of Early Modern Physics* (Berkeley, Los Angeles, London: University of California Press, 1977), p. 2.

9. This millenarian theme is thoroughly discussed in Charles Webster, *The Great Instauration: Science, Medicine, and Reform, 1626–1660* (New York: Holmes and Meier Publishers, 1976), chap. I, p. 31.

10. Bacon clearly encouraged this notion in his *Novum Organum*.

11. John Milton, *Complete Prose Works of John Milton*, ed. Don Wolfe (New Haven, 1953), 3: 296.

12. See James Jacob and Margaret Jacob, "The Anglican Origins of Modern Science . . . ," *Isis* 71 (1980): 251–267.

13. Cited from Joseph Glanvill, *Plus Ultra (1668) and Essays on Several Important Subjects (1676)*, in Barbara Shapiro, "Latitudnarianism and Science in Seventeenth-Century England," *Past and Present*, 40 (1968).

14. See W. H. Greenleaf, *Order, Empiricism, and Politics: Two Traditions of English Political Thought—1500–1700* (London: Oxford University Press, 1964), Chaps. 1–5 for a thorough discussion of this issue. In his "Cosmic Harmony and Political Thinking in Early Stuart England," *Transactions of the American Philosophical Society*, 69, Part 7 (1979), James Daly also suggests that Cosmic Harmony was used to justify theories of limited monarchy.

15. Daly, "Cosmic Harmony and Political Thinking," p. 61.

16. Charles Webster, "The Authorship and Significance of *Macaria*," *Past and Present*, 44 (1972).

17. James Harrington, *The Political Works*, ed. J. G. A. Pocock (Cambridge: Cambridge University Press, 1977), p. 656.

18. Peter Mathias, "Who Unbound Prometheus?" in Peter Mathias, ed., *Science and Society, 1600–1900* (Cambridge: Cambridge University Press, 1972), p. 75.

19. Cited in M. B. Hall, "Science in the Early Royal Society" in Maurice Crosland, ed., *The Emergence of Science in Western Europe* (New York: Science History Publications, 1976), p. 74.

20. John Donne, "First Anniversary" cited in Marjorie Nicolson, *Science and Imagination* (Ithaca: Cornell University Press, 1956), pp. 30–31. Nicolson's is the standard interpretation of Donne's response to the new science of the seventeenth century.

1: THE EXTENSION OF MECHANICAL AND
   MATHEMATICAL PHILOSOPHIES

1. The story of early modern skepticism and the ties between Protestantism, the Counterreformation, and the revival of Greek skepticism has been thoroughly and admirably told by Richard Popkin in his *History of Skepticism from Erasmus to Descartes* (The Hague: M. Nyhoff, 1959).

2. René Descartes, *Discourse on the Method* in *The Philosophical Works of Descartes*, trans. by E. Haldane and G. R. T. Ross (New York: Dover Publications, 1955), I: 83.

3. René Descartes, *Rules for the Direction of the Mind* in *The Philosophical Works of Descartes*, I: 4.

4. Descartes, *Philosophical Works*, 4.

5. Ibid., 4–5.

6. Ibid., 5.

7. Ibid., 57, 62, 66–68.

8. Preface to the first edition reprinted in Joseph Louis Lagrange, *Oeuvres* (Paris, 1887), II: xi–xii.

9. Descartes, *Rules for the Direction of the Mind* in *The Philosophical Works*, I: 7.

10. Cited in Alan G. R. Smith, *Science and Society in the 16th and 17th Centuries*, (New York: Science History Publications, 1972), p. 153.

11. Cited in Paul Hazard, *The European Mind: 1680–1715* (New York and Cleveland: The World Publishing Co., 1963, from 1935 French original), p. 307.

12. See John Herman Randall, Jr., "The Development of Scientific Method in the School of Padua," *Journal of the History of Ideas* 1 (1940): 177 ff.

13. Descartes, *Philosophical Works*, I: 14.

14. Benedictus Spinoza, *Selections*, ed. John Wild (New York: Charles Scribner's Sons, 1930), p. 56.

15. Cited in Mary B. Hesse, *Forces and Fields* (London: Thomas Nelson and Sons, Ltd., 1961), p. 160.

16. This topic is covered in Part IV of Descartes's *Principles of Philosophy*.

17. See Robert Kargon, *Atomism in England from Harriot to Newton* (Oxford: Oxford University Press, 1966) and Robert B. Lindsay, "Pierre Gassendi and the Revival of Atomism in the Renaissance," *American Journal of Physics*, 13 (1945): 235–342.

18. Robin Horton, "African Traditional Thought and Western Science," *Africa* (1967): 64.

19. See Trevor Aston, ed., *Crisis in Europe 1560–1660* (London: Routledge and Kegan Paul, 1965), for an introduction to the claim that this was a period of unusual instability.

20. See *De Motu Animalium*, 701B.

21. Contemporary report by Lynn White, Jr., in *Medieval Technology and Social Change* (Oxford: Oxford University Press, 1962), pp. 121, 122.

22. Silvio Bedini and Francis R. Maddison, "Mechanical Universe: The Astrarium of Giovanni de Dondi," *Transactions of the American Philosophical Society* 56, Part 5 (1966): 3–69.

23. Cited in G. H. Baille, *Clocks and Watches, An Historical Bibliography* (London, 1951), p. 3.

24. John Huizinga, *The Waning of the Middle Ages* (Garden City, N.Y.: Doubleday and Co., 1954), pp. 208–209.

25. See Alfred Chapuis, *De Horologus in Arte* (Lausanne, 1954), p. 12, and plates between pp. 16 and 17.

26. White, *Medieval Technology and Social Change*, p. 125.

27. Albert D. Menut and Alexander Deverney, eds., "Maestre Nicole Oresme; Le Livre du ciel et du monde, Book II," *Medieval Studies* 4 (1942): 170. My translation.

28. Derek Price, "On the Origin of Clockwork, Perpetual Motion Devices and the Compass," *Contributions from the Museum of History and Technology* Bulletin 218 (1959): 85–86.

29. See Arthur Koestler, *The Watershed* (Garden City: Doubleday and Co., 1960), p. 60.

30. Ibid.

31. Max Casper, *Kepler* (New York: Collier Books, 1962), p. 115.

32. Johannes Kepler, *Opera,* ed., Charles Frish (Frankfurt, 1859), II: 83–84.

33. See L. Defossez, *Les Savants du XVIIe siècle et la mésure du temps* (Laussane, 1946), pp. 58–60.

34. Ibid., p. 60.

35. For details of Kepler's work, see Robert Small, *An Account of the Astronomical Discoveries of Kepler* (Madison: University of Wisconsin Press, 1963), p. 151.

36. Bernard Fontenelle, *Conversations on the Plurality of Worlds,* trans. William Gardiner (London, 1715), pp. 11–12.

37. See Kargon, *Atomism in England from Harriot to Newton,* pp. 49–52.

38. Ibid., p. 52.

39. Cited in Marie Boas Hall, ed., *Nature and Nature's Laws: Documents of the Scientific Revolution,* (New York: Harper and Row, 1970), p. 311.

40. Ibid., p. 311.

41. Ibid.

42. Ibid., p. 312.

43. Ibid., p. 129.

44. Cited in Marie Boas, *The Scientific Renaissance* (New York: Harper and Brothers, 1962), p. 277.

45. Ibid., p. 279.

46. Julian Jaynes, "The Problem of Animate Motion in the Seventeenth Century." *Journal of the History of Ideas* 31 (1970): 223–224.

47. Lenora C. Rosenfield, *From Beast Machine to Man Machine: The Theme of Animal Soul in French Letters From Descartes to La Mettrie* (New York: Oxford University Press, 1940), p. 4.

48. Ibid.

49. Descartes, "Discourse of the Method of Rightly Conducting the Reason . . ." in the *Philosophical Works of Descartes,* I: 115–116.

50. Ibid., I: 116.

51. Ibid.

52. Marin Mersenne, *Harmonie universelle* (Paris, 1963; facsimile of 1636 ed.), II: 79. Translation mine.

53. René Descartes, *Oeuvres de Descartes*, ed. Victor Cousin (Paris, 1824), IV: 428. Translation mine.

54. Lenora Rosenfield, *From Beast Machine to Man Machine*, pp. 7–8.

55. In Descartes, *Philosophical Works*, II: 235.

56. Thomas Hobbes, *Leviathan*, ed. C. B. Macpherson (Harmondsworth, England: Penguin Books, 1968), p. 118.

57. Carolyn Merchant, *The Death of Nature: Women, Ecology, and the Scientific Revolution* (New York: Harper and Row, 1980), p. 3.

58. Ibid., p. 195.

59. Ibid., p. 193.

60. Ibid., p. 1.

61. Ibid., p. 193.

62. Rosenfield, *From Beast Machine to Man Machine*, pp. 59–60.

63. Ibid., p. 282.

64. Ibid., pp. 59–60. Translation mine.

65. Ibid., p. 54.

66. Thomas Hobbes, *The English Works of Thomas Hobbes*, ed. Sir William Molesworth (London, 1839), I: vii–ix.

67. See, for example, Marjorie Grene, "Hobbes and the Modern Mind," in M. Grene, ed., *The Anatomy of Knowledge* (London: Routledge and Kegan Paul, 1969), pp. 1–28.

68. Cited in Eugenio Garin, *Italian Humanism* (New York: Harper and Row, 1965), p. 32.

69. The story of Hobbes's conversion to geometry is told by his biographer John Aubrey in his *Brief Lives*. Corroborating evidence from Hobbes's correspondence has been discussed by G. R. DeBeer in "Some Letters of Hobbes," *Notes and Records of the Royal Society* 7 (1950): 205.

70. Cited in F. S. McNully, *The Anatomy of Leviathan* (New York: St. Martin's Press, 1968), p. 8.

71. Hobbes, *De Cive, or the Citizen*, ed. Sterling Lamprecht (New York: Appleton-Century-Crofts, 1949), p. 3.

72. Hobbes, *Leviathan*, Part I, Ch. 4, p. 105.

73. Hobbes, *The English Works of Thomas Hobbes*, I: 388. Emphasis mine.

74. Ibid.

75. Hobbes, *De Cive*, p. 15.

76. Ibid.

77. The current controversy over whether or not Hobbes's ethical doctrines "depend" upon his natural philosophy and its psychological

extension seems to have been given intense life by A. E. Taylor, "The Ethical Doctrine of Hobbes" in K. Brown, ed., *Hobbes Studies* (Oxford: Oxford University Press, 1965), 37 ff.

78. Hobbes, *De Cive,* p. 15.

79. Hobbes, *De Homine,* Ch. 10, Section 4, cited in Bernard Gert, ed., *Man and Citizen* (Garden City: Doubleday and Co., 1972), p. 41.

80. Hobbes, *Leviathan,* pp. 81–82.

81. Ibid., p. 82.

82. Ibid., p. 261.

83. Ibid., pp. 110–111.

84. See Joseph Weizenbaum, *Computer Power and Human Reason* (San Francisco: W. H. Freeman & Co., 1976) for a thoughtful assessment of the limitations of a digitized and instrumentalist view of human reason.

85. Hobbes, *Leviathan,* p. 151. Emphasis mine.

86. Alexandre Koyre, *Newtonian Studies* (Cambridge: Harvard University Press, 1965), p. 7.

87. Hobbes, *Leviathan,* pp. 114–115. Emphasis mine.

88. Ibid., p. 136.

89. Cited in Basil Willey, *The Seventeenth Century Background* (New York: Columbia University Press, 1934), p. 94.

90. Ibid., p. 28.

91. Ibid., p. 218.

92. Ibid.

93. David Hume, *Enquiries Concerning Human Understanding and Morals* (1777; reprint of 3d ed., Cambridge: Cambridge University Press, 1902), p. 165.

94. See Samuel I. Mintz, *The Hunting of Leviathan* (Cambridge: Cambridge University Press, 1969), pp. 9–10, for a concise introduction to Hobbes's early manuscript versions of *De Corpore* and *De Homine.*

95. Ibid., p. 10.

96. Hobbes, *De Cive,* p. 11.

97. Hobbes, *English Works,* I: 407.

98. Ibid., I: 391.

99. Hobbes, *Elements of Law,* chap. 7, cited in Richard Peters, ed., *Body Man and Citizen* (New York: Collier Books, 1962), pp. 9–10.

100. Hobbes, *Leviathan,* p. 119.

101. Hobbes, *De Cive,* p. 161.

102. Ibid., p. 11.

103. Michael Walzer, "On the Role of Symbolism in Political Thought," *Political Science Quarterly* 82 (1967): 201.

104. Locke, *The Second Treatise on Civil Government,* Section 4.

105. *Declaration of the Rights of Man and the Citizen* (1789).

106. Michael Walzer, "On the Role of Symbolism in Political Thought," p. 201.

107. Hobbes, *Leviathan,* pp. 238–239.

108. Ibid., p. 262.

109. Ibid., pp. 262–263.

110. Ibid., p. 264.

111. Ibid., p. 271.

112. Ibid., p. 164.

113. John Millar, cited in William C. Lehman, *John Millar of Glasgow: 1735–1801* (Cambridge: Cambridge University Press, 1960), p. 387.

114. Karl Marx, *The Manifesto of the Communist Party* (1848), Section I, "The Bourgeois and Proletarians."

115. See Alan Macfarlane, *The Origins of English Individualism: The Family, Property, and Social Transition* (Cambridge and New York: Cambridge University Press, 1979).

116. Cited in Samuel I. Mintz, *The Hunting of Leviathan,* pp. 55, 62.

117. See Margaret C. Jacob, *The Newtonians and the English Revolution: 1689–1720* (Ithaca: Cornell University Press, 1976), pp. 52–57.

118. Ibid., 66 ff.

2: EMPIRICIST POLITICAL SCIENCE IN THE SEVENTEENTH CENTURY

1. David Hume, *Essays, Moral and Political* (London, 1777), Essay XVI. Reprinted in Hume, *Essays, Moral, Political, and Literary* (Oxford: Oxford University Press, 1963), pp. 500–501.

2. See H. F. Russel-Smith, *Harrington and His Oceana* (Cambridge: Cambridge University Press, 1914), chap. 8, pp. 185–200.

3. James Harrington, *The Oceana and Other Works,* of James Harrington, Esq., Collected, Methodized, and Reviewed, with an Exact Account of His Life prefixed, by John Toland (London: A. Miller, 1737), p. 259.

4. Ibid., p. 516.

5. Ibid., p. 520.

6. Ibid., p. xviii.

7. John Adams to James Sullivan, May 26, 1776, cited in Charles Blitzer, ed., *The Political Writings of James Harrington: Representative Selections* (New York: Liberal Arts Press, 1955), p. xi.

8. Harrington, *Oceana,* p. 516.

9. Ibid., p. 516.

10. Ibid., p. 38.

11. Ibid., p. 38.

12. Ibid., p. 514. (A System of Politics.)

13. Ibid., p. 429. (The Art of Lawgiving.)

14. Ibid., p. 514. (Politicastor.)

15. Thomas More, *Utopia,* in *Famous Utopias of the Renaissance,* ed. F. R. White (New York: Henrich House, 1946), p. 117.

16. See Michael Downs, *James Harrington* (Boston: Twayne Publishers, 1977), p. 133.

17. Harrington, *Oceana,* p. 46.

18. Ibid.

19. Cited in John A. Wettergreen, "Chiron and Leviathan: James Harrington's Political Teaching," Ph.D. dissertation, Claremont Graduate School, 1970, p. iv.

20. James Harrington, *Oceana,* p. 41.

21. Ibid.

22. Ibid., p. 54.

23. Ibid.

24. Ibid., p. 47.

25. Ibid., pp. 54–55.

26. Ibid., p. 48.

27. Ibid.

28. See H. F. Russell-Smith, *Harrington and His Oceana,* chap. 8.

29. Harrington, *Oceana,* pp. 37, 38.

30. Cited in J. G. A. Pocock, *The Political Works of James Harrington* (Cambridge: Cambridge University Press, 1977), "Historical Introduction," p. 85.

31. Harrington, *The Prerogative of Popular Government,* in *Oceana,* pp. 265–266.

32. Charles Blitzer, editor's introduction to *The Political Writings of James Harrington: Representative Selections,* p. xii.

33. Felix K. Ralib, cited in Michael Downs, *James Harrington,* p. 14.

34. Pocock, *The Political Works of James Harrington,* p. 15.

35. Niccolò Machiavelli, *The Prince,* Ch. XV.

36. Blitzer, *The Political Writings of James Harrington,* p. 36.

37. Ibid., pp. 48, 49.

38. Ibid., p. 36.

39. Cited in *The Economic Writings of Sir William Petty.* ed. Charles Henry Hull (Cambridge: Cambridge University Press, 1899) I: xiii.

40. For a discussion of this work, published in 1674, see Robert Kargon, "William Petty's Mechanical Philosophy," *Isis* 56 (1965): 63–66.

41. See Emil Strauss, *Sir William Petty: Portrait of a Genius* (Glencoe, Ill.: The Free Press, 1954), p. 40.

42. Ibid., p. 41.

43. Ibid., p. 110.

44. Ibid., p. 113.

45. Ibid., p. 62.

46. Ibid., p. 63.

47. Ibid., p. 82.

48. Ibid., p. 93.

49. William Petty, *The Petty Papers: Some Unpublished Writings of Sir William Petty*, edited from the Bowood Papers by the Marquis of Lansdowne (London: Constable and Company, Ltd., 1927), I: 111.

50. Cited in Charles H. Hull, ed., *The Economic Writings of Sir William Petty*, I: ixi.

51. Ibid., I: 244.

52. *The Petty Papers*, II: 51.

53. *A Treatise of Taxes*, Chap. 2, in *The Economic Writings of Sir William Petty*, I: 42.

54. *Verbum Sapienti*, Chap. 1, in *The Economic Writings of Sir William Petty*, I: 105.

55. Ibid., I: 105–107.

56. Ibid., I: 107–108.

57. Ibid., I: 108.

58. Ibid.

59. Ibid., I: 21–22.

60. Ibid., I: 68.

61. Ibid., I: 69.

62. Ibid., I: 71.

63. See William Letwin, *The Origin of Scientific Economics* (Garden City: Doubleday, 1964), pp. 151–157, for an interesting analysis of Petty's *Treatise of Taxes*. Letwin focuses on this issue.

64. *The Economic Writings of Sir William Petty*, I: 36.

65. Ibid.

66. Ibid., I: 54–57.

67. Ibid., I: 61–64.

68. Ibid., I: 75.

69. Ibid., I: 94–95.

70. Petty to Southwell, n.d., cited in Lord Edmund Fitzmaurice, *The Life of Sir William Petty: 1623–1687* (London: John Murray, 1895), p. 158.

71. Emil Strauss, *William Petty*, p. 136.

72. See Peter Buck, "Seventeenth-Century Political Arithmetic: Civil Strife and Vital Statistics," *Isis* 68 (1977): 74.

73. Robert Southwell to William Petty, 4 October 1686, printed in *The Petty-Southwell Correspondence*, ed. the Marquis of Lansdowne (London: Constable and Co., Ltd., 1928), p. 235.

74. *The Economic Writings of Sir William Petty*, I: 110.

75. Ibid., I: 129.

76. Jonathan Swift, *A Modest Proposal for Preventing the Children of Poor People in Ireland from Being a Burden to Their Parents or Country; and for Making Them Beneficial to the Public* in John Hayward, ed., *Selected Prose Works of Jonathan Swift* (London: The Cresset Press, 1949), p. 429.

77. Ibid., pp. 430–434.

78. Ibid., p. 436.

3: THE RELIGIOUS IMPLICATIONS OF NEWTONIAN SCIENCE

1. See Richard S. Westfall, *Science and Religion in Seventeenth Century England* (New Haven: Yale University Press, 1958), passim.

2. Ibid., p. 73.

3. Ibid., p. 3.

4. See R. Olson, "On the Existence, Wisdom and Power of God . . ." in Fredrick Burwick, ed., *Organic Form* (Dordrecht: Riedel, 1987), pp. 1–48.

5. On the extent of the Hexameral tradition, see Frank Robbins, *The Hexameral Literature: A Study of the Greek and Latin Commentaries on Genesis* (Chicago: University of Chicago Press, 1912); on its importance for Medieval Science, see Bryan Stock, *Myth and Science in the Twelfth Century* (Princeton: Princeton University Press, 1972).

6. John Calvin, *The Institutes of Christian Religion*, cited in Paul H. Kocher, *Science and Religion in Elizabethan England* (San Marino, Calif.: The Huntington Library, 1953), p. 9. Emphasis mine.

7. See Robert Lenoble, *Mersenne: ou la naissance du méchanisme* (Paris, 1943).

8. See especially Gassendi's *syntagma philosophicum* in the first two volumes of his *Opera Omnia* (Lyons, 1658).

9. For a quick survey of seventeenth-century Catholic attitudes toward natural theology, see William Ashworth," Catholicism and Early Modern Science," in David Lindberg and Ronald Numbers, *God and Na-*

ture (Berkeley, Los Angeles, London: University of California Press, 1986), pp. 136–166.

10. Richard Hooker, *The Collected Works of That Learned and Judicious Divine, Mr. Richard Hooker* (Oxford: Oxford University Press, 1845), I: 215–216.

11. Ibid., I: 158.

12. Walter Charleton, *Physiologia Epicuro-Gassendo-Charletonians . . . .* (London, 1654), p. 128.

13. Barbara J. Shapiro, *Probability and Certainty in Seventeenth Century England* (Princeton: Princeton University Press, 1983), p. 58.

14. Ibid., p. 60.

15. Ibid.

16. Cited in Charles Coulston Gillispie, *The Edge of Objectivity* (Princeton: Princeton University Press, 1960), pp. 119–120.

17. "A Letter of Mr. Isaac Newton, Professor of Mathematics in the University of Cambridge; containing his New Theory about Light and Colors," *Philosophical Transactions* 80 (for 1671/72), reprinted in I. B. Cohen, ed., *Isaac Newton's Papers and Letters on Natural Philosophy* (Cambridge: Harvard University Press, 1958), p. 50.

18. Ibid.

19. Ibid., p. 53.

20. Ibid.

21. Ibid., p. 55.

22. Ibid., p. 57. Emphasis in last sentence mine.

23. Isaac Newton, *Mathematical Principles of Natural Philosophy*, "The System of the World," ed. Florian Cajori (Berkeley and Los Angeles: University of California Press, 1962), II: 547. All references to Newton's *Principia* are to be found in this widely available edition.

24. Newton, *Opticks* (New York: Dover reprint, 1952), p. 1. Emphasis mine.

25. Jean d'Alembert, *Elements de la philosophie* (1749) cited in Ernst Cassirer, *The Philosophy of the Enlightenment* (Boston: Beacon Press, 1955), pp. 46–47.

26. See, for example, I. B. Cohen, *Franklin and Newton: An Inquiry into Speculative Newtonian Experimental Science and Franklin's Work in Electricity as an Example Thereof* (Philadelphia: American Philosophical Society, 1956), passim.

27. A very thorough and fair analysis of this question based on a close reading of the precise way these terms were used by Newton was done by Alexandre Koyre in "L'Hypothèse et l'expérience Chez Newton," *Bulletin de la Société Française de Philosophie* 50 (1956): 57–79. Though my approach is very different, the basic argument that Newton

engaged in a form of self-deception which had remarkable public consequence is consistent with Koyre's analysis.

28. Cited in Thomas S. Kuhn, "Newton's Optical Papers," in I. B. Cohen, ed., *Isaac Newton's Papers and Letters,* p. 39.

29. Ibid.

30. A poignant example of Newton's tolerant attitude to bona fide confusion is described in I. B. Cohen's *Introduction to Newton's "Principia,"* (Cambridge: Harvard University Press, 1971), pp. 158–161.

31. I. B. Cohen, ed., *Newton's Papers and Letters,* pp. 118–119.

32. Alexandre Koyre, *Newtonian Studies* (Cambridge: Harvard University Press, 1965), p. 52.

33. I. B. Cohen, ed., *Newton's Papers and Letters,* p. 106.

34. Ibid.

35. See, for example, John Herschel, *Preliminary Discourse on the Study of Natural Philosophy* (London, 1830), p. 204.

36. Cited in Alexandre Koyre, *Newtonian Studies,* p. 62.

37. Thomas Reid, *Works* (Edinburgh: McLachlan and Stewart, 1863), p. 76. Emphasis mine.

38. I. Newton, *Principia,* I: 21.

39. Ibid., p. 42.

40. Ibid., p. 397.

41. Ibid., p. 398.

42. Ibid., p. 400.

43. Ibid., pp. 401–405.

44. Ibid., p. 408.

45. Ibid., p. 411.

46. Alexander Pope, *Epitaph,* intended for Sir Isaac Newton.

47. Jean d'Alembert, *Elements of Philosophy,* cited in Ernst Cassirer, *The Philosophy of the Enlightenment,* pp. 46–47.

48. See Morton Grosser, *The Discovery of Neptune* (Cambridge: Harvard University Press, 1962).

49. *The Impartial Philosopher* (London, 1749), p. 345, cited in Herbert Odom, "The Estrangement of Celestial Mechanics and Religion," *Journal for History of Ideas* 27 (1966): 540–541.

50. London, 1754, p. 51, cited in Odom, "Estrangement," p. 541.

51. I. Newton, "Four Letters from Sir Isaac Newton to Doctor Bentley . . . ," (London, 1756), p. 1.

52. Perry Miller, "Bentley and Newton," in I. B. Cohen, ed., *Isaac Newton's Papers and Letters,* pp. 273–274.

53. Bentley, *Confutation of Atheism . . . ,* Seventh Lecture, (London, 1692), p. 9.

54. Ibid., p. 33
55. Ibid., p. 34.
56. Ibid., p. 35.
57. Ibid.
58. Ibid., p. 28.
59. Ibid., p. 32.
60. Ibid.
61. Ibid.
62. William Whiston, *A New Theory of the Earth,* 2d ed. (London, 1708), p. 284.
63. Bentley, *Confutation,* pp. 5–6.
64. Samuel Clarke, *A Discourse Concerning the Unchangeable Obligations of Natural Religion, and the Truth and Certainty of the Christian Revelation* (London, 1706), pp. 19–23.
65. Henry Grove to Samuel Clarke, cited in John Dahm, *Science and Religion in Eighteenth Century England* (Ann Arbor: University Microfilms, 1969), p. 82.
66. Newton, "Four Letters," in Cohen, *Isaac Newton's Papers and Letters,* p. 303.
67. Bentley, *Confutation,* p. 39.
68. Ibid.; Newton, "Four Letters," in I. B. Cohen, ed., *Isaac Newton's Papers and Letters,* p. 292.
69. Bentley, *Confutation,* pp. 37–38.
70. Newton, *Opticks* (4th ed., 1730; reprint, New York: Dover, 1952), p. 402.
71. Colin McLaurin, *An Account of Sir Isaac Newton's Philosophical Discoveries* (London, 1748), p. 390, cited in Odom, "Estrangement," p. 542.
72. Ibid.
73. *A View of Sir Isaac Newton's Philosophy* (London, 1728), pp. 180–181, cited in David Kubrin, "Newton and the Cyclical Cosmos: Providence and the Mechanical Philosophy," *Journal of the History of Ideas* 28 (1967): 329.
74. Clarke to Leibniz, 29 October 1716, in Kubrin, "Newton and the Cyclical Cosmos," p. 329
75. Kubrin, "Newton and the Cyclical Cosmos," citing Clarke's first and fifth replies to Leibniz.
76. Cited in Margaret Jacob, *The Newtonians and the English Revolution: 1689–1720* (Ithaca: Cornell University Press, 1976), p. 124.
77. Ibid., p. 97.
78. Whiston, *Scripture Politicks* (London, 1717), p. 54, cited in James

Force, *William Whiston, The Honest Newtonian* (Cambridge: Cambridge University Press, 1985), p. 105. Emphasis mine.

79. See Herbert Odom, "Estrangement . . . ," passim.

80. Ibid., p. 535.

81. William Whewell, *Astronomy and General Physics Considered with Reference to Natural Theology* (London, 1833), p. vi.

82. Newton to Bentley, 10 December 1692, in I. B. Cohen, ed., *Papers and Letters,* p. 290.

83. James Force, "Beyond the Design Argument: Newton's 'Sleeping Argument' and the Newtonian Synthesis of Science and Religion," paper presented at the XVII International Congress of History of Science, Berkeley, California, 2 August 1985.

84. Robert Boyle, *The Works of the Honorable Robert Boyle in Five Volumes* (London, 1744), V: 526.

85. Isaac Newton, *Observations upon the Prophecies of Daniel and the Apocalypse of St. John* (London, 1733), p. 252. Emphasis mine.

86. Newton, Yahuda Ms. 1.1, fol. 3, cited in Frank Manuel, *The Religion of Sir Isaac Newton* (Oxford: Oxford University Press, 1974), p. 109.

87. Ibid., p. 124.

88. Ibid., pp. 115–116.

89. See Richard Popkin, "The Third Force in 17th Century Philosophy . . . ," *Nouvelles de la République des Lettres* (1983), I: 35–64.

90. Manuel, *The Religion of Isaac Newton,* p. 116.

91. Ibid., p. 120.

92. Ibid., p. 118.

93. Ibid., p. 120.

94. Ibid., pp. 120–121.

95. Newton, *Observations upon the Prophecies,* pp. 252–253.

96. Whiston, *A Supplement to the Literal Accomplishment of Scripture Prophecies* (London, 1725), p. 5.

97. Ibid., pp. 5–6.

98. Whiston, *The Genuine Nature of the Mosaic History of Creation* (London, 1707), cited in James Force, *William Whiston, the Honest Newtonian,* p. 95.

99. Whiston, *The Accomplishment of Scripture Philosophy* (London, 1708), p. 13.

100. Whiston, *A Supplement . . . ,* p. 8.

101. P. M. Heimann, "Voluntarism and Immanence: Conceptions of Nature in 18th Century Thought," *Journal of the History of Ideas,* 39 (1978): 282.

102. Ibid.

103. See John Yolton, *Thinking Matter, Materialism in 18th Century Britain* (Minneapolis: University of Minnesota Press, 1983), pp. 24–25, on the linkages between gravity and thought as properties of matter.

104. Ibid., p. 10.

105. John Locke, *An Essay Concerning Human Understanding,* IV. 3. 6. All references to Locke are from the sectioning of the 4th English edition, first published in 1704. This edition forms the basis for literally hundreds of subsequent editions, abridgments, and selections.

106. Ibid., II. 27. 5.

107. Ibid., II. 27. 6.

108. James O'Higgins, *Anthony Collins: The Man and His Works* (The Hague: Martinus Nijhoff, 1970), p. 73.

109. Ibid., pp. 69–76.

110. On Boscovich and his importance in Britain, see Robert Schofield, *Mechanism and Materialism* (Princeton: Princeton University Press, 1970), *passim.*

111. Roger Boscovich, *Theory of Natural Philosophy* (London, 1763; rpt. Chicago: The Open Court, 1922), p. 183.

112. Ibid.

113. Joseph Priestley, *Disquisitions Relating to Matter and Spirit* (London, 1777), p. 18, cited in Yolton, *Thinking Matter,* p. 113.

114. Cited in John Yolton, *Thinking Matter,* p. 113.

115. See Richard Olson, *Scottish Philosophy and British Physics* (Princeton: Princeton University Press, 1975), pp. 26–54.

116. Ibid., p. 28.

117. See James Force, "Secularization, The Language of God, and the Royal Society at the Turn of the Seventeenth Century," *History of European Ideas,* 2 (1981): 230.

118. See my forthcoming *Anglican Natural Theology and Its Enemies: 1592–1802* (Berkeley, Los Angeles, and Oxford: University of California Press).

4: LIBERALISMS AND SOCIALISMS:
THE IDEOLOGICAL IMPLICATIONS OF
ENLIGHTENMENT SOCIAL SCIENCE
I: THE TRADITION OF POLITICAL ECONOMY

1. Cited in Gary Wills, *Inventing America: Jefferson's Declaration of Independence* (New York: Random House, 1978), p. xxiii.

2. This definition is adapted from Joyce Appleby, *Economic Thought and Ideology in Seventeenth Century England* (Princeton: Princeton University Press, 1978), p. 5.

3. See Keith Michael Baker, *Condorcet: from Natural Philosophy to Social Mathematics* (Chicago: University of Chicago Press, 1975), Appendix A, p. 389.

4. Cited in Appleby, p. 78.

5. See Baker, *Condorcet*, p. 389.

6. William Letwin, *The Origins of Scientific Economics* (Garden City: Doubleday, 1964).

7. Peter Buck, "Seventeenth Century Political Arithmetic: Civil Strife and Vital Statistics," *Isis* 68 (1977): 67–84, and "People Who Counted: Political Arithmetic in the Eighteenth Century," *Isis* 73 (1982): 28–45.

8. Appleby, passim.

9. Ibid., pp. 187–188.

10. See Chapter 2 above. In addition, see Robert Kargon, "Sir William Petty's Mechanical Philosophy," *Isis* 56 (1965): 63–66.

11. Cited in Appleby, pp. 190–191.

12. See Joseph A. Shumpeter, *History of Economic Analysis* (New York: Oxford University Press, 1954), pp. 121–122.

13. See Chapter 2 above.

14. Cited in Appleby, p. 191.

15. Bacon, *Novum Organum,* Book I, aphorism II.

16. See Charles Webster, *The Great Instauration: Science, Medicine and Reform, 1626—1660* (London: Duckworth, 1975).

17. See Buck, "Seventeenth Century Political Arithmetic," p. 74.

18. Cited in Karl Pearson, *The History of Statistics in the 17th and 18th Centuries against the Changing Background of Intellectual, Scientific and Religious Thought* (London: Charles Griffin & Co., 1978), p. 70.

19. John Graunt, *Natural and Political Observations . . . Made upon the Bills of Mortality,* ed. Walter Willcox (Baltimore: The Johns Hopkins Press, 1939), p. 17.

20. Ibid., p. 78.

21. Ibid., p. 7.

22. Ibid., pp. 38–39.

23. Ibid., pp. 31, 33, 35.

24. Ibid.

25. Ibid., p. 54.

26. Ibid., pp. 57–61.

27. Ibid., pp. 67–69.

28. Ibid., pp. 69–70.

29. Ibid., pp. 71–76.

30. Ibid., p. 77.

31. Ibid.

32. Cited in Pearson, *The History of Statistics in the 17th and 18th Centuries*, p. 33.

33. Ibid., p. 80.

34. Graunt, *Natural and Political Observations*, pp. 78–79.

35. See G. S. Holmes, "Gregory King and the Social Structure of Pre-Industrial England," *Royal Historical Society Transactions* 26, 5th series (1976).

36. See Peter Buck, "People Who Counted," p. 36.

37. Ibid., pp. 33, 34, 37.

38. Ibid., p. 33.

39. Ibid., p. 39.

40. Ibid., pp. 40–42.

41. See C. C. Gillespie, *Science and Polity in France at the End of the Old Regime* (Princeton: Princeton University Press, 1980), pp. 50–73 for Lavoisier's work on munitions.

42. See Keith Michael Baker, *Condorcet*, pp. 10–12 for a short introduction to the special character of the Académie des Sciences.

43. Cited in ibid., p. 14.

44. Condorcet, *Vie de Turgot*, cited in Frank Manuel, *The Prophets of Paris* (New York: Harper and Row, 1962), p. 42.

45. Cited in Baker, *Condorcet*, p. 57.

46. See Gillespie, *Science and Polity*, for a super account of science in the Turgot Ministry.

47. Baker, *Condorcet*, p. 63.

48. Condorcet, *Mémoires sur l'instruction publique*, cited in ibid., p. 79.

49. Cited in ibid., p. 208.

50. Cited in ibid., p. 230.

51. Condorcet, *Mémories sur l'instruction publique*, cited in Baker, *Condorcet*, p. 79.

52. Richard Cantillon, *Essay on the Nature of Trade in General*, edited by Henry Higgs (London: The Royal Economic Society, 1931), pp. 13, 43.

53. Ibid., p. 3.

54. Ibid., pp. 5–7, 31.

55. Ibid., p. 15.

56. Ibid., pp. 15–17.

57. Ibid., p. 19.

58. Ibid., p. 23.

59. Ibid., p. 25.

60. Ibid., p. 27.

61. Ibid., p. 29.

62. Ibid., p. 119.

63. Ibid., p. 31.

64. Ibid., p. 43. For a description of Petty's calculation see Chapter 2, above.

65. Joseph Shumpeter, *History of Economic Analysis,* p. 219.

66. Cantillon, *Essay,* p. 37.

67. Ibid., p. 39.

68. Ibid., p. 47.

69. Ibid., pp. 47–55.

70. Ibid., p. 55.

71. Ibid., p. 57.

72. Ibid., p. 55.

73. Ibid., p. 83.

74. Ibid., p. 79.

75. Ibid., p. 67.

76. Ibid., pp. 79–81.

77. Ibid., p. 85.

78. Ibid.

79. Ibid.

80. Ibid., p. 87.

81. Ibid., pp. 87–93.

82. Ibid., p. 89.

83. Ibid.

84. Ibid., p. 91.

85. Ibid.

86. Ibid., pp. 227–233 for the extended development of this notion.

87. Ibid., p. 91.

88. Ibid., p. 131.

89. Ibid., p. 133.

90. Ibid., p. 159.

91. Ibid., p. 141.

92. Ibid., p. 151.

93. Ibid., p. 155.

94. Ibid.

95. Ibid., p. 161.

96. Ibid., p. 167.

97. Ibid., p. 171.

98. Ibid., p. 185.

99. Ibid., p. 195.

100. Ibid., p. 189.

101. Ibid., p. 201.

102. Ibid.

103. Ibid., p. 205.

104. Ibid., pp. 297–309.

105. Ibid., p. 215.

106. Ibid., p. 221.

107. Henry Higgs, *The Physiocrats* (London: McMillan & Co., 1897), p. 4.

108. See Elizabeth Fox-Genovese, *The Origins of Physiocracy* (Ithaca: Cornell University Press, 1976), p. 49.

109. Cited in Ronald Meek, *The Economics of Physiocracy, Essays and Traditions* (Cambridge: Harvard University Press, 1963), pp. 15–16.

110. Fox-Genovese, *The Origin of Physiocracy*, p. 76.

111. Cited in Meek, p. 108.

112. François Quesnay, "Fermiers," cited in Fox-Genovese, p. 115.

113. Ibid.

114. Cited in Meek, *The Economics of Physiocracy*, p. 17.

115. Quesnay, "Corn," in Meek, *The Economics of Physiocracy*, p. 81.

116. Ibid., p. 75.

117. Ibid., p. 77.

118. Ibid., p. 78.

119. Ibid., p. 99.

120. Quesnay, "Men," in Meek, *The Economics of Physiocracy*, p. 99.

121. Ibid., pp. 100–101.

122. Quesnay, "Tableau Economique," in Meek, *Economics of Physiocracy*, pp. 110–111.

123. See, for example, Joseph Shumpeter, *History of Economic Analysis*, pp. 239–243.

124. For a summary of the problems involved in interpreting the Tableau, see Meek, *The Economics of Physiocracy*, pp. 265–296.

125. Quesnay, "Tableau Economique," in Meek, *The Economics of Physiocracy*, pp. 110–111.

126. Quesnay, "Taxation," in Meek, *Economics of Physiocracy*, p. 102.

127. Ibid.

128. Ibid., p. 103.

5: THE IDEOLOGICAL IMPLICATIONS OF
ENLIGHTENMENT SOCIAL SCIENCE
II: THE SOCIOLOGICAL TRADITION

1. Anyone interested should begin by consulting Paul M. Spurlin's *Montesquieu in America, 1760–1801* (Baton Rouge: Louisiana State University Press, 1940).

2. Cited in Ronald Meek, *Social Science and the Ignoble Savage* (Cambridge: Cambridge University Press, 1976), pp. 233–234.

3. Cited in Gladys Bryson, *Man and Society: The Scottish Inquiry of the Eighteenth Century* (Princeton: Princeton University Press, 1945), p. 30.

4. Clarence Glacken, *Traces on the Rhodian Shore* (Berkeley and Los Angeles: University of California Press, 1967), details the environmentalist tradition from Greek Antiquity through Montesquieu.

5. Montesquieu, "Discourse on the Motives that Should Encourage Us in the Sciences," cited in J. Robert Loy, *Montesquieu* (Boston: Twayne Publishers, Inc., 1968), pp. 39–40.

6. Cited in George Havens, *The Age of Ideas* (New York: Henry Holt & Co., 1955), p. 132.

7. Montesquieu, *Spirit of the Laws,* trans. Thomas Nugent (New York: Hafner Press, 1949), p. lxvii.

8. Ibid., p. 1.

9. Ibid., p. 2.

10. Ibid.

11. Ibid., p. 6.

12. See Keith Baker, *Condorcet: From Natural Philosophy to Social Mathematics* (Chicago: University of Chicago Press, 1975), p. 222.

13. Turgot, "A Philosophical Review of the Successive Advances of the Human Mind," in Ronald Meek, *Turgot on Progress, Sociology, and Economics,* (Cambridge: Cambridge University Press, 1973), pp. 44–45.

14. Montesquieu, *Spirit of the Laws,* p. 113.

15. Helvetius to Montesquieu, cited in the editor's introduction to Montesquieu, *Spirit of the Laws,* p. xxvii. Doubt about the authenticity of the "Helvetius" letters to Montesquieu have been raised by R. Kroeber, "The Authenticity of the Letters on the *Esprit des lois* attributed to Helvetius," *Bulletin of the Institute for Historical Research, 24* (1951): pp. 19–43. For our purposes, the question of authorship is irrelevant.

16. Montesquieu, *Spirit,* p. 275.

17. Ibid., p. 113.

18. Ibid., p. 277.

19. Ibid., p. 278.

20. Ibid.

21. Ibid., p. 326.
22. Ibid., p. 316.
23. Ibid.
24. Ibid., pp. 316–317.
25. Ibid., p. 317.
26. Ibid.
27. Ronald Meek, *Social Science and the Ignoble Savage* (Cambridge: Cambridge University Press, 1976), p. 1.
28. Bernard Mandeville, "Fable of the Bees," cited in Basil Willey, *The Eighteenth Century Background* (New York: Columbia University Press, 1940), p. 96.
29. Montesquieu, *Spirit,* p. 317.
30. Ibid., pp. 296–297.
31. Ibid., pp. 298–299.
32. Ibid., 258.
33. Ibid., pp. 298–303.
34. Ibid., p. 302.
35. Ibid., p. 303.
36. Ibid., p. 302.
37. On the relations between Smith and Ferguson, see Ronald Meek, *Social Science and the Ignoble Savage,* pp. 99–130.
38. See H. H. Jogland, *Ursprunge und Grundlagen der soziologie bie Adam Ferguson* (Berlin, 1959).
39. See F. A. Hayek, *Studies in Philosophy, Politics, and Economics,* (Chicago: University of Chicago Press, 1967), chap. 6.
40. Adam Ferguson, *An Essay On the History of Civil Society,* ed. Duncan Forbes (Edinburgh: Edinburgh University Press, 1966), p. 2.
41. Ibid., pp. 3–6.
42. Ibid., p. 4.
43. Ibid., p. 5.
44. Ibid., p. 2.
45. Ibid., p. 122.
46. Ibid., p. 10.
47. Adam Ferguson, *Institutes of Moral Philosophy* (1766), cited in Gladys Bryson, *Man and Society* (Princeton: Princeton University Press, 1945), p. 35.
48. Ferguson, *Essay,* p. 15.
49. Ibid., p. 16–17.
50. Ibid., p. 10.
51. Ibid., p. 34.

52. Ibid., p. 38.
53. Ibid., p. 54.
54. Ibid., p. 19.
55. Ibid., p. 24.
56. Ibid., p. 22.
57. Ibid., p. 24.
58. Ibid., p. 17.
59. Ibid., p. 25.
60. Ibid., p. 119.
61. Ibid., p. 58.
62. Ibid., p. 17.
63. Ibid., p. 145.
64. Ibid., p. 60.
65. Ibid., p. 135.
66. Ibid., p. 128.
67. See, for example, Alan Swingewood, *A Short History of Sociological Thought* (New York: St. Martin's Press, 1984), pp. 22–24.
68. Ferguson, *Essay*, pp. 156–157.
69. Ibid., p. 157.
70. Ibid., pp. 158–159.
71. Ibid., p. 160.
72. Ibid., pp. 161–162.
73. Ibid., p. 182.
74. Ibid., pp. 180–181.
75. Ibid., pp. 181–182.
76. Ibid., p. 183.
77. Ibid., pp. 182–183.
78. Ibid., p. 186.
79. Ibid.
80. Ibid., pp. 183–189.
81. See George Elder Davie, *The Democratic Intellect: Scotland and Her Universities in the 19th Century* (Edinburgh: Edinburgh University Press, 1964).
82. Ferguson, *Essay*, p. 256.
83. Ibid., p. 255.
84. See Ronald Meek, "Smith, Turgot, and the Four Stages Theory" in R. Meek, *Smith, Marx, and After* (London: Chapman and Hall, 1977), pp. 18–32.
85. Joseph Shumpeter, *History of Economic Analysis,* p. 181.

86. Meek, *Smith, Marx, and After,* p. 3.

87. Adam Smith, *The Early Writings of Adam Smith,* ed. J. Ralph Lindgren (New York: Augustus M. Kelley, 1967), p. 66.

88. Ibid., p. 108. Emphasis mine.

89. See Vernard Foley, *The Social Physics of Adam Smith* (West Lafeyette, Ind.: Purdue University Press, 1976), for an extensive discussion of Smith's heterodox religious attitudes.

90. Smith, *Early Writings,* p. 49.

91. Smith, *Theory of Moral Sentiments,* Pt. III, Chap. 4, Par. 4, cited in the editors' introduction to Adam Smith, *An Inquiry into the Nature and Causes of the Wealth of Nations,* ed. R. H. Campbell and A. S. Skinner (Oxford: Oxford University Press, 1976), p. 9.

92. Ibid., p. 18.

93. Ibid., p. 7.

94. See Meek, *Smith, Marx, and After,* p. 183, for an introduction to the sources expressing Smith's ideas on law and government.

95. Cited in ibid., p. 178.

96. Ibid., p. 38.

97. Cited in Meek, *Social Science and the Ignoble Savage,* p. 123. Emphasis mine.

98. Francis Hutcheson, *An Inquiry Into the Original of Our Ideas of Beauty and Virtue,* 2d ed. (London, 1725), p. 164.

99. Cited in Meek, *Smith, Marx, and After,* p. 11.

100. Smith, *Wealth of Nations,* 1: 96.

101. Ibid., 2: 783–784.

102. Ibid., 788.

103. On Turgot's treatment of capital, see R. Meek, ed., *Turgot on Progress, Sociology, and Economics,* pp. 16–25.

104. Smith, *Wealth of Nations,* 1: 343.

105. Ibid., pp. 26–27.

106. Ibid., p. 422.

107. Ibid., p. 455.

108. Ibid., p. 456. Emphasis mine.

109. Ibid., p. 454. Emphasis mine.

110. Ibid., 2: 687.

111. Ibid.

112. Ibid., 1: 22

113. Ibid., 2: 687.

114. Ibid., 1: 31.

115. Ibid., p. 34.

116. See Frank Whitson Fetter, *The Economist in Parliament* (Durham, NC: Duke University Press, 1980), pp. 35–44.
117. Cited in ibid., p. 43.
118. Ibid., pp. 50–51.
119. Smith, *Wealth of Nations,* p. 85.
120. Ibid., p. 87.
121. Ibid., p. 89.
122. Ibid., p. 96.

6: THE IDEOLOGICAL IMPLICATIONS OF ENLIGHTENMENT SOCIAL SCIENCES III: SENSATIONALIST PSYCHOLOGY AND A NEW FOCUS ON EQUALITY, DISTRIBUTIVE JUSTICE, AND EDUCATION

1. Claude Adrien Helvetius to Montesquieu, cited in Destutt de Tracy, *A Commentary and Review of Montesquieu's Spirit of the Laws* (Philadelphia, 1811), p. 291.
2. Condorcet, cited in Keith Baker, *Condorcet: from Natural Philosophy to Social Mathematics* (Chicago: University of Chicago Press, 1975), p. 222.
3. Cited in Baker, *Condorcet,* p. 217. Emphasis mine.
4. Helvetius to Montesquieu, cited in de Tracy, *A Commentary,* p. 291.
5. William Godwin, *Political Justice* (1791), cited in Basil Willey, *The Eighteenth Century Background* (New York: Columbia University Press, 1940), p. 219.
6. Condorcet, cited in Baker, *Condorcet,* p. 223.
7. David Hartley, *Observations on Man, His Frame, His Duty, and His Expectations* (London, 1749), reprinted (Gainsville, Florida: Scholar's Facsimilies and Reprints, 1966), 1: 354.
8. Joseph Priestley, *Remarks on Reid, Beattie, and Oswald,* cited in Elie Halevy, *The Growth of Philosophical Radicalism,* trans. Mary Morris (London: Faber and Faber, Ltd., 1934), p. 9.
9. On the German reception of Hartley, see Robert Marsh, "The Second Part of Hartley's System," *Journal of the History of Ideas* 20 (1959): 54.
10. David Hartley, *Various Conjectures on the Perception, Motion, and Generation of Ideas* (Los Angeles: Augustan Reprint Society, 1959; rpt. of 1746 ed.), p. 54.
11. Ibid., pp. 54–55.
12. See Marsh, "The Second Part of Hartley's System," p. 64.
13. Hartley, *Observations on Man,* 1: 511.
14. Ibid., 512.

15. Ibid., 72.
16. Ibid., 6.
17. Ibid., 345.
18. Ibid., 345–346.
19. Ibid., 343.
20. Ibid.
21. Ibid., 8.
22. Ibid., 9.
23. Ibid., 5.
24. Ibid., 30–31.
25. Ibid., 35–36.
26. Ibid., 40.
27. Ibid., 59.
28. Ibid., 65.
29. Ibid., 67–72.
30. Ibid., 73–80.
31. Ibid., 80.
32. Ibid., 81.
33. Ibid.
34. Ibid., 82.
35. Ibid., 81.
36. Ibid., 83.
37. Ibid., 500.
38. Ibid., 371.
39. Ibid., 2: 306.
40. Ibid., 207.
41. Ibid., 1: 458–464.
42. Ibid., 464.
43. Ibid., 2: 292.
44. Ibid., 2: 263.
45. Ibid., 1: 273.
46. Ibid., 462.
47. Ibid., 281–282.
48. Ibid., 280.
49. Ibid., 315–321.
50. Etienne de Condillac, *La Logique: Logic,* ed. and trans. W. R. Albury (New York: Abaris Books, 1980), p. 151.
51. Ibid., pp. 127–129.

52. Ibid., editor's introduction, pp. 7–29.

53. Cited in ibid., p. 27.

54. Kingsley Martin, *French Liberal Thought in the 18th Century,* 3rd ed. (New York: Harper and Row, 1962), pp. 122–123.

55. See Irving L. Horowitz, *Claude Helvetius: Philosopher of Democracy and Enlightenment* (New York: Paine-Whitman Publishers, 1954).

56. Cited in D. W. Smith, *Helvetius: A Study in Persecution* (Oxford: Oxford University Press, 1965), p. 16.

57. Ibid., p. 40.

58. The entirety of Smith's *Helvetius* focuses on the fallout of the *De l'Esprit* affair.

59. Claude Adrien Helvetius, *Treatise on Man,* trans. W. Hooper (New York: Burt Franklin, 1969), 2: 279.

60. Ibid.

61. Ibid., 280.

62. Ibid., 279.

63. Fontenelle's perspective is well discussed in Frank Manuel, *The Eighteenth Century Confronts the Gods* (Cambridge: Harvard University Press, 1959), pp. 47–53.

64. Claude Adrien Helvetius, *De l'Esprit, or Essays on the Mind and Its Several Faculties* (New York: Burt Franklin, 1970; rpt. of 1809 ed.), p. xix.

65. Ibid., 39.

66. Ibid., 229.

67. Ibid., 124.

68. Ibid., 179.

69. Ibid., 135.

70. Ibid., 289.

71. Ibid., 57, 63, 124, 135, 146, 171, 185, 291, 336.

72. Ibid., 337.

73. Ibid., 149.

74. Ibid., 15.

75. Ibid., 146.

76. Helvetius, *Treatise on Man,* 2: 434–435.

77. Helvetius, *De l'Esprit,* p. 15.

78. Ibid., 16.

79. Ibid., 17–18.

80. Ibid., 18.

81. Ibid., 248–251, 261, 269–274, 278–283.

82. Ibid., 18–19.

83. Ibid., 19–25.
84. Helvetius, *Treatise on Man,* 2: 202.
85. Ibid., 88.
86. Ibid., 183.
87. Ibid., 88.
88. Ibid., 91.
89. Ibid., 95.
90. Ibid., 98.
91. Ibid., 103–105.
92. Ibid., 109.
93. Ibid., 113.
94. Ibid., 117, 124.
95. Ibid., 125–127.
96. Ibid., 198.
97. Ibid., 201.
98. Ibid., 202.
99. Ibid., 203–204.
100. Ibid.
101. Ibid., 207–208.
102. Helvetius, *De l'Esprit,* pp. 72–77.
103. Ibid., 71.
104. Ibid., 33.
105. Ibid., 126.
106. Helvetius, *Treatise on Man,* 2: 423–434.
107. Ibid., 439.
108. Helvetius, *De l'Esprit,* p. 172.
109. Ibid.
110. Ibid., 177.
111. Ibid., 183.
112. Helvetius, *Treatise on Man,* 2: 434–435.
113. Ibid., 435.
114. Ibid., 1: vi-vii.
115. Ibid., 2: 311.
116. For a handy guide to the literature on the role of the Enlightenment in fomenting the Revolution, see William F. Church, ed., *The Influence of the Enlightenment on the French Revolution* (Boston: D. C. Heath & Co., 1964).
117. See G. D. H. Cole, *Socialist Thought: The Forerunners, 1789–1850* (London: Macmillan, 1967), pp. 1–2.

118. See William B. Guthrie, *Socialism before the French Revolution* (London: Macmillan, 1907), p. 278.

119. See Gilbert Chinard, editor's introduction to Morelly, *Code de la nature; ou la veritable esprit de ses lois* (Paris, 1950), especially pp. 12–26. See also Richard Coe, *Morelly, Ein Rationalist auf dem Weg zun Sozialismus* (Berlin, 1961), passim.

120. Helvetius, *Treatise on Man,* 2: 423–432.

121. Cited in Keith Baker, *Condorcet: From Natural Philosophy to Social Mathematics,* p. 215.

122. Ibid., p. 218.

123. Ibid., p. 216.

124. Ibid., p. 217.

125. Ibid., p. 223.

126. Ibid., p. 225.

127. Ibid., p. 223.

128. See Dorothy Nelkin, "The Political Impact of Technical Expertise," *Social Studies of Science* 5 (1975): 35–54, and Dorothy Nelkin and Michael Pollak, *The Atom Besieged: Extraparliamentary Dissent in France and Germany* (Cambridge: MIT Press, 1981).

129. See Francisque Vial, *Condorcet et l'éducation démocratique* (Paris, 1902), and Manuela Albertone, "Enlightenment and Revolution: The Evolution of Condorcet's Ideas on Education" in Leonora C. Rosenfield, ed., *Condorcet Studies* (Atlantic Heights, N.J.: Humanities Press, 1984), 1: 131–144.

130. See J. Salwyn Shapiro, *Condorcet and the Rise of Liberalism* (New York: Octagon Books, 1963), pp. 198–199.

131. Ibid., p. 199.

132. See Baker, *Condorcet,* pp. 293–294.

133. Cited in Shapiro, *Condorcet and the Rise of Liberalism,* p. 203.

134. Ibid., p. 200.

135. Ibid., p. 204.

136. Ibid., pp. 204–205.

137. Ibid., pp. 209–210.

138. P. S. G. Cabanis, "Travail sur l'éducation . . ." cited in L. Pearce Williams, "Science, Education and the French Revolution," *Isis* 44 (1953): 316.

7: SCIENCE AND AESTHETICS: THE BEAUTIFUL TO THE SUBLIME, CA. 1600–CA. 1800

1. John Locke, *Some Thoughts Concerning Education* (London, 1693), section 165, p. 207.

2. Cited in Douglas Bush, *Science and English Poetry* (New York: Oxford University Press, 1950), p. 40.

3. Cited in W. B. Stanford, *Enemies of Poetry* (London: Routledge and Kegan Paul, 1980), p. 45.

4. Ibid., p. 46.

5. See Joel E. Spingarn, *A History of Literary Criticism in the Renaissance*, 2d ed. (New York: Columbia University Press, 1908), pp. 24–27.

6. Paraphrase of Menturno in ibid., p. 118.

7. John Dennis, *Grounds of Criticism in Poetry* (1704) cited in Richard P. Cowel, ed., *The Theory of Poetry in England: Its Development in Doctrines and Ideas from the Sixteenth Century to the Nineteenth Century* (London: Macmillan & Co., 1914), pp. 55–56.

8. Francis Hutcheson, *An Inquiry Concerning Beauty, Order, Harmony, and Design*, ed. Peter Kivy (The Hague: Martinus Nyhoff, 1973), pp. 40, 55.

9. Aristotle, *Poetics*, 1460a 20.

10. Ibid., 1460b 25–1460b 30.

11. Ibid., 1460b 35–1461a 4, 1461b 12–1461b 15.

12. Ibid., 1451a 37–1451b 7.

13. Ibid., 1448b 4–1448b 19.

14. Ibid., 1449b 24–1449b 28.

15. Samuel H. Butcher, *Aristotle's Theory of Poetry and Fine Art* (New York: Dover Publications, 1951; rpt. of 4th ed., 1907), p. 254.

16. Aristotle, *Poetics*, 1458a 18–1458a 22.

17. Ibid., 1459a 6–1459a 10.

18. J. E. Spingarn, "Jacobean and Caroline Criticism," in *The Cambridge History of English Literature* (Cambridge: Cambridge University Press, 1911), 7: 266.

19. See Clarence DeWitt Thorpe, *The Aesthetic Theory of Thomas Hobbes* (New York: Russell and Russell, Inc., 1964), pp. 7–8, 302–307.

20. Sir Philip Sidney, *Defense of Poetry*, ed. Lewis Soens (Lincoln: University of Nebraska Press, 1970), p. 12.

21. Cited in Thorpe, *The Aesthetic Theory of Thomas Hobbes*, p. 97.

22. Ibid., p. 99.

23. Ibid.

24. Hobbes, *The Answer to Davenant*, cited in ibid., p. 108.

25. Ibid.

26. John Locke, *An Essay Concerning Human Understanding*, 3.10.34.

27. Cited in Thorpe, *The Aesthetics of Thomas Hobbes*, p. 109, n. 76.

28. Ibid., p. 97.

29. Ibid., p. 105.

30. Thomas Hobbes, *The Whole Art of Rhetoric,* cited in ibid., p. 134.

31. Cited in Thorpe, p. 140.

32. Ibid., p. 141.

33. Ibid., p. 243.

34. John Dennis, *The Usefulness of the Stage,* excerpted in Henry Hitch Adams and Baxter Hathaway, eds., *Dramatic Essays of the Neoclassic Age* (New York: Columbia University Press, 1950), pp. 203–204.

35. Ibid., p. 204.

36. Edmund Burke, *A Philosophical Enquiry into the Origin of our Ideas of the Sublime and the Beautiful,* ed. with an introduction and notes by J. T. Boulton (London: Routledge & Kegan Paul, 1958), pp. xix–xx.

37. Ibid., p. xx.

38. Thorpe, *The Aesthetic Theory of Thomas Hobbes,* p. 136.

39. Ibid., p. 143.

40. John Dennis, *The Sounds of Criticism in Poetry,* cited in ibid., p. 250. Emphasis mine.

41. Editor's introduction to Burke, p. lvii.

42. Ibid., p. lxxxii.

43. Burke, *A Philosophical Enquiry,* p. 1.

44. Ibid., pp. 13–14.

45. Ibid., p. 17.

46. Ibid., p. 129.

47. Ibid., p. 15.

48. Ibid., pp. 130–131.

49. Ibid., p. 18.

50. Ibid., p. 26.

51. Ibid., p. 27.

52. Ibid., p. 38.

53. Ibid., pp. 40–41.

54. Ibid., p. 39.

55. Ibid., p. 43.

56. Ibid., p. 40.

57. Ibid., p. 44.

58. Ibid., p. 46.

59. Ibid., p. 47.

60. Ibid., p. 49.

61. Ibid., p. 134.

62. Ibid., p. 136.

63. Ibid., p. 137.

64. Ibid., p. 141.

65. Ibid., p. 140.

66. Editor's introduction to Burke, pp. cvii–cxvii.

67. See Uvalde Price, *An Essay on the Picturesque, as Compared with the Sublime and the Beautiful* (London, 1794).

68. Burke, p. 60.

69. Ibid., p. 61.

70. Ibid., p. 63.

71. Ibid., pp. 60–61.

72. Cited in Basil Willey, *The Eighteenth Century Background*, pp. 227–278, 280.

73. Burke, pp. 92, 105.

74. Ibid., pp. 93–95.

75. Ibid., pp. 94–95.

76. Ibid., p. 96.

77. Ibid., p. 98.

78. Ibid., p. 101.

79. Ibid.

80. Ibid., pp. 104–105.

81. Ibid., pp. 110–111.

82. Ibid., pp. 113–117.

83. Ibid., p. 139.

84. Editor's introduction to Burke, p. lxxiv. For a more extended discussion of this notion, see Erika von Erhardt-Siebold, "Harmony of the Senses in English, German and French Romanticism," *PMLA* 47 (1932): 578ff.

8: SCIENCE, TECHNOLOGY, AND THE INDUSTRIAL
REVOLUTION: THE CONFLATION OF
SCIENCE AND TECHNICS

1. On the role of monasticism in stimulating Western technological development see Lynn White, Jr., "What Accelerated Technological Progress in the Western Middle Ages," in A. C. Crombie, ed., *Scientific Change* (New York: Basic Books, 1963), pp. 272–291.

2. Hugh of St. Victor, *The Didascolicon*, trans. from the Latin with an introduction by Jerome Taylor (New York: Columbia University Press, 1961), p. 9.

3. Ibid., p. 56.

4. Marie Boas Hall, "Oldenburgh, The *Philosophical Transactions,* and Technology" in John Burke, ed., *The Uses of Science in the Age of Newton* (Berkeley, Los Angeles, London: University of California Press, 1983), p. 21. See also Derick De Solla Price, *Science Since Babylon* (New Haven: Yale University Press, 1975), chap. 3.

5. A. R. Hall, *The Science of Ballistics in the Seventeenth Century* (Cambridge, 1952), p. 1.

6. Richard Westfall, "Robert Hooke, Mechanical Technology, and Scientific Investigation," in John Burke, ed., *The Uses of Science in the Age of Newton,* p. 89.

7. See Peter Mathias, "Who Unbound Prometheus? Science and Technical Change, 1600–1800," in Peter Mathias, ed., *Science and Society 1600–1700* (Cambridge: Cambridge University Press, 1972), pp. 54–55 for a summary of this histographic position. Also see G. N. von Tunzelmann, "Technical Progress in the Industrial Revolution," in Roderick Flood and Donald McCloskey, eds., *The Economic History of Britain since 1700* (Cambridge: Cambridge University Press, 1981), esp. 1: 147–151.

8. C. C. Gillispie, "The Natural History of Invention," *Isis* 48 (1957): 402–403.

9. David Hume, *Enquiries Concerning the Human Understanding and Concerning the Principles of Morals* (Oxford, 1902), p. 76. Cited in A. L. Donovan, *Philosophical Chemistry in the Enlightenment* (Edinburgh: 1975), p. 60.

10. William Cullen, cited in Donovan, *Philosophical Chemistry,* p. 60.

11. Cited in A. E. Musson and Eric Robinson, *Science and Technology in the Industrial Revolution* (Toronto: University of Toronto Press, 1969), pp. 5–6.

12. See G. N. von Tunzelmann, "Technical Progress during the Industrial Revolution," in Roderick Floud and Donald McCloskey, eds., *The Economic History of Britain since 1700, Vol. I, 1700–1860* (Cambridge: Cambridge University Press, 1981), p. 149.

13. Gillispie, "The Natural History of Industry," p. 402.

14. Ibid., p. 400.

15. Ibid., p. 405.

16. See E. L. Jones, ed., *Agricultural and Economic Growth in England, 1650–1815* (London: Methuen, 1967) for a good survey of the causes and character of the Agricultural Revolution.

17. Taken from J. H. Plumb, *England in the Eighteenth Century* (Harmondsworth, England: Penguin Books, 1950), p. 82.

18. On Lavoisier's experimental farm and its impact, see Douglas McKie, *Antoine Lavoisier* (New York: Collier Books, 1956), chaps. 17 and 18.

19. A. L. Donovan, *Philosophical Chemistry,* p. 69.

20. See James Small, *A Treatise on Ploughs and Wheeled Carriages* (Edinburgh, 1784).

21. Cited in Donovan, *Philosophical Chemistry*, p. 69.

22. Cited in Mathias, "Who Unbound Prometheus?" pp. 74–75, 76 n. 1.

23. Ibid., pp. 75–76.

24. See A. E. Musson and Eric Robinson, *Science and Technology in the Industrial Revolution* (Manchester: University of Manchester Press, 1969), esp. pp. 73 and 141.

25. Ibid., p. 108.

26. Ibid., p. 73, n. 6.

27. See R. J. Forbes, "Power to 1850," p. 164, and H. W. Dickinson, "The Steam Engine to 1830," pp. 179–181, in Charles Singer, et. al., *A History of Technology, Vol. IV, The Industrial Revolution c. 1750–c. 1850* (Oxford: Oxford University Press, 1958).

28. "An Experimental Examination of the Quantity and Proportion of Mechanic Power Necessary to be Employed in Giving Different Degrees of Velocity to Heavy Bodies from a State of Rest," *Phil. Trans.,* 66 (1776): 450–475 and "New Fundamental Experiments upon the Collision of Bodies," *Phil. Trans.,* 72 (1782): 337–354.

29. "Observations on the Graduations of Astronomical Instruments with an Explanation of the Method Invented by the late Mr. Henry Hindley of York, Clock Maker, to divide Circles into any Given Number of Parts," *Phil. Trans.,* 76 (1786): 1–47.

30. John Smeaton, *Reports of the Late John Smeaton, F. R. S.* (London: M. Taylor, 1837), 1: x–xi.

31. John Farey, *A Treatise on the Steam Engine* (1827), p. 296, cited in Musson and Robinson, *Science and Technology*, p. 69.

32. Smeaton's experiments are nicely summarized in A. Stowers, "Watermills c1500–c1850," in Charles Singer, et al., *A History of Technology*, 4: 189–213.

33. Ibid., p. 204.

34. Musson and Robinson, *Science and Technology*, pp. 71–72.

35. Ibid., p. 74.

36. Archibald and Nan Clow, *The Chemical Revolution* (London: The Batchworth Press, 1952), pp. 611–615, cited in Robert E. Schofield, *The Lunar Society of Birmingham: A Social History of Provincial Science and Industry in Eighteenth Century England* (Oxford: Oxford University Press, 1963), pp. 3–4.

37. Neil McKendrick offers an excellent catalog and characterization of those who are skeptical of the scientific element in eighteenth-century technological innovation. See "The Role of Science in the Industrial Revolution: A Study of Josiah Wedgwood as a Scientist and Industrial Chemistry," in Mikulas Teich and Robert Young, eds., *Changing Per-*

spectives in the History of Science (Dordrecht, Holland: D. Reidel Publishing Co., 1973).

38. Cited in Neil McKendrick, "The Role of Science in the Industrial Revolution," p. 292, from A. R. Hall, "What Did the Industrial Revolution in Britain Owe to Science?" in N. McKendrick, *Historical Perspectives, Studies in English Thought and Society in Honor of J. H. Plumb* (London, 1974).

39. Schofield, *The Lunar Society of Birmingham,* p. 170.

40. McKendrick, "The Role of Science in the Industrial Revolution," pp. 303–304.

41. The results of this study were read to the Royal Society in December of 1780, but were never published. See Schofield, *The Lunar Society of Birmingham,* pp. 170–171, for a summary description.

42. McKendrick, "The Role of Science in the Industrial Revolution," p. 309, n. 159.

43. Ibid., p. 309

44. Ibid., pp. 303–304.

45. Cited in Schofield, *The Lunar Society of Birmingham,* pp. 172, 173.

46. Ibid., p. 173.

47. Ibid., pp. 160–161.

48. See McKendrick, "The Role of Science," pp. 317–318.

49. Ibid., p. 318.

50. On the spread of scientific literacy in the eighteenth century, see Margaret Jacob, *The Cultural Meaning of the Scientific Revolution* (New York: Alfred Knopf, 1988), pp. 120–131, 144–151.

51. A. R. Hall, *The Historical Relations of Science and Technology* (London: Imperial College of Science and Technology, 1963), pp. 127–128.

52. Cited in McKendrick, "Josiah Wedgwood and Factory Discipline," in David Landes, ed., *The Rise of Capitalism* (New York: Macmillan Co., 1966), p. 67.

53. Ibid.

54. Ibid., p. 73.

55. Ibid., p. 74.

56. McKendrick, "The Role of Science," pp. 306–307.

57. On the role of machines in the early textile revolution, see, for example, Abbott Payson Usher, "The Textile Industry, 1750–1830," pp. 230–245, in Melvin Kransberg and Carroll Pursell, eds., *Technology in Western Civilization, I* (New York: Oxford University Press, 1967).

58. A. and N. L. Clow, "The Chemical Industry: Interaction with the Industrial Revolution," in Charles Singer, et al., *A History of Technology,* 4: 244–247.

59. See, for example, Donovan, *Philosophical Chemistry,* pp. 78–83.

60. See A. Clow and N. L. Clow, "The Chemical Industry," pp. 246–247.

61. See Musson and Robinson, *Science and Technology in the Industrial Revolution,* pp. 252–254.

62. The *Journal de physique* 26 (1885): 231–235 and in the *Annals de chimie* 2 (1788): 151–190, and 6 (1790): 204–240.

63. Musson and Robinson, *Science and Technology in the Industrial Revolution,* pp. 256–259.

64. Ibid., pp. 289–290.

65. Ibid., pp. 321–322.

66. Cited in A. and N. L. Clow, "The Chemical Industry," p. 248.

9: ROMANTIC REACTIONS TO SCIENTIZED POLITICS AND PRODUCTION

1. Joseph de Maistre, *The Works of Joseph de Maistre,* selected, translated, and introduced by Jack Lively (New York: Schocken Books, 1971), p. 111.

2. Ibid., p. 114.

3. Ibid., pp. 59–60.

4. Ibid., p. 118.

5. Ibid., pp. 65–66.

6. Ibid., p. 47.

7. Ibid.

8. Ibid., p. 62.

9. Ibid., pp. 66–67.

10. Ibid., p. 49.

11. Ibid., p. 80.

12. Ibid., pp. 45–46.

13. Ibid., p. 126, emphasis mine.

14. Ibid., p. 127.

15. Ibid., p. 175.

16. Ibid., p. 150.

17. Cited in ibid., p. 156.

18. Ibid.

19. Ibid., p. 102.

20. Ibid., p. 109.

21. Ibid., p. 128.

22. Ibid., p. 179.

23. Ibid., p. 180.

24. Ibid., p. 166.

25. Ibid., p. 112.
26. Edmund Burke, *Reflections on the Revolution in France* (Garden City, New York: Anchor Books, 1973), p. 103.
27. Ibid., pp. 105, 168.
28. Ibid., p. 212.
29. Ibid., p. 38.
30. Ibid., pp. 169–170.
31. Ibid., p. 94.
32. Ibid., p. 156.
33. Ibid., p. 174.
34. Ibid., p. 73.
35. Ibid., pp. 71–72.
36. Ibid., p. 73.
37. Ibid., pp. 173–174.
38. Ibid., p. 103.
39. Ibid., p. 47.
40. Ibid., p. 45.
41. Ibid., p. 94.
42. Ibid., p. 77.
43. Ibid., p. 91.
44. Ibid., p. 89.
45. Ibid., p. 75.
46. Ibid., p. 189.
47. Ibid., pp. 213–214.
48. Ibid., p. 67.
49. Ibid., p. 147.
50. Ibid., p. 125.
51. Ibid., p. 44.
52. Ibid., pp. 51, 70.
53. Ibid., p. 259.
54. William Blake, *The Complete Writings of William Blake* (London: Oxford University Press, 1972), p. 97.
55. Ibid., emphases mine. I have also changed the order in which these lines are usually printed. Blake's original order is uncertain.
56. Ibid.
57. Ibid., p. 98.
58. Ibid., p. 191.
59. Ibid., pp. 149–150.

60. Ibid., p. 685.

61. Ibid., pp. 701–702.

62. Ibid., p. 700.

63. Ibid., pp. 635–636.

64. Cited in John Purkis, *A Preface to Wordsworth* (New York: Charles Scribner's Sons, 1970), p. 117.

65. William Wordsworth, *Preface to the Lyrical Ballads,* reprinted in *The Oxford Anthology of English Literature* (New York: Oxford University Press, 1973), 2: 604–605.

66. Cited in Purkis, *A Preface to Wordsworth,* p. 63, emphasis mine.

67. Ibid., pp. 61–62.

68. Ibid., p. 50.

69. Ibid., p. 58.

70. Wordsworth, *The Excursion,* VIII. 167–176, 180–185.

71. Ibid., 122–126.

72. Cited in Herbert L. Sussman, *Victorians and the Machine* (Cambridge: Harvard University Press, 1968), p. 103.

73. See William Blake, "The French Revolution" in Geoffrey Keynes, ed., *Blake, Complete Writings, with Variant Readings* (London: Oxford University Press, 1972), pp. 143–145; and Wordsworth, A Letter to the Bishop of Llandaff on the Extraordinary Avowal of His Political Principles" (1793, unpublished).

Bibliographical Essay

INTRODUCTION

During the nineteenth and early twentieth centuries there was an important historiographical tradition that sought to interpret Western culture as shaped and given its "progressive" character largely by the development of science. Indeed, in 1936, just as this tradition was being almost universally rejected, George Sarton, founder of the History of Science Society, laid out its central dogma with breathtaking clarity in *The Study of the History of Science* (Cambridge: Harvard University Press, 1936), p. 5:

> *Definition*: Science is systematized positive knowledge, or what has been taken as such at different ages and in different places.
> *Theorem*: The acquisition and systematization of positive knowledge are the only human activities which are truly cumulative and progressive.
> *Corollary*: . . . If we wish to explain the progress of mankind, then we must focus our attention on the development of science and its applications.

Many of the classics of this genre, including Ernest Renan's *L'Avenir de la science* (Paris, 1848–1849), Emil du Bois Reymond's *Kulturegeschichte und Naturwissenschaft* (Leipzig, 1878) and John William Draper's *A History of the Intellectual Development of Europe* (New York: Harper & Bros., 1862), are now interesting primarily as historical curiosities; but some of the late and less doctrinaire offshoots remain outstanding if used with some caution. Preserved Smith's two-volume *A History of Modern Culture* (New York: Holt, Rinehart & Winston, Inc., 1934) and John Herman Randall's *Making of the Modern Mind* (Boston: Houghton Mifflin Co., 1926) are two of the best. Another excellent work that seems to accept a triumphal progressive view without its positivist grounding is Ernst Cassirer, *The Philosophy of the Enlightenment* (Princeton: Princeton University Press, 1951, translated from the German edition of 1932).

In the aftermath of two unprecedentedly destructive world wars and consequent fears of nuclear annihilation, as well as in the face of severe environmental degradation and the increasing enrichment of the few in the face of the impoverishment of the many, it has been impossible for most recent historians to accept the "triumphal progressivist" views associated with Positivism. As a consequence, general historians have tended to shy away from acknowledging the centrality of science to broader cultural concerns except as it is mediated through technology. So when Thomas Kuhn was asked to discuss "the relations between History and History of Science" for *Daedalus* in 1971 (Spring, 1971, pp. 271–304), he reported that for over twenty years there had been "only very tenuous links" between the history of science and other kinds of history.

Since the mid-1960s there has been a growing concern among social critics and philosophers regarding the seemingly dominant and oppressive role played by science in modern Western culture. Jacques Ellul weighed in from the conservative Catholic position with *The Technological Society* (New York: Alfred A. Knopf, 1964, from the French, 1954 original). The continental radical socialist position is represented in Jurgen Habermas's *Knowledge and Human Interest* (Boston: Beacon Press, 1970), in Herbert Marcuse's *One Dimensional Man* (Boston: Beacon Press, 1964) and in William Leiss's *The Domination of Nature* (New York: George Braziller, 1972); and a more Romantic radical position has been espoused by Theodore Roszak in *Where the Wasteland Ends* (New York: Doubleday & Co., 1972). More recently, a strong tradition of feminist analyses, represented by Sandra Harding's *The Science Question in Feminism* (Ithaca: Cornell University Press, 1986) has emerged. Important and sometimes disturbing defenses of science and its centrality for modern culture have been offered by C. P. Snow, *The Two Cultures: And a Second Look* (Cambridge: Cambridge University Press, 1963), Jacques Monod, "On Values in a Age of Science" in *The Place of Values in a World of Facts: The Fourteenth Nobel Symposium* (Stockholm: Swedish Academy of Science, 1964), B. F. Skinner, *Beyond Freedom and Dignity* (New York:

Vintage Books, 1971) and Paul Kurtz, *The Transcendental Temptation: A Critique of Religion and the Paranormal* (Buffalo: Prometheus Books, 1986), as well as by a host of less innovative authors.

A few of the major contemporary social critics of science have historical interests and training (Roszak, for example, holds the Ph.D. degree in history from Princeton and teaches in the History Department at California State University at Hayward), and the New Critical perspectives have begun to inform a number of detailed studies of relatively narrow topics by historians of science. But as Kuhn rightly observed, the key role of science in Western culture has not returned as a major theme in main-line historiography; and very few works have attempted any revisionist responses to the triumphalist synthetic works of John Randall or Harry Elmer Barnes's *An Intellectual and Cultural History of the Western World,* 3 vols. (New York: Reynal and Hitchock, Inc., 2d ed. 1941). J. Bronowski and Bruce Mazlish did produce *The Western Intellectual Tradition from Leonardo to Hegel* (New York: Harper and Row, 1960) for use in the Humanities Program at MIT, but it returned to a nauseatingly celebratory stance regarding the scientific tradition.

The present work is intended to fill this lacuna; but there are a handful of closely related works that deserve to be mentioned. The British journalist and TV personality turned popular historian, James Burke, developed a perverse but highly entertaining and thought-provoking series of programs for the BBC on the ideological implications of scientific change. The associated text, *The Day the Universe Changed* (London: The British Broadcasting System, 1985), is a nicely illustrated reinterpretation—elementary and occasionally outrageous—of the cultural roles of science from a historical relativist perspective. For an introduction to more scholarly perspectives on the topics selected for attention by Burke, see the anthology of readings assembled by John Burke (no relation), *Science and Culture in the Western Tradition* (Scottsdale, Arizona: Gorsuch Scarisbrick, Publishers, 1987). Another closely related collection of readings is Richard Olson, *Science as Metaphor* (Belmont, California: Wadsworth Publishing Co., 1971). John Marks's, *Science and the Making of the Modern World* (London: Heinenan, 1983) was designed for use as a textbook for a science and society course taught at the Polytechnic of North London. It is useful but very superficial. Another, more scholarly work with substantial linkages with *Science Deified* is Colin A. Russell, *Science and Social Change in Britain and Europe: 1700–1900* (New York: St. Martin's Press, 1983). Russell, however, spends much more time discussing social influences on scientific activities and on technologically mediated social impacts of science, leaving relatively little space for the relations between science and ideology. One work that appeared too late for me to take advantage of it stands out above those mentioned so far; this is Margaret Jacob's *The Cultural Meaning of the Scientific Revolution* (New York: Alfred A. Knopf, 1988). Though Ms. Jacob and I often have very different readings of science/culture interrelations, we both tend to emphasize ideological implications of scientific activity, and her "social" and my "intellectual" orientations complement one another.

Most modern social interpreters of science and its cultural roles renounce all conceptions of science which, like the positivist vision, assert its "objectivity" and value neutrality. Anyone interested in the general themes of this work should explore some answers to the question, "What is science?" At a very simple level and with great good humor, Garwin McCain and Erwin M. Segal's *The Game of Science* (Pacific Grove, Calif.: Brooks/Cole Publishing Co., 5th ed. 1988) provides a marvelous introduction to the nature of science. One of the most accessible updated versions of positivist understandings of science is Rudolph Carnap, *An Introduction to the Philosophy of Science,* edited by Martin Gardner (New York: Basic Books, 1966). Within most elements of the scientific community, a modification of positivist views grounded in Karl Popper's focus on "falsifiability" and explored in Karl R. Popper, *Conjectures and Refutations* (London: Routledge and Kegan Paul, 1968) now prevails. One of the most intriguing expressions of this currently dominant view—because it was implicitly accepted in 1987 by the U.S. Supreme Court as a legal definition of *Science*—is to be found in a twenty-seven-page pamphlet, "*Amicus Curiae* Brief of 72 Nobel Laureates, 17 State Academies of Science, and 7 other Scientific Organizations, in Support of Appelees," #85–1513 in the Supreme Court of the United States, October Term, 1986, in the matter of Edwin W. Edwards, governor of Louisiana, et al., *V. Don Aquillard.* (Caplin and Drysdale, chartered; One Thomas Circle, N.W., Washington D.C. 20005).

Recent developments in the philosophical tradition of science studies are summarized in A. F. Chalmers, *What is This Thing Called Science? An Assessment of the Nature and Status of Science and Its Methods* (St. Lucia, London, and New York: University of Queensland Press, 1976).

In 1962, Thomas Kuhn's *The Structure of Scientific Revolutions* (Chicago: University of Chicago Press, 1962) focused the attention of philosophers of science on the social-psychological dimensions of the pursuit of science; and since that time, increasing numbers of students of science have concentrated on what are now called "social studies of science." This entire movement is nicely characterized in Steve Woolgar, *Science: The Very Idea* (London and New York: Tavistock Publications and Ellis Horwood Ltd., 1988); and David Wade Chambers has collected an excellent selection of founding documents edited for student use in *On the Social Analysis of Science* (Victoria: Deakin University Press, 1984.) A more scholarly introduction can be found in K. D. Knorr-Cetina and M. Mulkay, eds., *Science Observed: Perspectives on the Social Studies of Science* (London: Sage, 1983).

CHAPTERS 1 AND 2

Before seriously exploring science as a factor in the shaping of European culture, one should probably have at least a basic orientation to European history and history of science for the period in question. No treatment of either topic is likely to suit everyone's pallet, but a few titles in each field deserve special recognition.

For my money the best elementary single-volume orientation to European history during the period covered by this work is Fernand Braudel's *Capitalism and Material Life: 1400–1800* (New York: Harper and Row, 1973, from 1967 French original). This work is totally without scholarly apparatus and is essentially a popular summary of the themes dealt with in greater detail in Braudel's three-volume *Civilization and Capitalism, 15th–18th Century* (English trans., New York: Harper and Row, 1981–1984). Though it is now dated, no comprehensive work has appeared to replace E. A. Wrigley's extremely useful *Population and History* (New York: McGraw Hill, 1969). Myron Gutmann's *Toward the Modern Economy: Early Industry in Europe: 1500–1800* (New York: Alfred A. Knopf, 1988) updates and corrects Braudel on economic issues, but alas, it is not very gracefully written. T. K. Rabb's *The Struggle for Stability in Early Modern Europe* (New York: Oxford University

Press, 1975) is one of a number of good introductions to the notion of a "crisis of authority" in early modern Europe which I assume in this work.

For the history of science, Richard S. Westfall, *The Construction of Modern Science: Mechanisms and Mechanics* (New York: John Wiley and Sons, 1971) and Thomas L. Hankins, *Science and the Enlightenment* (Cambridge: Cambridge University Press, 1985) together serve the function of Braudel for European social and economic history. For those who want more details, these should be supplemented by John Heilbron, *Elements of Early Modern Physics* (Berkeley: University of California Press, 1982) on natural philosophy and institutional factors in early modern science, by G. S. Rousseau and Roy Porter, eds., *The Ferment of Knowledge; Studies in the Historiography of Eighteenth Century Science* (Cambridge: Cambridge University Press, 1980) for the 18th century, and Robert G. Frank, Jr., *Harvey and the Oxford Physiologists: A Study of Scientific Ideas and Social Interaction* (Berkeley, Los Angeles, London: University of California Press, 1980) on central biological developments in the seventeenth century.

There are many works centering on the so-called "mechanical" philosophy. Of these, Marie Boas Hall's early classic, "The Establishment of the Mechanical Philosophy," *Osiris* 10 (1952): 412–541, remains valuable as an introduction though it undoubtedly overemphasizes Robert Boyle and underemphasizes Pierre Gassendi and the French. Rom Harré's *Matter and Method* (London: Macmillan, 1964) provides an elementary historical *and* philosophical analysis. Derek de Solla Price's "Automata and the Origins of Mechanism and Mechanistic Philosophy," reprinted in his *Science since Babylon* (New Haven: Yale University Press, 1975), is an excellent introduction to the artifactual background. Robert H. Kargon's *Atomism in England from Hariot to Newton* (Oxford: Clarendon Press, 1966) emphasizes the revival of atomism in England, whereas Robert Lenoble, *Mersenne, ou la naissance du mécanisme* (Paris: Vrin, 1943) does a similar if more limited job for France. Lisa Sarasohn links both French and English atomism and mechanism with moral issues in "Motion and Morality: Pierre Gassendi, Thomas Hobbes, and the Mechanical World View," *Journal of History of Ideas* 46 (1985): 363–379.

The *esprit géométrique* has received much attention, and Isabel Knight discusses it admirably at the beginning of her *The Geometric Spirit: The Abbé de Condillac and the French Enlightenment*

(New Haven: Yale University Press, 1968). The broader question of why mathematics became so central to European culture in the early modern period has been addressed from a variety of interesting perspectives. My own is closely related to that expressed by Richard Popkin in *The History of Skepticism from Erasmus to Spinoza* (Berkeley, Los Angeles, London: University of California Press, 1977). Popkin sees the reputation of Euclidean certitude as a kind of life raft in a sea of doubt about all other kinds of knowledge. Alexandre Koyré saw the fascination with mathematical reason as linked to a non-occult tradition of Renaissance neo-Platonism in "Galileo and Plato" in Philip P. Wiener and Aaron Noland, eds., *The Roots of Scientific Thought* (New York: Basic Books, 1957). Frances Yates, "The Hermetic Tradition in Renaissance Science" in *Art, Science, and History in the Renaissance,* ed. Charles Singleton (Baltimore: The Johns Hopkins University Press, 1967), was the leader of a group who views the intensification of interest in the Occult and "mathematical magic" as central to the fascination with mathematics. John Heilbron, "Introductory Essay" to Wayne Schumacher and John Heilbron, eds., *John Dee on Astronomy* (Berkeley, Los Angeles, London: University of California Press, 1978), is the current standard-bearer for those who see the commercial utility of "mixed" or "applied" mathematics for navigational, accounting, and engineering purposes as most important.

René Descartes is dealt with in every history of modern philosophy and mentioned in every treatment of early modern intellectual history; yet I know of no satisfactory discussion of the range of his interests and influences. The articles "Descartes," "Descartes: Mathematics and Physics," and "Descartes: Physiology," in *The Dictionary of Scientific Biography,* Vol. 4 (New York: Charles Scribner's Sons, 1971), are good places to start; and two collections of articles, Norman Kemp Smith's *New Studies in the Philosophy of Descartes* (New York: Garland, 1966) and Thomas M. Lennon, ed., *Problems in Cartesianism: Studies in the History of Ideas* (Montreal: McGill-Queens University Press, 1982), will guide one into the literature. Richard Popkin's *History of Skepticism from Erasmus to Spinoza,* cited above, contains the most compelling discussion of the context of Cartesian thought that I know. The long-range impact of Descartes's ideas has been explored by Aram Vartanian in his *Diderot and Descartes* (Princeton: Princeton University Press, 1953), by Paul Muoy, *Le Développement de la physique cartésienne* (Paris: Vrin, 1934), and by Leonora Cohen Rosenfield, *From Beast*

Machine to Man Machine: Animal Soul in French Letters from Descartes to LaMettrie, 2d ed. (New York: Octagon Books, 1968).

The relative shortage of intelligent and readable treatments of Descartes and his cultural significance is balanced by a huge and fascinating literature on Hobbes. Though he often argues for a more subtle influence of scientific ideas on Hobbesian arguments than I tend to seek, J. W. N. Watkins's, *Hobbes' System of Ideas: A Study in the Political Significance of Philosophical Theories* (London: Hutchinson, 1965) stands as a model whose successes I would like to be able to attain. One additional work that emphasizes the relations between Hobbes's natural science and his political theory is Thomas A. Spragens, Jr., *The Politics of Motion: The World of Thomas Hobbes* (Lexington: University of Kentucky Press, 1973). Comments on the radical uses of Hobbesian ideas are scattered through Christopher Hill's *The World Turned Upside Down: Radical Ideas during the English Revolution* (Harmondsworth: Penguin, 1975), but the best short discussion of this theme is Michael Walzer, "On the Role of Symbolism in Political Thought," *Political Science Quarterly,* 82 (1967): 191–204. The immediate negative response to Hobbes was first explored at length by Samuel Mintz in *The Hunting of Leviathan: Seventeenth-Century Reactions to the Materialism and Moral Philosophy of Thomas Hobbes* (Cambridge: Cambridge University Press, 1962). The now standard analysis of Hobbes's centrality for liberal ideology is Colin McPherson, *The Political Theory of Possessive Individualism* (Oxford: Clarendon Press, 1962.) For an analysis that sees Hobbes as the successful founder of a modern conservative political tradition, see Reinhart Koselleck, *Critique and Crisis: Enlightenment and the Pathogenesis of Modern Society* (Cambridge: MIT Press, 1988, from the 1959 German original). Quentin Skinner, "Thomas Hobbes and his Disciples in France and England," *Comparative Studies in Social History* 8 (1966): 153–167, has looked across the political spectrum at those who took Hobbes seriously from the beginning.

W. H. Greenleaf, *Order, Empiricism, and Politics: Two Traditions of English Political Thought* (Oxford and London: University of Hull Publications, 1964), was one of the first authors to isolate two distinctive scientific traditions of political discourse based on methodological disagreements. James Harrington, who was among the first to launch an explicitly political attack in the name of empiricism against Hobbesian mathematicism, has recently been the focus of a number of studies by J. A. G. Pocock, whose editor's introductory essay to *The Political Works of James Harrington* (Cambridge: Cambridge University Press, 1977) is superb. William Graig Diamond, "Natural Philosophy in Harrington's Political Thought," *Journal of the History of Philosophy*, 16 (1978): 387–398, focuses especially on the linkages between Harrington and the English tradition of alchemical studies. Though now outdated and clearly overly enthusiastic, H. F. Russell-Smith's *Harrington and His Oceana* (Cambridge: Cambridge University Press, 1914) suggests how important Harrington was in Colonial America. William Petty is discussed in Chapter 2 solely to explore his links to Hobbes and the mechanical philosophy. See Robert Kargon, "William Petty's Mechanical Philosophy," *Isis* 55 (1965): 63–66, and the elegant biography by Emil Strauss, *Sir William Petty: Portrait of a Genius* (Glencoe: The Free Press, 1954). Petty will be treated at greater length in connection with Chapter 4.

CHAPTER 3

Scholarly understandings of the relationships between science and religion have undergone tremendous changes over the past century. Dominated during the late nineteenth century by a sense that science and religion were somehow inevitably in conflict with one another—see John William Draper, *History of the Conflict between Religion and Science* (New York: D. Appleton, 1874) and Andrew Dickson White, *A History of the Warfare of Science with Theology in Christendom* (New York: D. Appleton, 1896)—attitudes shifted dramatically in the late 1930s with Robert K. Merton's argument that the rise of science was closely linked to the emergence of "Puritan" values in "Science, Technology, and Society in Seventeenth Century England," *Osiris* 4 (1938): 360–632. A modified and updated version of the Merton thesis is meticulously documented in Charles Webster, *The Great Instauration: Science, Medicine, and Reform: 1626–1660* (London: Duckworth, 1975). There has been a constant string of Catholic rebuttals to Merton, of which Stanley L. Jaki's *The Road of Science and the Ways to God* (Chicago: University of Chicago Press, 1978) is a good representative. The state of the discussion became increasingly complex during the late 1950s with the publication of Richard S. Westfall's seemingly nonpartisan *Science and Religion in Seventeenth-Century England* (New Haven: Yale University Press, 1958). I disagree with many of Westfall's assumptions and conclusions, but there is no question that his work has shaped almost all subsesquent discussions of British science and religion. For a general introduction to current understandings of the interactions among various sciences and various segments of the Christian community, there is now one single work to be recommended without reservation, David C. Lindberg and Ronald L. Numbers, eds., *God and Nature: Historical Essays on the Encounter between Christianity and Science* (Berkeley, Los Angeles, London: University of California Press, 1986). For basic background to the political, social, and religious contexts for the growth of Anglican natural theology, see Conrad Russell, *The Crisis of Parliaments: English History: 1509–1660* (London: Oxford University Press, 1971), Charles Wilson, *England's Apprenticeship: 1603–1773* (London: Longmans, 1965), Henry R. McAdoo, *The Spirit of Anglicanism* (London: A. & C. Blade, 1965), and G. R. Cragg, *From Puritanism to the Age of Reason* (Cambridge: Cambridge University Press, 1950).

In the late 1960s Barbara Shapiro initiated a radical challenge to the Merton thesis. In her view, now widely held with some modifications, seventeenth-century British science was most intimately connected, not with Puritanism, but with liberal, or "Latitudinarian," Anglicanism. See her *John Wilkins, 1614–1672: An Intellectual Biography* (Berkeley and Los Angeles: University of California Press, 1969). The state of the Latitudinarianism/science link in 1980 is well expressed in James Jacob and Margaret Jacob, "The Anglican Origins of Modern Science: The Metaphysical Foundations of the Whig Constitution," *Isis* 71 (1980): 251–267. I have explored the natural theological tradition within Latitudinarianism from a different perspective in "On the Nature of God's Existence, Wisdom and Power: The Interplay between Organic and Mechanistic Imagery in Anglican Natural Theology," in Frederick Burwick, ed., *Approaches to Organic Form* (Dordrecht: D. Reidel Publishing Co., 1987), pp. 1–48.

The special place of Richard Hooker in the Anglican tradition is discussed in J. S. Marshall, *Hooker and the Anglican Tradition* (London: Adam and Charles Black, 1962); Boyle's role is best treated in a number of articles and books by James Jacob, of which my favorite is "Boyle's Atomism and the Restoration Assault on Pagan Naturalism," *Social Studies of Science* 8 (1978).

Charles Raven's extensive discussion of John Ray's natural theology in *John Ray* (Cambridge: Cambridge University Press,

1942) is unfortunately much too adulatory; and John Locke's role as a natural theologian has not yet been adequately studied, though Richard Ashcraft has approached the issue in "Faith and Knowledge in Locke's Philosophy," in J. W. Yolton, ed., *John Locke: Problems and Perspectives* (Cambridge: Cambridge University Press, 1969). Latitudinarian uses of one aspect of Newtonian religious ideas have been treated in Margaret C. Jacob's very influential *The Newtonians and the English Revolution: 1689–1720* (Ithaca: Cornell University Press, 1976).

Isaac Newton's life and science have been treated in literally thousands of books and articles during the past ten years. Good, but very elementary, treatments are to be found in E. N. da C. Andrade, *Sir Isaac Newton, His Life and Work* (Garden City: Doubleday Anchor Books, 1954) and I. Bernard Cohen, *The Birth of a New Physics* 2d ed. (New York and London: W. W. Norton, 1985). An excellent and marvelously illustrated elementary appreciation of Newton's work and his cultural importance produced to celebrate the 300th anniversary of the publication of the *Principia* is John Fauvel, Raymond Flood, Michael Shortland, and Robin Wilson, eds., *Let Newton Be: A New Perspective on His Life and Works* (Oxford, New York, and Tokyo: Oxford University Press, 1988). The most comprehensive biography to date is Richard S. Westfall, *Never At Rest: A Biography of Isaac Newton* (Cambridge: Cambridge University Press, 1980).

The best single account of Newton's—as opposed to Newtonian—religion is Frank E. Manuel, *The Religion of Isaac Newton* (Oxford: Clarendon Press, 1974), though it should be complemented by James Force and Richard Popkin, eds., *Essays on the Theology of Isaac Newton* (Dordrecht: Martinus Nijhoff, 1990). There were a number of Newtonian religious traditions: one, incorporated into traditional Anglican natural theology, is nicely introduced in Perry Miller's "Bentley and Newton" in I. B. Cohen, ed., *Isaac Newton's Papers and Letters on Natural Philosophy* (Cambridge; Harvard University Press, 1958) and linked to political issues by Steven Shapin, "Of Gods and Kings: Natural Philosophy and Politics in the Leibniz-Clarke Disputes," *Isis* 72 (1981): 187–215; another, linked to traditions of prophecy interpretation, is nicely protrayed in James E. Force, *William Whiston: Honest Newtonian* (Cambridge: Cambridge University Press, 1985); and a third, which links Newtonian ideas with Deism, is discussed by Richard Westfall in "Isaac Newton's *Theologiae Gentilis Origines Philo-*

sophicae," in Warren Wager, ed., *The Secular Mind, Transformations of Faith in Modern Europe* (New York: Holmes and Meier, Inc., 1982).

Conservative reactions to the Newtonian natural theological tradition are generally unsympathetically but effectively presented in John Redwood, *Reason, Ridicule and Religion: The Age of Enlightenment in England, 1660–1750* (Cambridge: Harvard University Press, 1976). One fascinating and influential anti-Newtonian tradition of natural philosophy and natural theology is discussed by C. B. Wilde in "Hutchinsonian Natural Philosophy and Religious Controversy in Eighteenth-Century Britain," *History of Science* 18 (1980): 1–24.

Freethinking and atheistic developments growing out of Anglican natural theology are well discussed in James O'Higgins's *Anthony Collins: The Man and His Works* (The Hague: Martinus Nijhoff, 1970), Margaret Jacob, *The Radical Enlightenment: Pantheists, Freemasons, and Republicans* (London: Allen and Unwin, 1981), Robert Sullivan, *John Toland and the Deist Controversy* (Cambridge: Harvard University Press, 1982), and James Force, "The Origins of Modern Atheism," *Journal of the History of Ideas* 50 (1989): 153–162.

CHAPTER 4

For a general overview of the history of economic doctrines, nothing has yet superceded Joseph A. Shumpeter, *A History of Economic Analysis* (New York: Oxford University Press, 1954); and for an overview of the dramatic economic changes that occasioned the intense English debates over economic matters in the mid-seventeenth century, Barry Emmanuel Supple's *Commercial Crisis and Change in England: 1600–1642* (Cambridge: Cambridge University Press, 1959) remains a classic. The development of fascination with statistical information and analyses, which is so important for the growth of political economy is explored in Karl Pearson, *The History of Statistics in the Seventeenth and Eighteenth Centuries against the Changing Background of Intellectual, Scientific, and Religious Thought* (London: Charles Griffin & Co., 1978), and in G. S. Holmes, "Gregory King and the Social Structure of Pre-Industrial England,: *Royal Historical Society Transactions* 26 (1976).

Two excellent books survey the growth of economic theorizing in seventeenth-century England: William Letwin, *The Origin of Scientific Economics* (Garden City: Doubleday, 1964), and Joyce Oldham Appleby, *Economic Thought and Ideology in Seventeenth-Century England* (Princeton: Princeton University Press, 1978). Both, in contrast to my approach, tend to emphasize the extent to which even putatively scientific theorizing was developed to serve special economic interests, though Appleby is also sensitive to the ways that theory authorizes beliefs and behaviors that do not obviously serve the interests of the actors involved. A very different understanding of how theory came to mirror economic realities in the early modern period is suggested by Mark Blaug in "Economic Theory and Economic History in Great Britain: 1650–1776," *Past and Present,* 68 (1964): 111–116, and Peter Buck's "Seventeenth-Century Political Arithmetic: Civil Strife and Vital Statistics" *Isis* 68 (1977): pp. 67–85, presents an aggressively antiestablishment interpretation of the ideological function of early scientific economic ideas.

Early social scientific theorizing developed against the background of at least three different traditions of social discourse. Its relation to the Scholastic tradition, often associated with the term *moral economy,* is analyzed in Leonard Bauer and Herbet Mathis, "From Moral to Political Economy: The Genesis of Social Sciences," *History of European Ideas* 9 (1988): 125–143. The persistence of a moral economy tradition well into the early nineteenth century is discussed by E. P. Thompson, "The Moral Economy of the English Crowd in the Eighteenth Century," *Past and Present* 50 (1971): 76–136, and in Raymond de Roover, "Scholastic Economics: Survival and Lasting Influence from the 16th Century to Adam Smith," *Quarterly Journal of Economics* 69 (1955): 161–190. The second, Mercantilist, tradition, is explored in D. C. Coleman, ed., *Revisions in Mercantilism* (London: Oxford University Press, 1969). A good, short summary article is R. W. K. Hinton, "The Mercantile System in the Time of Thomas Mun," *Economic History Review* 7 (1955). The third, Humanist, tradition associated with the terms *prudence* and *practical reason* has recently been well discussed in Eugene Garver, *Machiavelli and the History of Prudence* (Madison: University of Wisconsin Press, 1987).

William Petty's role in the emergence of political economy is treated briefly in Tony Aspromourgos, "The Life of William Petty in Relation to His Economics: A Tercentenary Interpretation," *History of Political Economy* 20 (1988): 337–356, and at more length in

Alessandro Roncaglia, *Petty: The Origins of Political Economy* (Cardiff: Cardiff University Press, 1985). On the relations between Petty and John Graunt, see P. D. Groenewegen, "Authorship of *Natural and Political Observations upon the Bills of Mortality*," *Journal of the History of Ideas* 28 (1967). Debates about the role of expertise in government policymaking are dealt with for France in Keith Baker, *Condorcet, from Natural Philosophy to Social Mathematics* (Chicago: University of Chicago, 1975) and by Peter Buck, "People Who Counted: Political Arithmetic in the Eighteenth Century," *Isis* 73 (1982): 28–45.

Richard Cantillon is periodically rediscovered. A short older appreciation is J. J. Spengler, "Richard Cantillon: First of the Moderns," *Journal of Political Economy* 62 (1954): 281–95, 406–24. A more complete presentation is A. E. Murphy, *Richard Cantillon: Entrepreneur and Economist* (Oxford: Clarendon Press, 1986).

On physiocratic doctrines, Elizabeth Fox-Genovese, *The Origins of Physiocracy* (Ithaca: Cornell University Press, 1976) and Ronald Meek, *The Economics of Physiocracy* (Cambridge: Harvard University Press, 1963) are excellent. Meek's *Economics and Ideology and Other Essays* (London: Chapman and Hall, 1967) and "Smith, Turgot, and the 'Four Stages' Theory," *History of Political Economy* 3 (1971): 9–27 are also very insightful.

CHAPTER 5

Among the general histories of sociological thought which give substantial emphasis to pre-Comtean developments, Alan Swingewood, *A Short History of Sociological Thought* (New York: St. Martin's Press, 1984) is good. A similar job is done for early anthropology by Marvin Harris, *The Rise of Anthropological Theory: A History of Theories of Culture* (New York: Crowell, 1968). The central role of contact with non-European cultures for the emergence of the sociological imagination is briefly surveyed by Richard Cole, "Sixteenth-Century Travel Books as a Source of European Attitudes toward Non-White and Non-Western Cultures," *Proceedings of the American Philosophical Association* 116 (1972): 59–67, while Ronald Meek treats this issue brilliantly and at greater length in *Social Science and the Ignoble Savage* (Cambridge: Cambridge University Press, 1976). The broad environmentalist tradition that shaped Montesquieu's analyses is treated by Clarence Glacken,

Traces of the Rhodian Shore: Nature and Culture in Western Thought from Ancient Times to the End of the Eighteenth Century (Berkeley and Los Angeles: University of California Press, 1967).

J. Robert Loy, *Montesquieu* (New York: Twayne Publishers, Inc., 1968) is a useful brief biography of Montesquieu. Alan Baum is, I think, basically wrong and ahistorical in viewing Montesquieu as an Ideal-type theorist who was misunderstood by his Scottish emulators; but his *Montesquieu and Social Theory* (Oxford: Pergamon Press, 1979) is nonetheless filled with valuable information regarding the background to and reception of Montesquieu's work. In spite of attempts to modernize his views, C. J. Beyer's claim in *Montesquieu et l'esprit cartesien* (Bordeaux: Académie de Bordeaux, 1955) that Montesquieu retained his Cartesian scientific attitudes throughout his active career remains convincing to me. On the basic conservatism of Montesquieu's views see M. Hulling, *Montesquieu and the Old Regime* (Berkeley, Los Angeles, London: University of California Press, 1976), and on the use of Montesquieu's work within the subsequent conservative tradition, see C. P. Courtney, *Montesquieu and Burke* (Oxford: B. Blackwell, 1963). On the "liberal" uses of Montesquieu's works, see Paul M. Spurlin, *Montesquieu in America, 1760–1801* (Baton Rouge: Louisiana State University Press, 1940).

The development of "philosophical history" in Scotland is surveyed in Gladys Bryson's classic *Man and Society: The Scottish Inquiry in the Eighteenth Century* (Princeton: Princeton University Press, 1945), and Louis Schneider has collected and commented on selections from a number of key figures in *The Scottish Moralists on Human Nature and Society* (Chicago: University of Chicago Press, 1967). William C. Lehman has written outstanding critical biographies of three of the central figures of Scottish sociology in *Adam Ferguson and the Beginnings of Modern Sociology* (New York: Columbia University Press, 1930), *John Millar of Glasgow: 1735–1801, His Life and Thought and His Contributions to Sociological Analysis* (Cambridge: Cambridge University Press, 1960), and *Henry Home, Lord Kames, and the Scottish Enlightenment* (The Hague: Martinus Nijhoff, 1971).

A more recent and less worshipful treatment of Ferguson can be found in David Kettlery, *The Social and Political Philosophy of Adam Ferguson* (Columbus: The Ohio State University Press, 1965); and the importance of Ferguson in the development of German sociology is to be found in Herta H. Jogland, *Urspringe und*

Grundlagen der Soziologies bei Adam Ferguson (Berlin: Duncker and Humblot, 1959). Peter Bowles approaches an important topic virtually ignored in my discussions in "John Millar, the Four Stages Theory, and Women's Position in Society," *History of Political Economy* 16 (1984): 619–638.

There is probably no historical figure with the possible exception of Karl Marx who has been the subject of as much horrendously biased scholarship as Adam Smith. C. R. Fay's *Adam Smith and the Scotland of His Day* (Cambridge: Cambridge University Press, 1956) is a reliable biography for the most part. J. Ralph Lindgren's *The Social Philosophy of Adam Smith* (The Hague: Martinus Nijhoff, 1973) is in my judgment by far the best introduction to his work, for it dispels the myth that there was a significant element of natural theological belief in his notion of the invisible hand, and it deals intelligently with Smith's methodological commitments. T. D. Campbell's *Adam Smith's Science of Morals* (London: Allen and Unwin, 1971) offers a less relativist view of Smith's beliefs which is only marginally less compelling than Lindgren's. David Reisman has emphasized the sociological element in *Adam Smith's Sociological Economics* (New York: Barnes and Noble, 1976), and Ronald Meek has commented in very interesting ways on both Smith's lectures on jurisprudence and his impact on subsequent popular economic liberalism in *Smith, Marx, and After* (London: Chapman and Hall, 1977).

On the Continent the spread of sociological ways of thinking prior to Comte is discussed in G. A. Wells, *Herder and After, A Study in the Development of Sociology* (The Hague: Martinus Nijhoff, 1959).

CHAPTER 6

The overall roles of sensationalist and associationist psychology in the history of psychology were first discussed at length in G. S. Brett, *History of Psychology*, abridged by R. S. Peters (Cambridge: The MIT Press, 1965, from 1912–1921 original). Though Brett's approach is now seriously dated, no satisfactory alternative exists. The basic argument of this chapter—i.e., that sensationalist/associationist psychology played a central role in the development of liberal and radical social doctrines—I met first in Kingsley Martin's *French Liberal Thought in the 18th Century*, 3rd ed. (New York:

Harper and Row, 1962), though it is also strongly held throughout Elie Halévy, *The Growth of Philosophical Radicalism* (London: Faber and Faber, 1934, from 1929 original).

Two figures central to the growth of psychology in the eighteenth century are glossed over quickly here. On the importance of Locke and *An Essay Concerning Human Understanding,* one should consult John Yolton, ed., *John Locke, Problems and Perspectives* (Cambridge: Cambridge University Press, 1969), and *Thinking Matter, Materialism in 18th-Century Britain* (Minneapolis: University of Minnesota Press, 1983). On Hume and associationism, see especially John Bricke, *Hume's Philosophy of Mind* (Princeton: Princeton University Press, 1980) and Eugene F. Miller, "Hume's Contribution to Behavioral Science," *Journal of the History of Behavioral Sciences* 7 (1971): 154–168.

One of the very few articles to argue that Newtonian influences on psychology were pernicious is Daniel Robinson, "The Scottish Enlightenment and Its Mixed Bequest," *Journal of the History of Behavioral Sciences* 22 (1986): 171–197.

On Hartley's special appeals to Newtonian ideas and methods, see C. U. M. Smith, "David Hartley's Newtonian Neurophysiology," *Journal of the History of the Behavioral Sciences* 23 (1987): 123–136. A broader brief view of Hartley's place in eighteenth-century culture is in Chapter 8, pp. 136–154, of Basil Willey, *The Eighteenth Century Background* (New York: Columbia University Press, 1940). The political implications of early British associationism are nicely suggested in Isaac Kramnick, "Eighteenth Century Science and Radical Social Theory: The Case of Joseph Priestley's Scientific Liberalism," *Journal of British Studies* 25 (1986): 1–30.

In the late eighteenth and early nineteenth centuries, important political and social consequences were drawn from associationist principles by the "utilitarians" whose intellectual leader was Jeremy Bentham. On early utilitarianism, see Douglas Long's *Bentham on Liberty: Jeremy Bentham's Idea of Liberty in Relation to His Utilitarianism* (Toronto: University of Toronto, 1977) and David Lyons, *In the Interests of the Governed: A Study of Bentham's Philosophy of Utility and Law* (Oxford: Clarendon Press, 1963).

The most frequently cited work on the associationist/sensationalist tradition in France is still François Picavet, *Les Ideologues: essai sur l'histoire des idées et des théories scientifiques, philosophiques, religieuses etc. en France depuis 1789* (Paris: Felix Alcan, 1891). W. R. Albury provides an excellent short introduction to Condillac in his "Introduction" to *Entienne Condillac: La Logique* (New York: Abaris Books, 1980). Isabel Knight's *The Geometric Spirit: The Abbé de Condillac and the French Enlightenment* (New Haven: Yale University Press, 1968) is more thorough.

Claude Helvetius is given a worshipful, but still interesting, treatment from a socialist perspective in Irving L. Horowitz, *Claude Helvetius: Philosopher of Democracy and Enlightenment* (New York: Paine-Whitman, 1954). D. W. Smith's *Helvetius, A Study in Persecution* (Oxford: Oxford University Press, 1965) is, in spite of its title, a critical study with marvelous insights into the political infighting among French savants. Robert Weygant provides a short intriguing glimpse at the contemporary impact of Helvetius in "Helvetius and Jefferson: Studies of Human Nature and Government," *Journal of the History of the Behavioral Sciences* 9 (1973): 29–41. For Helvetius's role in prerevolutionary socialist thought, see Gilbert Chinard's editor's introduction to Morelly, *Code de la nature; ou la veritable esprit de ses lois* (Paris: Raymond Clavreuil, 1950).

Condorcet is given an interesting, elementary, heroic treatment in J. Salwyn Shapiro, *Condorcet and the Rise of Liberalism* (New York: Octagon Books, 1963), while the standard scholarly treatment is now Keith Michael Baker, *Condorcet: From Natural Philosophy to Social Mathematics* (Chicago: University of Chicago Press, 1975). His role in revolutionary and educational activities is emphasized in Manuela Albertone's "Enlightenment and Revolution: The Evolution of Condorcet's Ideas on Education," in Leonora C. Rosenfield, ed., *Condorcet Studies* I (Atlantic Heights, N.J.: Humanities Press, 1984, pp. 131–144), and in Robert Palmer, *The Improvement of Humanity: Education and the French Revolution* (Princeton University Press, 1985). David Williams explores an intriguing aspect of associationist implications in "Condorcet, Feminism, and the Egalitarian Principle," *Studies in 18th-Century Culture* 5 (1976): 151–163.

A handy guide to the massive literature on the relation of scientistic intellectuals to the French Revolution is to be found in William F. Church, ed., *The Influence of the Enlightenment on the French Revolution* (Boston: D. C. Heath and Co., 1964). Though a number of recent detailed studies of individual ideologues have to some extent superceded it, Charles Hunter Van Duzer's *The Contribution of the Ideologues to French Revolutionary Thought* (Baltimore: Johns Hopkins University Studies in Historical and Political Science, #53, 1935) remains a useful introduction.

CHAPTER 7

I know of no really satisfactory general discussions of aesthetic theories and their relationships to scientific developments during the period covered by this work. For the seventeenth century, the essays by Stephen Toulmin, Douglas Bush, James Ackerman, and Claude Palescoe in Hadley Howell Rhys, ed., *Seventeenth-Century Science and the Arts* (Princeton: Princeton University Press, 1961) at least begin to cover music, the visual arts, and literature; but the emphasis is on content rather than aesthetic theory. For the eighteenth century, Walter Jackson Bate's *From Classic to Romantic: Premises of Taste in Eighteenth Century England* (New York: Harper & Row, 1946) is very useful. Chapter 7, "The Fundamental Problem of Aesthetics," pp. 175–360, in Ernst Cassirer, *The Philosophy of Enlightenment* (Princeton: Princeton University Press, 1951, from German 1932 original), focuses on French and German developments which are virtually ignored here.

On the Classical and Aristotelian background to early modern aesthetic theories, see especially Samuel H. Butcher, *Aristotle's Theory of Poetry and Fine Art* (New York: Dover Publications, 1951, rpt. of 4th ed., 1907) and J. E. Spingarn, *A History of Literary Criticism in the Renaissance,* 2d ed. (New York: Columbia University Press, 1908).

On science and music theory, see especially Jacob Opper, *Science and the Arts: A Study in Relationships from 1600–1900* (Rutherford, N.J.: Fairleigh-Dickinson University Press, 1973) which, in spite of its title, focuses almost exclusively on music, and Georgia Cowert, *The Origins of Modern Musical Criticism: French and Italian Music, 1600–1750* (Ann Arbor, Michigan: University of Michigan Research Press, 1981). On the visual arts, Stephanie Ross, "Painting the Passions: Charles Le Brun's *Conference sur l'expression,*" *Journal of the History of Ideas* 45 (1984): 25–47 deals with a key issue, and Michael Leverey, *Rococo to Revolution: Major Trends in 18th-Century Painting* (New York: Frederick Praeger, 1966) provides a useful survey. For architecture, Anthony Blunt's *Art and Architecture in France, 1500–1700,* 2d ed. (Baltimore: Penguin Books, 1953) and *Artistic Theory in Italy: 1450–1600* (Oxford: Clarendon Press, 1956) provide the best introduction to classical styles, whereas Elizabeth MacDougall and F. H. Hazelhurst, eds., *The Formal Garden* (Washington, D.C.: The Dunbarton Oaks Trustees, 1974) discusses the classical style in gardening. Wendy Hilton extends to the dance the discussion of the impact of mathematical ideas on aesthetics in *Dance of Court and Theater: The French Noble Style: 1690–1725* (Princeton: The Princeton Book Co., 1981) as does Ivor Forbes in *Le Ballet de l'Opera de Paris* (Paris: Theatre National de l'Opera, 1976).

For all of the arguments of this chapter, Clarence Dewit Thorpe, *The Aesthetic Theory of Thomas Hobbes, with Special Reference to the Psychological Approach to English Literary Criticism* (New York: Russell and Russell, Inc., 1964) has been absolutely central. Charles Henant, "Hobbes on Fancy and Judgement," *Criticism* 18 (1976): 615–626, George Dillon, *The Art How to Know Men; A Study of Rationalist Psychology and Neo-Classical Dramatic Theory* (Ph.d. Dissertation, University of California at Berkeley, 1969) and James Engell, *The Creative Imagination* (Cambridge: Harvard University Press, 1981) serve only to amplify the importance of Hobbesian psychologism for British aesthetic theory.

The scientific element in Edmund Burke's aesthetic theory is dealt with briefly in Vincent M. Bevalaqua, "Two Newtonian Arguments Concerning Taste," *Philological Quarterly* 47 (1968): 385–590, and in vastly greater depth in Martin Kallick, *The Association of Ideas and Critical Theory in Eighteenth-Century England: A History of a Psychological Method of English Criticism* (The Hague: Mouton, 1970).

CHAPTER 8

For the traditional view that science was of marginal significance in supporting technical innovation until well into the nineteenth century, see A. R. Hall, *The Science of Ballistics in the Seventeenth Century* (Cambridge: Cambridge University Press, 1952), Charles Coulston Gillispie, "The Natural History of Invention," *Isis* 48 (1957): 152–170, G. N. von Tunzelmann, "Technical Progress in the Industrial Revolution," in Roderick Flood and Donald McCloskey, eds., *The Economic History of Britain since 1700,* I: 1700–1860 (Cambridge: Cambridge University Press, 1981), or Richard Westfall, "Robert Hooke, Mechanical Technology, and Scientific Investigation," in John Burke, ed., *The Uses of Science in the Age of Newton* (Berkeley, Los Angeles, London: University of California Press, 1983).

For the contrary view (which I share) that science was very important in stimulating technical innovations as early as the late seventeenth century, see especially Peter Mathias, "Who Unbound Prometheus? Science and Technical Change, 1600–1800," in Peter Mathias, ed., *Science and Society 1600–1900* (Cambridge: Cambridge University Press, 1972), and Neil McKendrick, "The Role of Science in the Industrial Revolution: A Study of Josiah Wedgwood as a Scientist and Industrial Chemist" in Micklas Teich and R. M. Young, eds., *Changing Perspectives in the History of Science* (London: Heineman, 1973).

The best surveys of the role of science during the Industrial Revolution are A. E. Musson and Eric Robinson, *Science and Technology in the Industrial Revolution* (Toronto: University of Toronto Press, 1969) and A. E. Musson, ed., *Science, Technology, and Economic Growth in the Eighteenth Century* (London: Methuen & Co. Ltd., 1972), while Part 3 of C. C. Gillespie, *Science and Polity in France at the End of the Old Regime* (Princeton: Princeton University Press, 1980), offers a French perspective to augment Musson and Robinson's focus on England.

On the roles of science in the agricultural revolution of the late seventeenth and early eighteenth century, see G. E. Mingay, *The Agricultural Revolution* (London: Adam & Charles Black, 1977), E. L. Jones, ed., *Agricultural and Economic Growth in England, 1650–1815* (London: Methuen, 1967), and the sections on William Cullen in Arthur Donovan, *Philosophical Chemistry in the Scottish Enlightenment* (Edinburgh: University of Edinburgh, 1975).

On the societies that served to encourage the application of science to technical problems, see especially Robert Schofield, *The Lunar Society of Birmingham: A Social History of Provincial Science and Industry in Eighteenth-Century England* (Oxford: Oxford University Press, 1963), D. Hudson and K. W. Lockhurst, *The Royal Society of Arts: 1754–1954* (London, 1954), and S. B. Donkin, "The Society of Civil Engineers: Smeatonians," *Newcomen Society Transactions* 17 (1936–37).

Archibald and Nan Clow's *The Chemical Revolution* (London: the Batchworth Press, 1952) is the classic on science and the chemical industry in eighteenth-century Britain. This should be supplemented for France by John Graham Smith, *The Origins and Development of the Heavy Chemical Industry in France* (Oxford: Oxford University Press, 1979).

On science and the development of power technologies, see R. L. Hills, *Power in the Industrial Revolution* (Manchester: Manchester University Press, 1970), articles by A. Stowers, R. J. Forbes, and H. W. Dickinson in Charles Singer, et al., eds., *A History of Technology, Vol 4: The Industrial Revolution ca. 1750–1850* (Oxford: Oxford University Press, 1958), and E. Robinson and A. E. Musson, *James Watt and the Steam Revolution* (New York: Kelley, 1969).

Neil McKendrick's "Josiah Wedgwood and Factory Discipline" in David Landes, ed., *The Rise of Capitalism* (New York: The Macmillan Co., 1966) provides an introduction to the application of scientific attitudes to the management of technical enterprises, as does S. Pollard, *The Genesis of Modern Management* (Cambridge: Harvard University Press, 1965).

CHAPTER 9

One of the most eloquent brief portrayals of Conservatism and Romanticism as blanket renunciations of the roles of science in eighteenth-century culture is Isaiah Berlin, "The Counter-Enlightenment" in *Against the Grain: Essays in the History of Ideas* (Harmondsworth, England: Penguin Books Ltd., 1979). This should be supplemented by Harold E. Pagliaro, ed., *Irrationalism in the Eighteenth Century: Studies in 18th-Century Culture,* vol. 2 (Cleveland: The Press of Case Western Reserve University, 1972).

Continental conservative reactions to the French Revolution are outlined in P. H. Beik, "The French Revolution Seen from the Right, Social Theories in Motion: 1789–1799," *Transactions of the American Philosophical Society* 46 (1956), and in Richard Fargher, "The Retreat from Voltaireanism" in *The French Mind: Studies in Honor of Gustave Rudler* (Oxford: Clarendon Press, 1952).

On de Maistre in particular, see J. C. Murray, "The Political Thought of Joseph de Maistre," *Review of Politics* 11 (1949): 63–86, Elisha Greifer, "Joseph de Maistre and the Reaction against the 18th Century," *American Political Science Review* 55 (1961): 591–98, and E. D. Watt, "Locked in: De Maistre's Critique of French Lockeanism," *Journal of the History of Ideas* 32 (1971): 129–133.

Burke's political ideas are explored in Alfred Cobban, *Edmund Burke and the Revolt against the Eighteenth Century* (New York: Barnes and Noble, 2d ed., 1960) and C. P. Courtney, *Montesquieu and Burke* (Oxford: Basil Blackwell, 1963), who acknowledges the importance of a sociological perspective in Burke's thought.

For the relationship between pre-Romantic and Romantic poetry and the triple revolutions of the eighteenth century, see especially Meyer H. Abrams, *The Mirror and the Lamp: Romantic Theory and the Critical Tradition* (London: Oxford University Press, 1953), Ernest Tuveson, *The Imagination as a Means of Grace* (Berkeley and Los Angeles: University of California Press, 1960), and Marjorie Hope Nicholson, *Mountain Gloom and Mountain Glory: The Development of the Aesthetics of the Infinite* (Ithaca: Cornell University Press, 1959).

On Wordsworth, see James A. W. Hefferman, *Wordsworth's Theory of Poetry: The Transforming Imagination* (Ithaca: Cornell University Press, 1969) and on Blake, see Peter Fisher, *The Valley of Vision: Blake as Prophet and Revolutionary* (Toronto: University of Toronto Press, 1961) and G. E. Bentley, *William Blake: The Critical Heritage* (London: Routledge and Kegan Paul, 1975).

Index

Designer:	U.C. Press Staff
Compositor:	Prestige Typography
Text:	10/12 Auriga
Display:	Helvetica
Printer:	Thomson-Shore, Inc.
Binder:	Thomson-Shore, Inc.

DATE DUE